Build you:

HIGH-END AUDIO EQUIPMENT

Build your own

HIGH-END AUDIO EQUIPMENT

Elektor Electronics (Publishing)
P.O. Box 1414
Dorchester
England DT2 8YH

British Library Cataloguing in Publication Data
A catalogue record for this book is available from the British Library

ISBN 0 905705 40 8

First published in the United Kingdom 1995

Printed in the Netherlands by Tulp BV, Zwolle

Contents

WARNING: electricity is dangerous

The projects in this book are, to the best of the Publishers' knowledge and belief, both accurately described and safe. Nevertheless, great care must always be taken when assembling electronic circuits that carry mains voltages: the Publishers can not accept responsibility for any accidents that may occur.

Because electricity is dangerous, its use, application and transmission are governed by rules, regulations and guidance. These are laid down in numerous laws, British Standards and recommendations of the Institution of Electrical Engineers (IEE). Some of these may be obtained from your local electricity showroom, but most, if not all, should be available for reference in your local library.

Introduction

After peaking in the 1970s, it appeared for a time in the early 1980s that the public's interest in hi-fi had all but disappeared. There was a feeling that the techology had moved as far as it could and that better sound reproduction was becoming a matter of spending disproportionate large amounts of money. The introduction of the compact disc, like the compact audio cassette invented by Philips, changed all that. Hi-fi equipment is once again a best-seller in all electronics retail shops.

Top-of-the-range hi-fi apparatus is now termed 'high-end equipment' and the name is a good indication of the prices charged for it. For those who can not, or will not, pay these prices, there is another solution offered in this book: build your own (at considerable cost savings). But this book is aimed not only at this sector of the market, but also at the many enthusiasts who want to be able to experiment and make modifications to their 'high-end equipment'.

Build your own High-end Audio Equipment deals with building and testing preamplifiers, power amplifiers, headphone amplifiers, loudspeaker systems and appropriate test equipment. There is also a chapter dedicated to an overview of PC-controlled AF Measurement Systems. The book is largely, but by no means entirely, an anthology of articles that have appeared in *Elektor Electronics*, the International Electronics Magazine.

The track layouts of all printed-circuit boards used in the projects are given in an Appendix to facilitate the making of the relevant board. Where possible, the drawings of these boards, are given in true size, but in some cases, owing to the book format, this was not possible. In these cases, the drawings are illustrated reduced to 70.7%, which is the conversion factor from A3 to A4 paper size. This makes it possible to make photocopies of these reduced drawings in full size merely by setting the photocopier to A4→A3 (or to 141.4%).

Part 1
Voltage amplifiers

Chapter 1

Preamplifier I

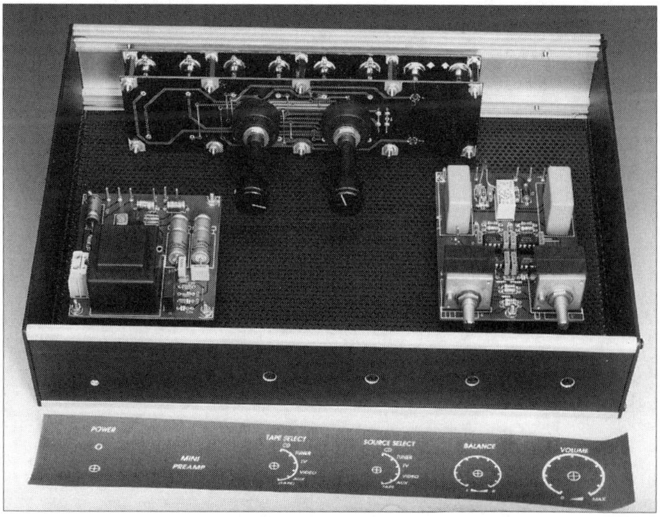

The two major properties of this design are simplicity and quality. Simplicity is achieved by omitting such superfluous facilities as tone control, mono/stereo selection, rumble filters, noise filters, and so on. Such measures also improve the quality, which may be further enhanced by the use of the best available components.

The changes that have taken place in the audio world over the past decade are reflected in design philosophy. In the past, there were two clear camps in audio engineering: one that advocated full control of frequency, the use of various filters, and so on; and the other which wanted the minimum of controls. Nowadays, what is the use of a 33/78 input and a rumble filter in the absence of a record player? Why have a noise filter when available signals sources do not produce noise? And what is the use of a mono/stereo selector? Moreover, the quality of current signal sources and the recording quality of compact disks surely make tone control and equalizers superfluous?

When all these facilities are omitted, what is left? Only the basic functions: input selection, volume control, balance control, and perhaps a separate selector for record out. These functions require relatively few components and this is an aspect that audio purists have always seen as a great plus point. After all, what is not there can not cause noise or distortion.

The design

Figure 1.1. shows that the omissions mentioned in the previous paragraph result in a fairly simple circuit. The input signals enter via phono sockets K_1–K_{12}. Each of the inputs is individually terminated by R_1–R_{12}. Switch S_1 selects the record out signal, which is applied to output sockets K_{13} and K_{14} via R_{13} and R_{14} respectively. Switch S_2 functions as the standard input selector.

The signals at poles A and B of S_2 (left-hand and right-hand channels respectively) are applied to a second terminating resistor, R_{15} (R_{16}). The overall terminating impedance of the selected input has the standard value of 47 kΩ. The signals are then applied to a buffer stage, IC_1 (IC_2), which is arranged as a unity gain amplifier. Since the NE5534 is not inherently stable at unity gain, a compensating capacitor, C_1 (C_2), is connected between pins 5 and 8.

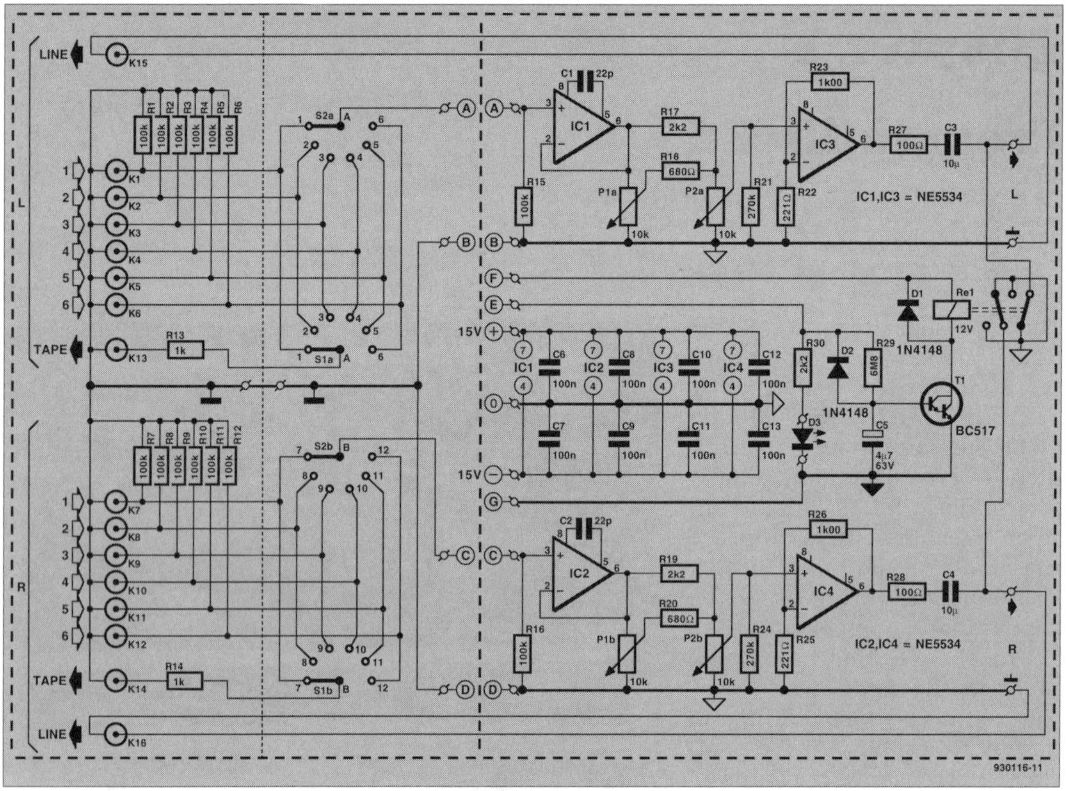

Fig. 1.1. Circuit diagram of Preamplifier I.

The output of the buffer is applied to a voltage amplifier, IC_3 (IC_4), via balance control P_1 and volume control P_2. The amplification of IC_3 (IC_4) is set to ×5.5 with R_{22}–R_{23} (R_{25}–R_{26}). This ensures that, in spite of the losses in the balance control, a nominal output level of 1 V is obtained for an input of 250 mV.

The outputs of IC_3 and IC_4 are applied to output sockets K_{15} and K_{16} respectively via contacts of relay Re_1. Delay stage T_1 arranges for the relay to be energized a few seconds after the supply has been switched on. This ensures that any switch-on noise is kept away from the outputs. On/off indication is provided by D_3-R_{30}.

The design of the power supply follows the same philosophy as the amplifier: no complex circuit where simple ones will do. Circuits IC_5 and IC_6 are voltage regulators. The delay stage derives its own supply directly from the secondary of the mains transformer via D_4, D_5, R_{31}, C_{20}.

Enhancement measures

Quality starts with the constituent components. For example, there are many types of phono input socket available, but for best results only gold-plated ones will do. Good-quality rotary switches are also a must, but these can be difficult to obtain. The type of potentiometer used for the balance and volume controls has a very real effect on the quality of the amplifier. Again, for best results, use a top quality component, such as an Alps type. Capacitors C_3 and C_4 should be polypropylene types. Note, however, that the printed circuit board can accommodate less expensive passive components as well.

There is a wide choice of integrated circuits as shown in the parts list. This does not

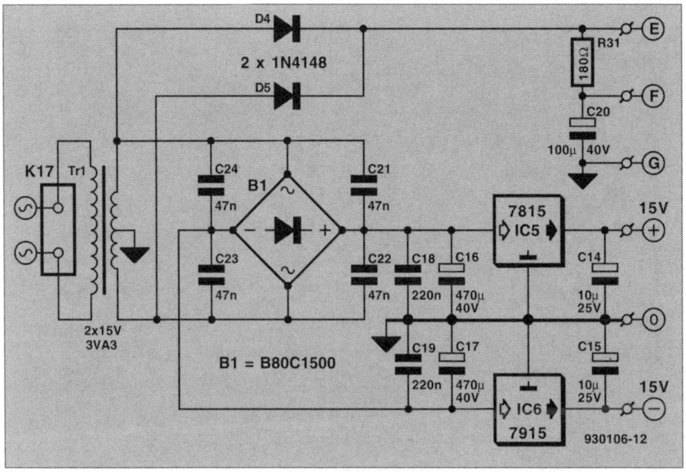

Fig. 1.2. Circuit diagram of the power supply.

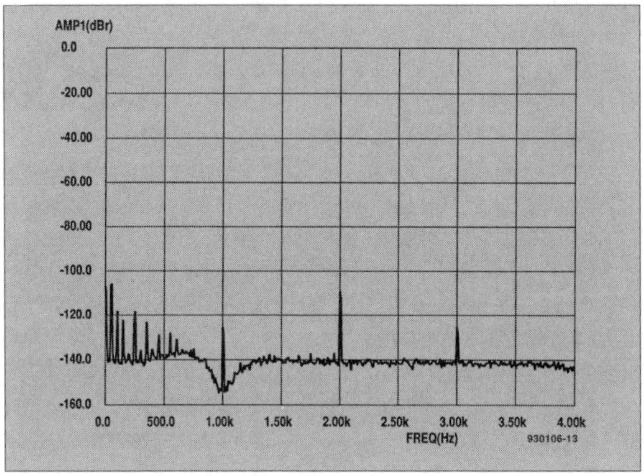

Fig. 1.3. THD characteristic of the preamplifier.

mean that the NE5534 used in the prototype is not a satisfactory choice, but there are other, more expensive, types that may meet an individual need better. It should be borne in mind, however, that a more expensive device does not necessarily provide better aural quality.

Of course, there are different yardsticks for the buffer and the amplifier. For the buffer, low noise and high input impedance are prime requirements, whereas for the amplifier a good gain-bandwidth product and a low output impedance are important. The slew rate reflects much about the quality of an op amp, but its importance in top quality audio equipment must not be exaggerated: other parameters may be just as important.

Since few people will be able to try out all the op amps in the parts list, the editor's recommendations are the SSM2131 for IC_1 and IC_2 (the OPA627 is also excellent, but perhaps rather expensive for this application), and the OPA637 for IC_3 and IC_4. In the later case, the LT1028 and OP37 are good second choices. Bear in mind that all types which are not stable at unity gain require special compensation when they are used as buffers. This compensation varies from one type to another and is not always wholly satisfactory. It is, therefore, best if the recommended is not, or can not, be used, to choose another that is stable at unity gain. Compensating capacitors C_1 and C_2 should be omitted when such types are used.

Construction

Before construction is started, cut or saw the printed circuit board in Fig. 1.4. into four along the lines indicated. Populating the four boards should prove straightforward.

Interconnect E, F, G,'–' and '+' on the supply board and the amplifier board with appropriate lengths of flexible circuit wire.

Sandwich the other two boards together with the aid of spacers as shown in Fig. 1.5. and interconnect them with short lengths of bare wire. The interconnections between the boards near K_{13} and K_{14} are via R_{13} and R_{14}. Note that resistors R_1–R_{12} are soldered directly to the terminals of the phono sockets. When the sandwich has been completed,

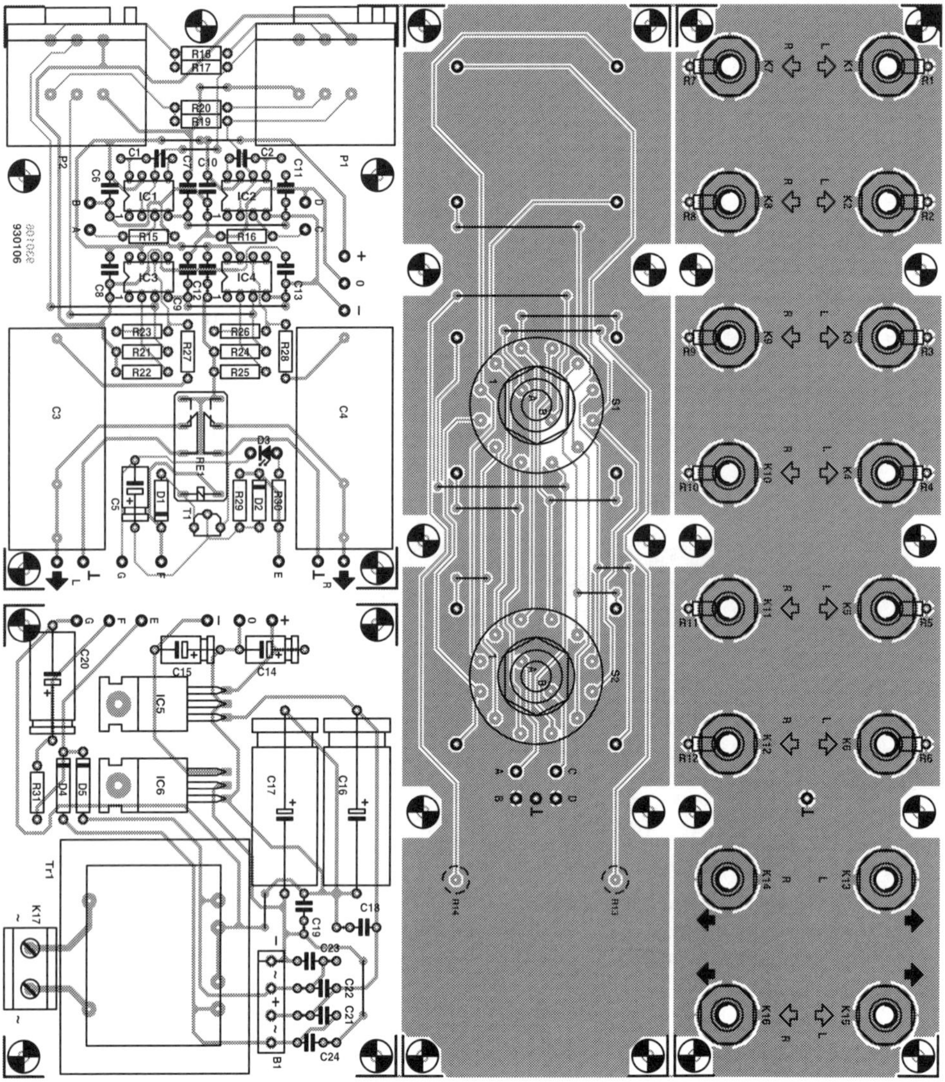

Fig. 1.4. Printed-circuit board (component layout) shown at 70.7% (=A3→A4).
The track layout is given on page 215.

connect A, B and C, D to the amplifier board via two short lengths of screened cable.

Connect the outputs of he amplifier board to output sockets K_{15} and K_{16} via two lengths of screened cable.

The boards can then be built into a suitable enclosure. It is the intention that the sandwiched boards are mounted on the inside rear panel and that the spindles of the rotary switches are extended to the front panel—see Fig. 1.6. A suggested front panel suitable for a number of enclosures is shown in Fig. 1.7. It should not prove difficult to adapt it to a particular enclosure.

Fig. 1.5. The board for the sockets and that for the rotary switches are assembled into a 'sandwich' with the aid of suitable spacers.

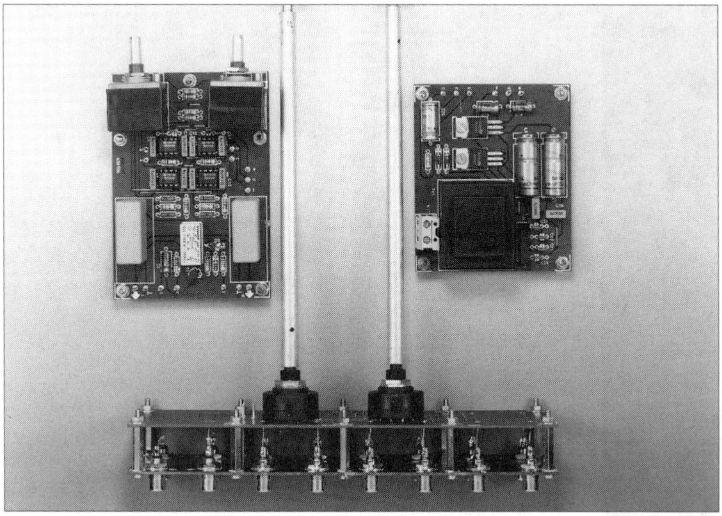

Fig. 1.6. The spindles of S_1 and S_2 must be extended to protrude through the front panel.

Some parameters

- The input sensitivity is 250 mV into 47 kΩ for an output of 1 V into 100 Ω.
- Channel separation is 82 dB at 1 kHz.
- Noise suppression and total harmonic distortion (THD) are shown in Fig. 1.3.

Parts list

Resistors:

R_1–R_2, R_{15}, R_{16} = 100 kΩ
R_{13}, R_{14} = 1 kΩ
R_{17}, R_{19}, R_{30} = 2.2 kΩ
R_{18}, R_{20} = 680 Ω
R_{21}, R_{24} = 270 kΩ
R_{22}, R_{25} = 220 Ω, 1%
R_{23}, R_{26} = 1 kΩ, 1%
R_{27}, R_{28} = 100 Ω
R_{29} = 6.8 MΩ
R_{31} = 180 Ω

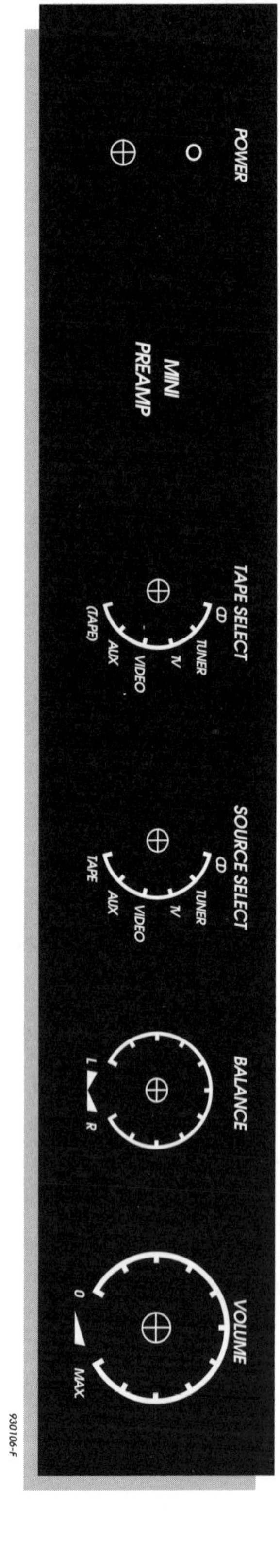

Potentiometers:
P_1 = 10 kΩ, linear, stereo
P_2 = 10 kΩ, logarithmic, stereo

Capacitors:
C_1, C_2 = 22 pF
C_3, C_4 = 10 µF, polypropylene
C_5 = 4.7 µF, 63 V
C_6–C_{13} = 100 nF
C_{14}, C_{15} = 10 µF, 25 V
C_{16}, C_{17} = 470 µF, 40 V
C_{18}, C_{19} = 220 nF
C_{20} = 100 µF, 40 V
C_{21}, C_{24} = 47 nF, ceramic

Semiconductors:
D_1, D_2 = 1N4148
D_3 = LED, 3 mm
D_4, D_5 =1N4003
B_1 = B80C1500
T_1 = BC517

Integrated circuits:
IC_1–IC_4 = see text:
NE5534 (bipolar)
SSM2131 (FET)
SSM2134 (bipolar)
OP27 (bipolar)
OP37 (bipolar)
OPA627 (FET)
OPA637 (FET)
LT1028 (bipolar)
LT1115 (bipolar)
TLE2027 (bipolar)
TLE2037 (bipolar
AD743 (FET)
AD745 (FET)
LT1007 (bipolar)
LT1037 (bipolar)
LM627 (bipolar)
LM637 (bipolar)
IC_5 = 7815
IC_6 = 7915

Miscellaneous:
K_1–K_{16} = phono socket for board mounting
K_{17} = 2-way terminal block, pitch 7.5 mm
S_1, S_2 = 2-pole, 6-position rotary switch for board mounting
Re_1 = 12 V miniature relay with 2 change-over contacts for board mounting
Tr_1 = mains transformer, secondary 2×15 V, 3.3 VA
Enclosure as appropriate
Mains socket for panel mounting
Mains on/off switch

Fig. 1.7. Suggested front panel layout. (Reduced to 66.7%).

Chapter 2

Preamplifier II

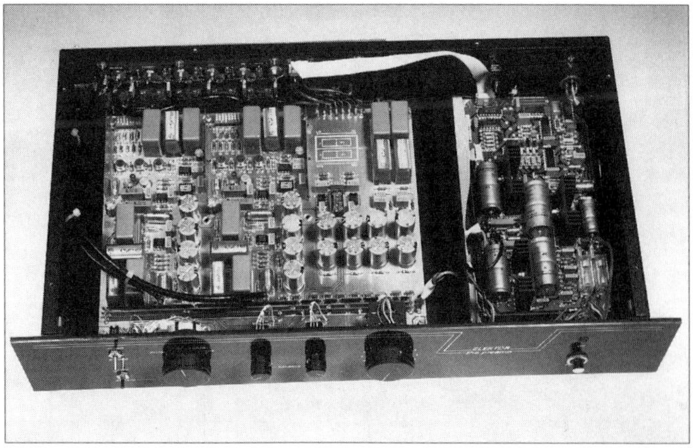

Music lovers and audiophiles are interested only in audio equipment in which the quality of the reproduced sound has been the prime consideration rather than the provision of a range of interesting, but strictly unnecessary, facilities, such as complex tone controls and remote operation. It is interesting to note that there is a certain connection between the price and the number of operating controls. As the quality and, consequently, the price rise, the number of controls decrease. This may not always be strictly true, but there is a strong definite trend towards it. At the top, there is equipment designed by and for purists from which anything that has no direct bearing on the sound quality has been omitted. Such equipment is geared to

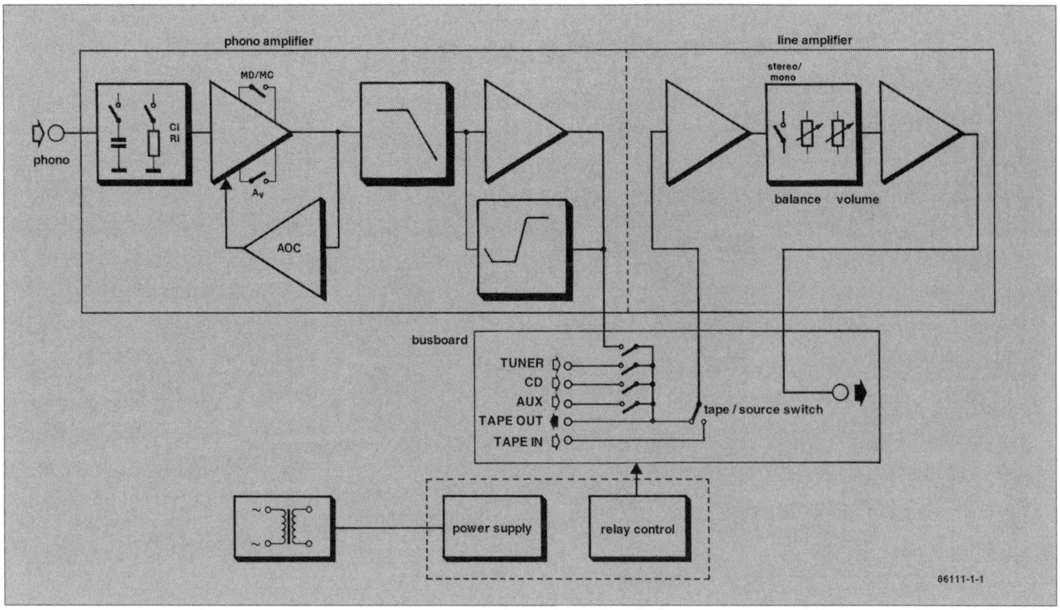

Fig. 2.1. Block diagram of preamplifier II.

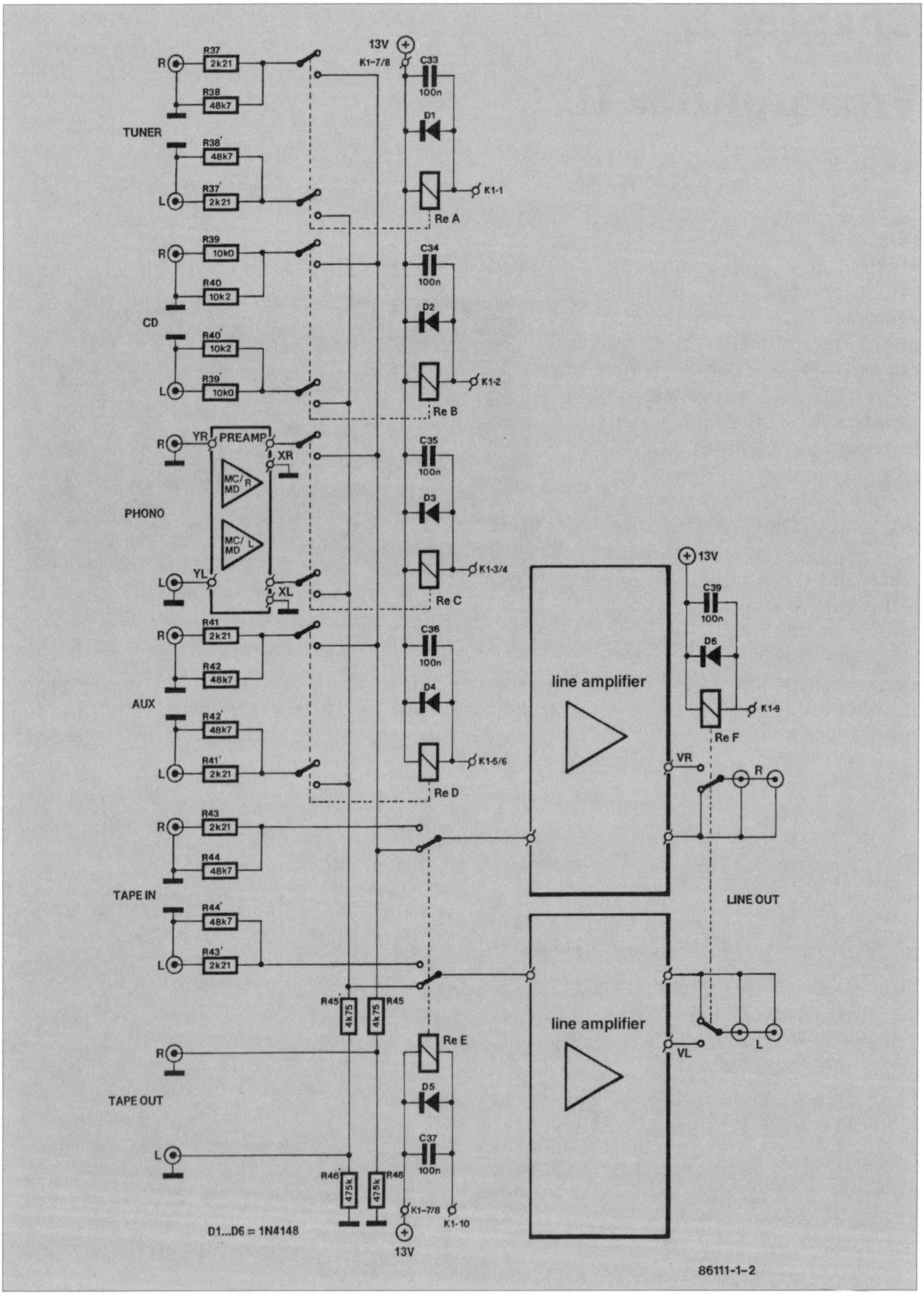

Fig. 2.2. Diagram of the input and relay circuits.

the utmost perfection of the reproduced sound. Often, the preamplifiers of this kind of equipment have only an on/off switch, input selector and volume control. The preamplifier described in this chapter belongs to this class of equipment, although it has three more controls than mentioned: mono/stereo; tape source and balance.

Basic design

The layout of the preamplifier is shown in the block diagram in Fig. 2.1. Each of the three sections in dashed lines is located on a separate printed-circuit board. That at the top is the preamplifier proper, which is, of course, a stereo setup, although only one channel is shown. The section underneath it is the busboard which contains the input and output connectors, the various selectors, and associated parts. The third board contains the power supply with the exception of the mains transformer, which is fitted in a separate case, and the relay control circuits.

The relay control circuits enable selection of various modes of operation to be made as close as possible to the relevant input.

The preamplifier is almost entirely DC-coupled; where this proved difficult or impossible, high quality polypropylene capacitors, which obviate the usual drawbacks of capacitor coupling, are used. The characteristics of these capacitors and the special semiconductors used will be discussed later.

The preamplifier proper consists of two parts: a phono amplifier that can work with either a moving-coil (MC) or a magneto-dynamic (MD) cartridge, and a line amplifier with inputs for TUNER, CD (compact disc), AUX(iliary) and TAPE. Many constructors may decide to omit the phono section. This section is rather special, because it is not the usual combination of MD and MC preamplifiers, but a single stage whose amplification can be set to suit both types of cartridge.

The input stage of the phono amplifier offers the facility of terminating the pickup cartridge used into the correct capacitance and resistance: an indispensable feature in this class of amplifier.

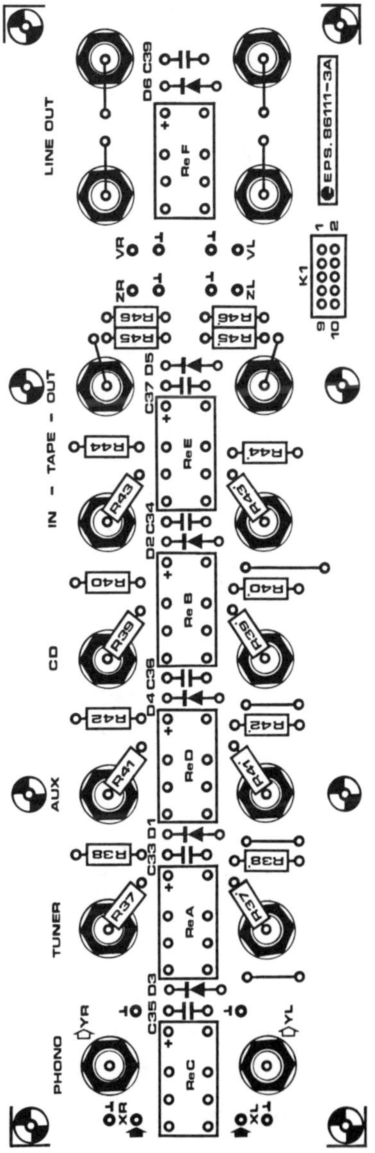

Fig. 2.3. The busboard.
(Reduced to 70.7% - A3→A4)
(Track layout on page 217)

The voltage gain, A_V, of the second stage can be arranged not only to accommodate either an MD or an MC element, but also—in two steps—to suit the output voltage of these elements. The active offset correction (AOC) stage ensures that the offset voltage at the output of the linear amplifier remains negligibly small at all times without the need for any adjustments.

The final stage in this section provides the necessary deemphasis for record reproduction. The deemphasis characteristic is within 0.1 dB of the relevant requirements of the IEC (International Electrotechnical Commission) and the RIAA (Record Industry Association of America). The corresponding preemphasis characteristic—see Fig. 2.4—is used by all major broadcasting organizations, virtually the whole of the recording industry and is recognized as such by organizations like the AES (Audio Engineering Society), the RIAA, and the NARTB (National Association of Radio and Television Broadcasters).

The linear line amplifier contains the volume control, the balance control and the stereo/mono selector.

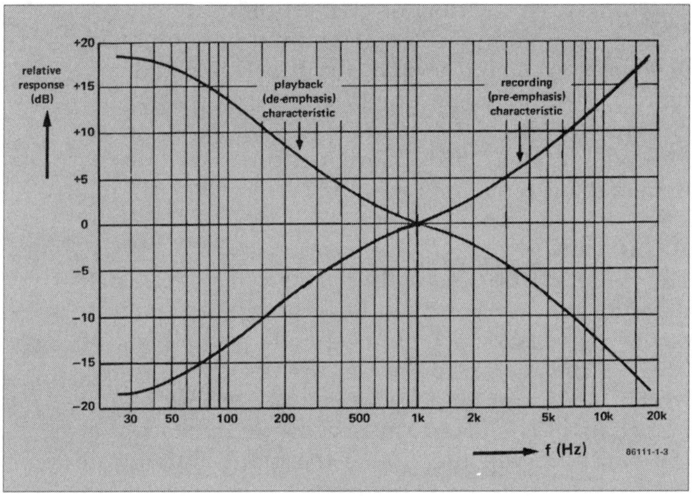

Fig. 2.4. The IEC/RIAA recommended recording and playback characteristics.

Busboard

The busboard, the component layout of which is shown in Fig. 2.3 (for the track layout see Appendix), contains not only all input and output connectors and relays, but also the voltage dividers required for level matching. Its circuit diagram is shown in Fig. 2.2. At the left are all the inputs, and at the right all the outputs. The relays (with associated capacitors and freewheeling diodes) take care of all the switching. The relays are controlled from the relay control section on the power supply board.

Since a CD player provides a much higher output voltage than, say, a tuner or a tape recorder, the CD input is attenuated by voltage divider R_{39}-R_{40}. The voltage dividers on the other inputs merely serve to reduce near-end crosstalk even further and are, strictly speaking, not necessary. Inputs not in use are taken to ground via resistors R_{38}, R_{40}, R_{42} and R_{44}.

The relays are miniature PCB types which must be of the best quality obtainable to ensure that no unnecessary resistances are introduced in the (low-level) signal paths. It is recommended to use, if at all possible, phono connectors and relays with gold-plated contacts.

Relay control

The relays on the busboard are controlled by a number of driver stages on the power supply board, which have been designed to ensure almost noiseless switching operations. The circuit diagram of the relay control section is shown in Fig. 2.5.

When the supply is first switched on, the output relay is energized after a slight delay; when the supply is switched off, the output relay is deactuated immediately, however.

Source selection. When either the input selector or the tape monitor switch is operated, the output relay is deenergized before the relevant change is effected, and reactuated only after a new input or tape position has been selected.

The non-used contacts of the source selector, S_2, are logic high via resistors R_{16}–R_{19}, while the selected input contact is logic low via the switch wiper. The switching arrangement is passed on to inputs A_0–A_3 of comparator IC_6, wherc it is compared with the situation at pins B_0–B_3. Because of the time delay introduced by R_{25}-C_{30}, R_{26}-C_{31}, R_{27}-C_{32} and R_{28}-C_{33} respectively, the two compared quantities will differ by a few microseconds. This

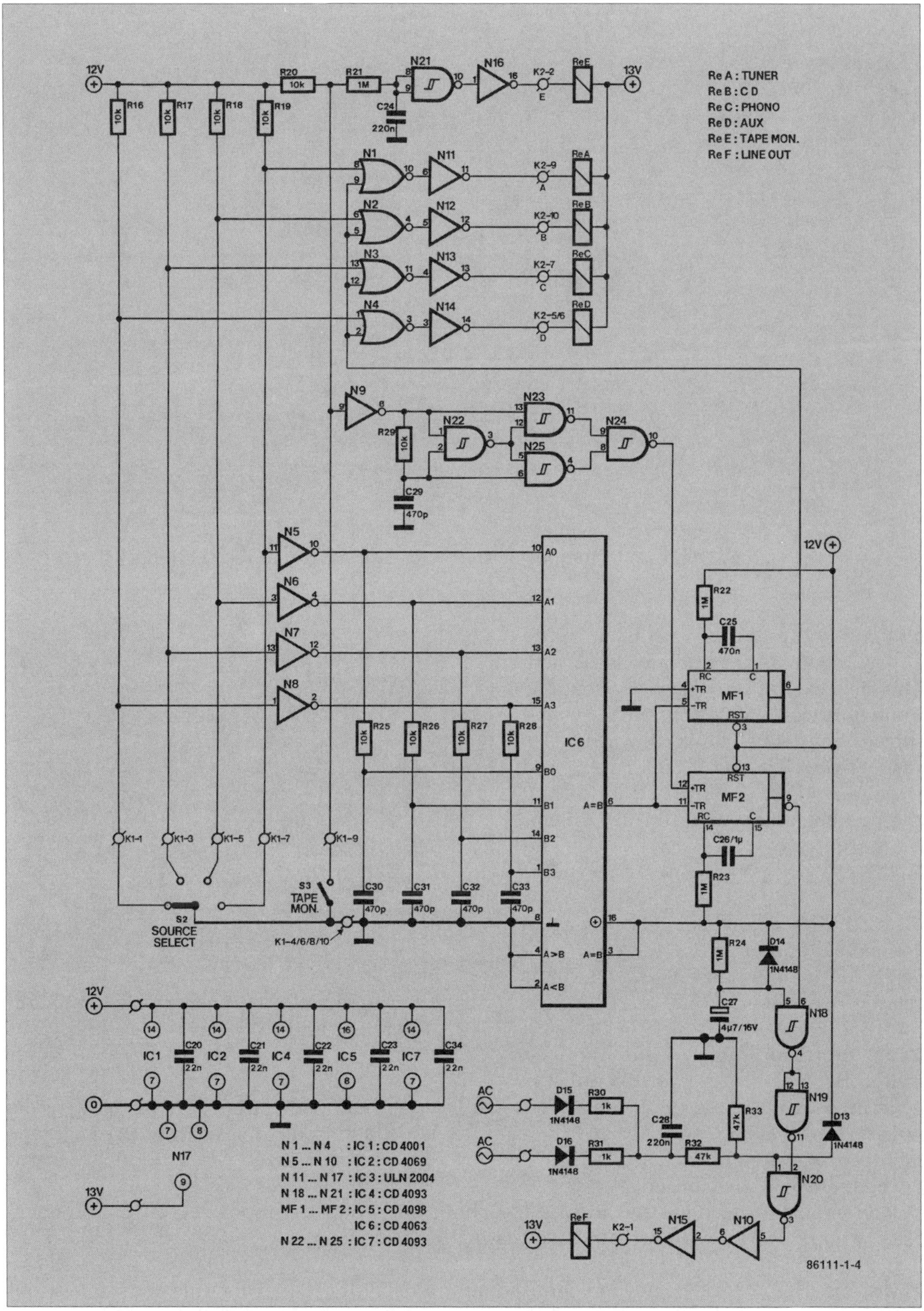

Fig. 2.5. Diagram of the relay control circuits.

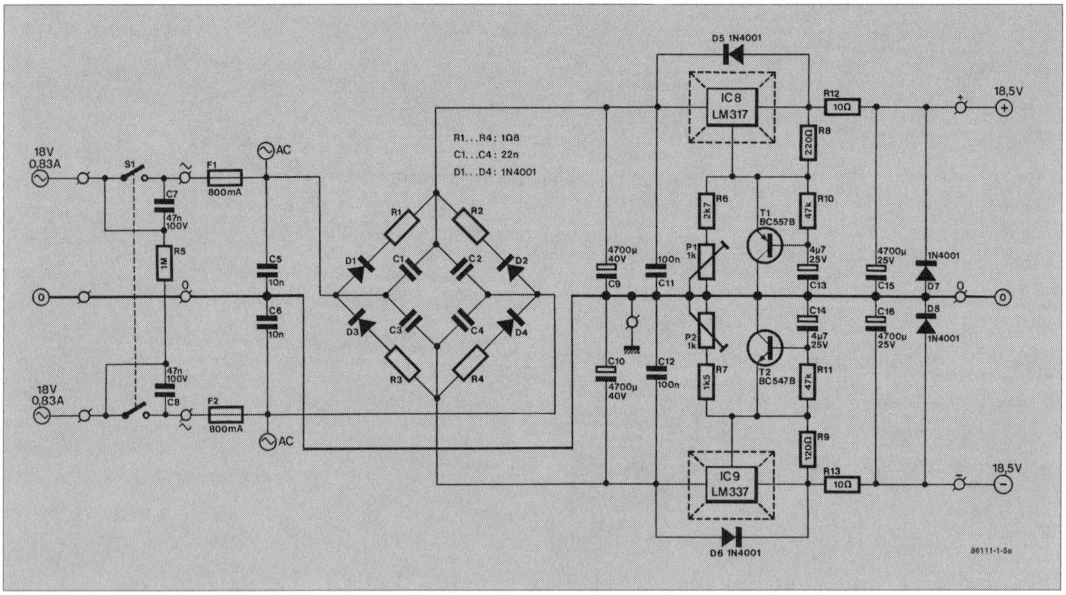

Fig. 2.6. Diagram of the power supply circuit for the audio sections.

will cause the output, pin 6, of IC_6 to go low momentarily when S_2 is turned. This negative pulse triggers monostable multivibrators MMV_1 and MMV_2, which introduce delays of 0.5 s and 1 s respectively. If both are triggered simultaneously, the selected input and the line out relay, Re_F, are disconnected instantly by N_1, N_2, N_3 or N_4 and N_{18}-N_{19}-N_{20}-N_{10}-N_{15} respectively. After the delay provided by MMV_1 has elapsed, the newly selected input is connected, and after the delay generated by MMV_2 has passed, the LINE OUT relay is reenergized.

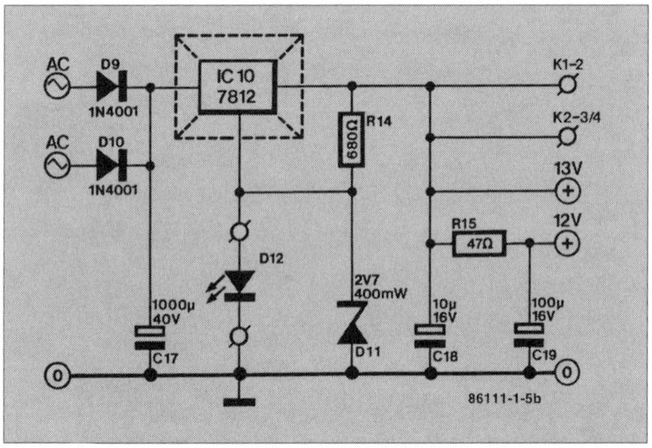

Fig. 2.7. Diagram of the power supply circuit for the relays and relay control circuits.

Tape monitor. When tape monitor switch S_3 is closed, a positive pulse is generated with the aid of N_9, delay network R_{29}-C_{29}, and XOR gate N_{22}-N_{23}-N_{24}-N_{25}. This pulse triggers MMV_2, so that LIN OUT relay Re_F is deenergized. After a delay R_{21}-C_{24}, tape monitor relay Re_E is energized. The LINE OUT relay is reenergized when the delay introduced by MMV_2 has elapsed.

It is important to note that during the above operation the input relays remain energized, and the connection with TAPE OUT is not broken.

Power on. The LINE OUT relay is energized after a delay R_{24}-C_{27}. This time constant is just a little longer than the time required by the power supply to attain full output. Diode D_{14} ensures that C_{27} is discharged rapidly when the power is switched off.

Power failure. The secondary voltage of the mains transformer is rectified by D_{15} and D_{18} and smoothed to some extent by C_{28}. It is then roughly halved by voltage divider R_{32}-R_{33} to provide a suitable input (<12 V) to pin 1 of N_{20}. Diode D_{13} affords protection against noise peaks. Because of the very short time constant R_{32}-R_{33}-C_{28} (about 20 ms), the LINE OUT relay is deenergized the instant the power is switched off or fails.

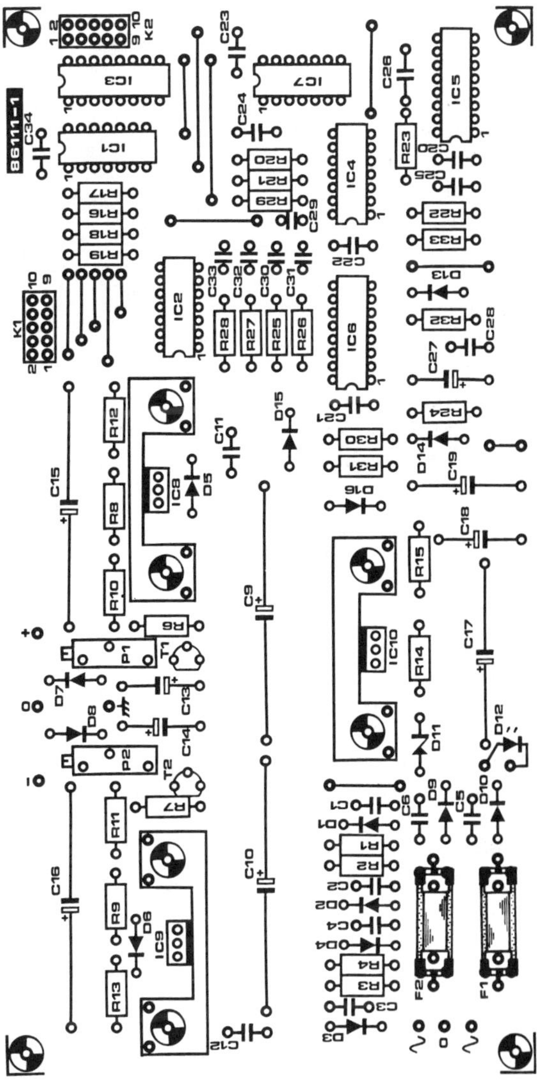

Fig. 2.8. Printed-circuit board (component layout) for the power supplies and relay control circuits. (Reduced to 70.7%: A3→A4)
Track side on page 219.

Power supply

The power supply is rather more extensive than is usual with this type of equipment; this is because of the requirement for different voltages for th audio sections, the relays, and the relay control.

The supply for the audio sections—see Fig. 2.6—provides a symmetrical voltage of ±18.5 V. Everything feasible has been done to reduce hum and other noise to a minimum, and the circuit therefore contains components not often found in power supplies.

The mains transformer has two secondary windings, each providing 18 V at 1 A. It is not housed in the preamplifier enclosure, but in a separate case; this reduces hum in the preamplifier to an absolute minimum.

The mains on/off switch, S_1, is debounced by C_7 and C_8. Noise peaks are shorted to ground by C_5 and C_6.

Resistors R_1–R_4 in series with rectifiers D_1–D_4 limit current peaks at switch-on. Capacitors C_1–C_4 effectively suppress the internal noise of the rectifiers.

Reservoir capacitors C_9 and C_{10} are shunted by foil capacitors C_{11} and C_{12} to improve the suppression of r.f. noise.

Voltage regulators IC_8 and IC_9 stabilize the ±18 V lines. Their action is enhanced by transistors T_1 and T_2, which act as variable zener diodes. Presets P_1 and P_2 enable the output voltage to be set to the exact level.

Networks R_{12}-C_{15} and R_{13}-C_{16} are low-pass filters with a very low cut-off frequency which ensure the almost complete elimination of any noise from the supply lines.

The supply for the relays and the relay control circuits—see Fig. 2.7—is fairly simple. The output voltage of regulator IC_{10} is increased slightly by connecting the ground pin to earth via diode D_{12}. This LED also functions as the on/off indicator. Zener diode D_{11} provides a safety precaution which ensures correct operation if for some reason the LED breaks down.

The components layout of the printed-circuit board for the power supply and he relay control circuits is shown in Fig. 2.8 (see Appendix for track layout). Note that voltage regulators IC_8, IC_9 and IC_{10} must be mounted on a suitable heat sink.

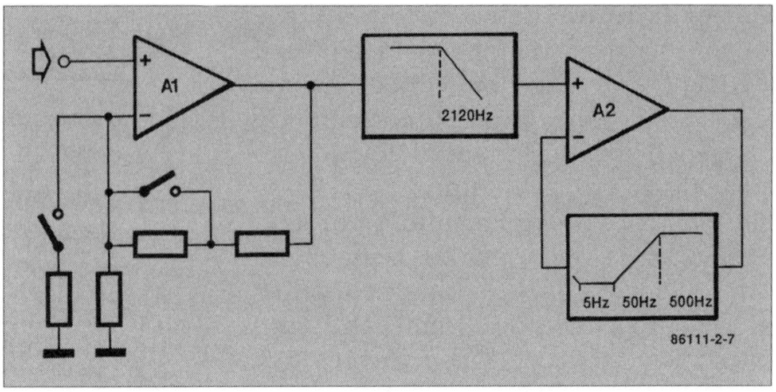

Fig. 2.8. The dual nature of the phono amplifier becomes clear from this block diagram.

Design considerations

As stated earlier, the phono section (if used) of the preamplifier can work from either a magneto-dynamic (MD) or moving-coil (MC) cartridge. The main difference between these two types is the level of output voltage they provide: that of an MC type is 100–400 μV and that of an MD type is 2–5 mV. In the design of an RIAA (IEC) equalization circuit suitable for use with both types of cartridge, there are two choices: one in which one or more stages of amplification can be switched on or off depending on which type of cartridge is used, or one with variable (or switch-selected) amplification. The second choice has been adopted in the present preamplifier.

This choice makes heavy demands on the input stage because it is not easy to achieve a good signal-to-noise ratio with MC cartridges owing to the combination of low output voltage and small hum resistance of these elements. At very low hum impedances (a few ohms), the noise of the input stage tends to drown the signal. Fortunately, careful design and the use of specially selected components can reduce the noise factor of the input stage to a very low value.

Another requirement of a universal input stage is that its input capacitance and resistance can be varied: some MC cartridges must be terminated into 47 Ω, others into 100 Ω, whereas MD cartridges need a much higher terminating resistance of around 47 kΩ. The input capacitance is particularly important when MD cartridges are used, because it affects the frequency response of these elements in the 10–20 kHz range.

The RIAA (IEC) deemphasis characteristic is obtained by a well-tried combination of a passive low-pass filter and an active low-frequency correction section as shown schematically in Fig. 2.9. Input stage A_1 raises the level of the signal from the cartridge: its amplification is matched to the type of cartridge and the input voltage by switched resistors.

The output of A_1 is passed through a passive low-pass section which has a cut-off frequency of 2120 Hz. The signal is then applied to amplifier A_2, the negative feedback loop of which contains a low-frequency equalization section with cut-off frequencies of 500 Hz, 50 Hz and 5 Hz.

Circuit of phone amplifier

In Fig. 2.10, the input capacitance and resistance are selected by DIL switch S_2. Capacitors C_5, C_6 and C_7 are needed to prevent any d.c. from reaching the sensitive MC cartridge. All capacitances in the signal path are formed by parallel combinations of a polypropylene and a polyester capacitor; this will be reverted to later.

The input amplifier consists of three dual transistors Type MAT-02. These are low-noise, carefully matched devices with a low offset voltage drift. Current source T_4 provides the d.c. bias for the transistors. The voltage drop across an LED is used as the reference voltage. The three transistors are connected in parallel because the base resistance of the input transistor generates most of the thermal noise when the signal source has a low output impedance: the thermal noise here is, therefore, reduced by nearly 67%. Another

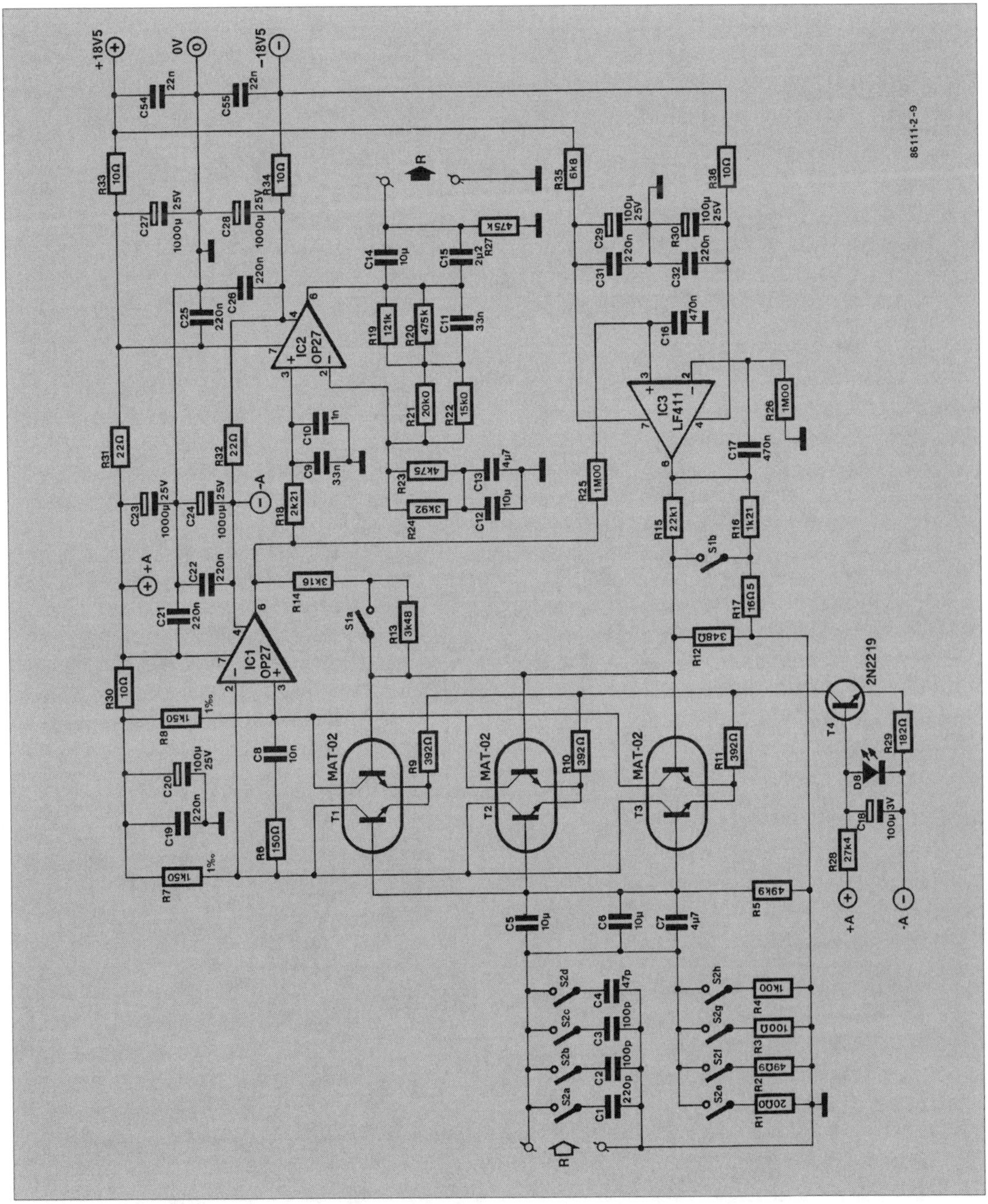

Fig. 2.10. Circuit diagram of the phono amplifier.

type of noise, the Schottky noise, is determined largely by the collector current of the input transistor. Generally, the Schottky noise diminishes when the collector current increases—up to a limit. The collector current here is set at 1 mA per transistor; ideally, it should have been 3 mA (according to the manufacturers), but this value would give difficulties with the active offset control, which will be discussed later. A value of 1 mA is a good compromise resulting in excellent signal-to-noise ratios.

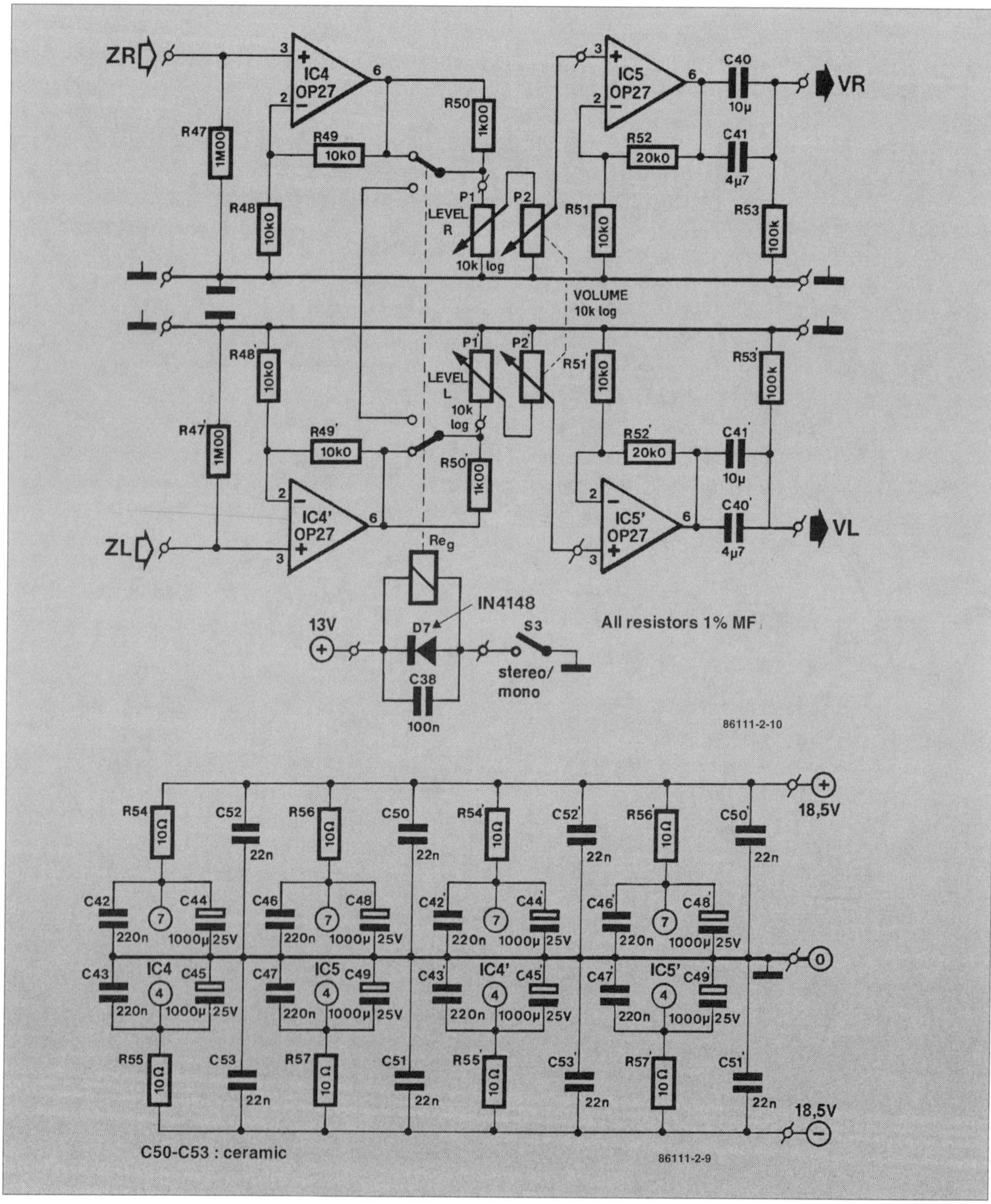

Fig. 2.11. Circuit diagram of the line amplifier.

Fig. 2.12. Close-up of the three supply rails on the mother board.

The value of input capacitors C_5–C_7 also has an effect on the amount of noise. The total value should be of the order of 100–200 µF for a negligible contribution to the noise at the lowest frequencies. Since electrolytic capacitors can not be used in the signal path (because of their poor performance), an empiric compromise was found between the dimensions of the capacitors and their noise contribution.

All the measures to reduce the noise to an absolute minimum are for the benefit of MC cartridge users; if only MD pick-ups are used, one MAT-02 is sufficient. The collector current of this single transistor can be reduced to, say, 500–600 µA by increasing the value of R_{29}.

The input stage forms one half of a differential amplifier; the other half is IC_1. This op amp raises the level of the difference signal at the collectors of the dual transistors.

Operational amplifier IC_1 is a high-quality device that is not cheap, but which gives excellent performance. The negative feedback loop of this stage, R_{12}-R_{13}-R_{14}-R_{17}, contains two switch sections, S_{1a} and S_{1b}, with which the input sensi-

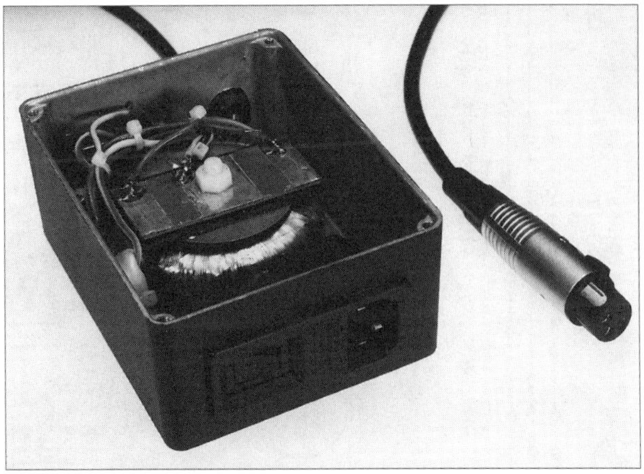

Fig. 2.13. The mains transformer must be housed in a separate metal enclosure.

tivity can be set to 0.1 mV, 0.2 mV, 2 mV and 4 mV. This arrangement enables optimum matching of the dynamic range and the signal-to-noise ratio to the output signal of the cartridge. Note the low values of R_{12} and R_{17} which enable the noise at the inverting input

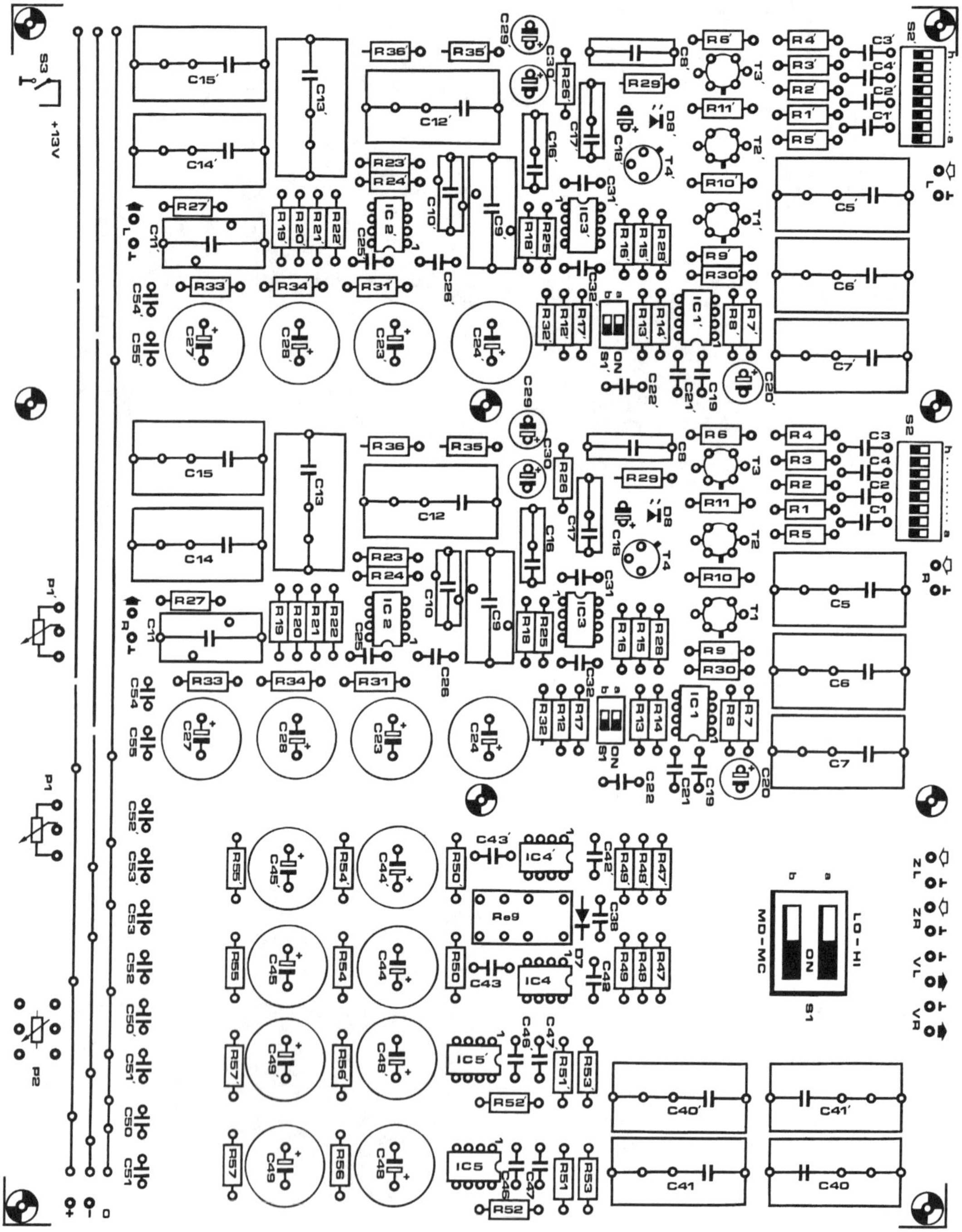

Fig. 2.14. Component layout of the motherboard (reduced to 70.7%: A3→A4).
For track side and overlay, see pages 221, 223.

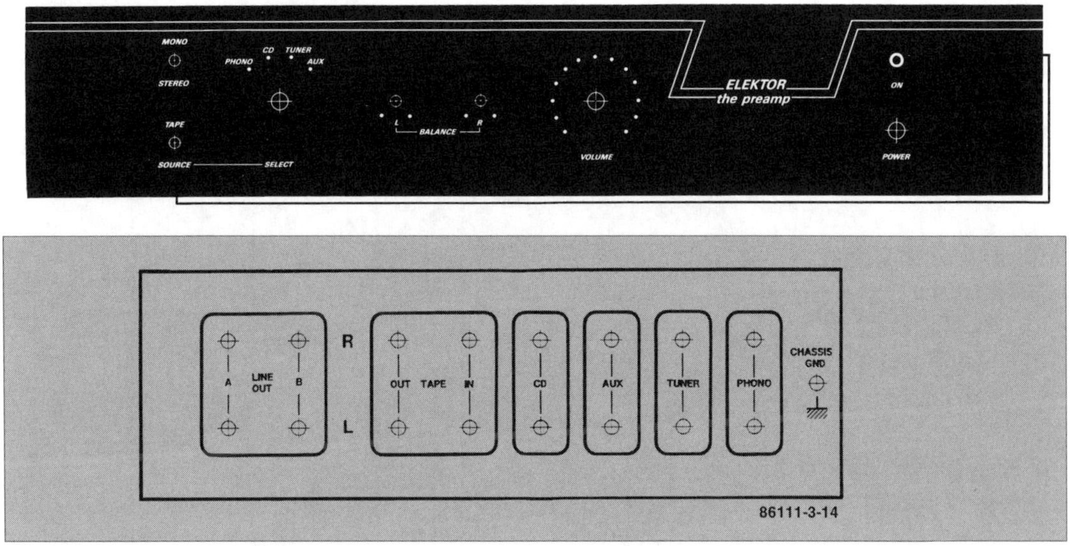

Fig. 2.15. Suggested front and rear panel layouts (reduced to 75% of true size).

of IC_1 to be kept to a minimum.

The difference in value between R_5 and R_{12}-R_{17} results in a relatively large, unwanted offset voltage. This is particularly troublesome when an MC cartridge is used, because the gain of the stage is then 40–46 dB. This problem is resolved by integrator IC_3, which provides active offset correction. The output of IC_1 is first taken through a low-pass filter, R_{25}-C_{16}, which has a cut-off frequency of 0.3 Hz, and then integrated by IC_3. The d.c. level at the bases of the dual transistors is set by IC_3 via R_{15} or R_{16}, depending on the position of S_{1b}, to a value which results in zero output from IC_1.

Since the supply voltage to a Type LF411 should not exceed 30 V, R_{35} has been inserted into the positive supply line to reduce the supply to IC_3 to about +10 V. This creates no problems, since the output of the op amp is always negative when the offset is being adjusted: the output current then flows via the negative supply line.

The current required from IC_3 is fairly large, mainly because of the low value of R_{17} (even a small potential difference across this resistor requires a fairly large current). The output current must be 6–8 mA to keep the output voltage at zero. This explains why the collector current of the dual transistors is arranged at 1 mA: higher values would necessitate an even larger current through R_{12}-R_{17}. Increasing the values of the feedback resistors would result in increased noise in the input stage.

The passive part of the deemphasis circuit is formed by R_{18}-C_9-C_{10}; the capacitors are 1% polystyrene types. The output of IC_1 is fed to the non-inverting input of a second Type OP-27 op amp. The negative feedback loop of this stage contains the low-frequency correction section of the deemphasis circuit. All resistances in the loop are formed by two 1% resistors in parallel; strictly speaking, this is not necessary, but it is done to enable constructors to make up the exact values with other combinations of resistors.

Capacitors C_{12} and C_{13} limit the d.c. amplification of the op amp to unity. Automatic offset correction in this stage was decided against because of the requirement to suppress low-frequency (<5 Hz) components. Although the nominal gain of IC_2 is only 14 dB, the frequency-selective networks cause an additional gain of 20 dB for signals below 50 Hz.

Strictly speaking, the coupling capacitors in the output circuit of IC_2 are not necessary, because the automatic offset correction at the input stages works so well that there is no

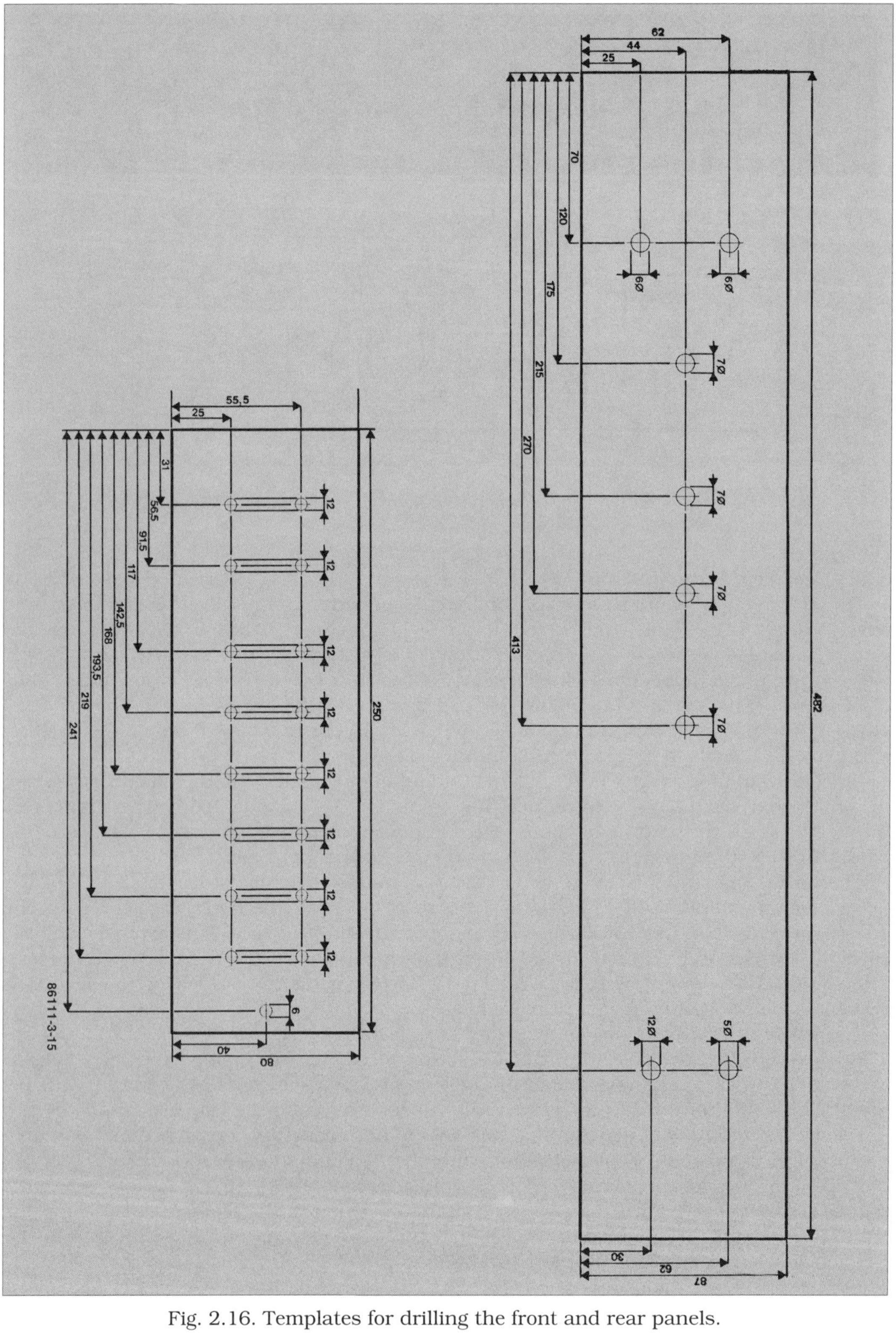

Fig. 2.16. Templates for drilling the front and rear panels.

discernible d.c. at the output of IC_2.

The supply lines to the different stages are decoupled separately. The relevant network consists in each case of two 1000 μF electrolytic capacitors, each shunted by a 200 nF ceramic capacitor for better high-frequency performance. Each electrolytic capacitor is connected in series with a low-value resistor to further improve the decoupling.

Line amplifier

The quality of the line amplifier is particularly important for the faithful reproduction of compact discs, when a good dynamic range, large bandwidth and minimal distortion are essential.

The dynamic range is determined by the maximum supply voltage to the ICs, which has been made as high as feasible at ±18.5 V. This value makes possible an undistorted output voltage of about 12 V, which is ten times as high as the nominal output level of 1.2 V (to give about 20 dB headroom).

Since the noise produced by op amps is very low, and the gain of the devices is only about 14 dB, the signal-to-noise ratio is of the order of 100 dB. With the headroom of about 20 dB, this gives a total dynamic range of around 120 dB.

As stated earlier, the voltage dividers at the inputs merely serve to reduce crosstalk and hardly attenuate the wanted signal. Only the CD input is provided with an attenuator to lower the input signal by about 6 dB. The reason for this is that the majority of CD players provide a fairly high output signal—of the order of 1 V. The closer the CD player output is to the nominal sensitivity of the preamplifier, the smaller the likelihood of overload during peak levels. The signal-to-noise ratio is not affected by the attenuator (bear in mind that the output level control on th CD player performs the same function).

The circuit of the line amplifier—see Fig. 2.11—consists of two OP-27 op amps per channel. The two-stage arrangement has the benefit of a greater dynamic drive range, because, since the volume control is normally nowhere near fully open, and, like the balance control, is fitted between the two amplifiers, the first op amp can deliver a greater undistorted signal without overloading the second op amp. This set-up has the further advantage that both controls are isolated from the inputs and the output.

The first stage, IC_4, has a gain of 6 dB. Its output is connected to the second amplifier via a mono/stereo selector (and, as stated earlier, the volume and balance controls). The selector is actually a relay contact to avoid long signal paths to a conventional switch. Resistor R_{50} ensures that the op amps do not short-circuit each other's output in the mono mode.

Each channel has its own individual (mono) balance control, P_1' (P_1'), but the volume control, P_2 is of the customary stereo type. The individual balance controls enable the output signal to be set at maximum level. Moreover, good-quality stereo balance controls are difficult to obtain.

The second stage, IC_5, has a gain of about 10 dB, resulting in the line amplifier delivering an output signal of 1.2 V for an input of 200 mV. The output is provided with d.c. blocking capacitors; again, these are not strictly necessary, since the op amps have no offset problems. But it could just happen that one of the signal sources delivers a d.c. component and this would be amplified together with the signal.

The output terminals are connected to the op amps via relay contacts; the relay action is delayed at switch-on, but is immediate on switch-off. The output terminals are also disconnected briefly when the inputs are switched to prevent annoying clicks and plops. As in the phono amplifier, the supply lines are thoroughly decoupled. Each op amp is supplied via a separate 10 Ω resistor, and individually decoupled by two 1000 μF electrolytic capacitors, each of which is shunted by a 200 nF ceramic capacitor for improved performance.

The earth tracks of the two channels on the printed-circuit board are kept separate and combined only at the main earth rail. This arrangement further improves the channel separation figures.

Choice of components

It is important to use only high-quality components to ensure optimum performance. All resistors should be metal-film types with a tolerance of 1%, although that of R_7 and R_8 should preferably be 0.1%. If these can not be obtained, select a pair of 1% resistors that are identical in value, or very nearly so, with the aid of a digital multimeter.

All op amps are Type OP-27, while the dual transistors at Type MAT-02. Do not use Type OP-37s in the line amplifier because this type has offset compensation only for gains greater than 14 dB.

All capacitors in the signal paths are metallized polypropylene types. Frequency-determining capacitors in the RIAA (IEC) compensation section are 1% polystyrene types. Electrolytic capacitors in the power supply are all types for PCB mounting. Decoupling capacitors shunting electrolytic capacitors may be polypropylene or ceramic types.

It is advisable to use silver- or gold-plated phono input sockets: these guarantee freedom from oxidation and consequent contact potentials between plug and socket.

The relays on the bus board must, of course, be of prime quality. The components lists shows four possible types. The excellent SDS type is, unfortunately, polarized and its coil connections are exactly the reverse of the others; if this type is used, its coil connections must, therefore, be reversed.

The volume control potentiometers must be of the highest quality; in the prototype, a stereo version from Alps was used with excellent results.

The balance potentiometers are rather less critical, but should still be of very good quality; they should definitely not be carbon type, but conductive plastic or cermet. Bourns or Spectrol types are recommended.

The switches are not critical components, since they only switch direct voltages to the relays.

A few tips to bring the total cost down somewhat. The OP-27 may be replaced by a 5534, which is a lot cheaper and still a good-quality device, but it may give offset problems. The MAT-02 may be replaced by an LM394, but the overall quality will come down slightly. In this context, if it is unlikely that MC pick-ups will be used, only one MAT-02 is required per channel as explained earlier.

Construction

The mains transformer, which can be a laminated or a toroidal type, must be mounted in a separate, aluminium case—see Fig. 2.13. The mains cable should enter the case at one end, while the three-core cable carrying the secondary voltage should emerge from the other end. The three-core cable must be terminated into a suitable socket that mates with a corresponding three-pin plug at the rear of the preamplifier enclosure. This arrangement is essential to keep any hum away from the preamplifier circuits.

Next, the power supply board should be completed. Fit the voltage regulators on adequate heat sinks, which can be secured to the board with self-tapping screws. When the board is finished, mount it at the right-hand side of the enclosure. Do not forget a screen between it and the mother board. The secondary voltage from the mains transformer is taken to the board via the double-pole mains on/off switch. At this stage, he mains on/off indicator, D_{12}, should be connected to the supply board.

The earth connection on the supply board is then connected to the enclosure via a short length of heavy-duty cable.

The supply can then be switched on to test whether the direct voltages are present; if so, they should be set to ±18.5 V with presets P_1 and P_2.

The busboard can be completed fairly quickly. First, screw all the phono sockets to the

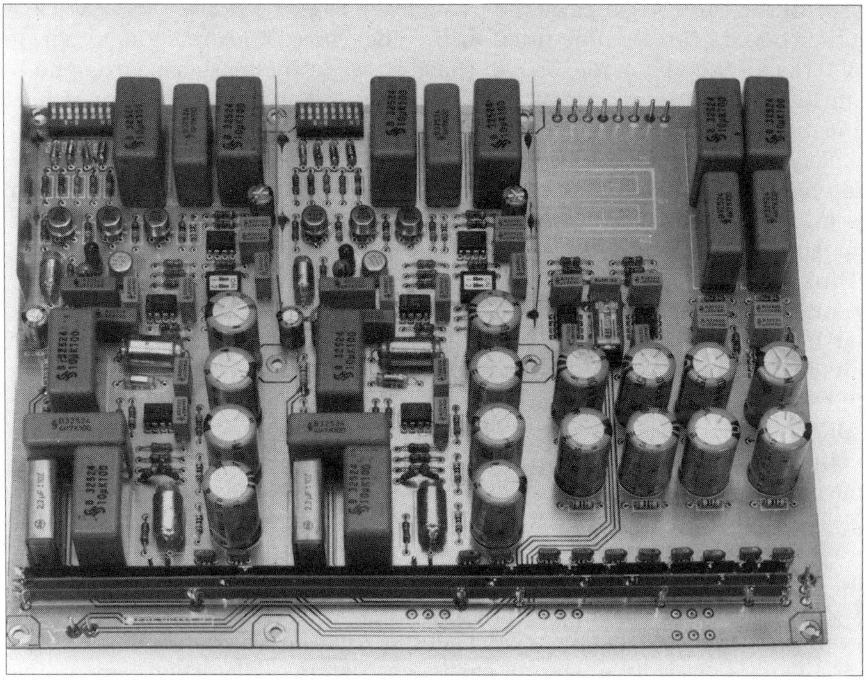

Fig. 2.17. Completed prototype board.

board (inputs at the track side). Tighten them by hand and then solder them lightly to the board: this prevents them coming loose when later the corresponding plugs are withdrawn and plugged in again. Then tighten the socket nuts with a suitable spanner; do not over-tighten them. After that, all other components, including the relays, can be fitted on to the board. Some resistors are soldered directly to the centre terminal of a socket.

The connections between socket and board at the tape and line outputs are made with short lengths of flexible, insulated circuit wire. The remaining connections are provided with soldering pins to make them easily accessible during the remainder of the work.

Remove any resin from the board with a brush dipped into white spirit or alcohol, and then seal the track side with a suitable plastic spray. Take care that no spray gets into the sockets or relays. This cleaning and insulating of the board reduces the risk of crosstalk to a minimum.

The board is then mounted to the rear panel of the enclosure with the aid of insulated spacers: this prevents any possibility of the tracks or sockets touching the enclosure.

The earth connection adjacent to the sockets must be connected securely to the enclosure to become the case earth: this same point should be connected to the central earth point on the supply board via a short length of heavy-duty cable.

The mother boards should be completed in the following order: resistors, capacitors, mechanical parts, semiconductors and integrated circuits—do not use sockets for the ICs.

Three supply rails have to be provided at the front of the boards. To do this, first fit soldering pins in all the holes; then cut narrow strips of brass or tin sheet, and solder these to the pins a few millimetres above the board—see Fig. 2.12 and 2.17.

Next, fit soldering pins to all interconnecting points. At this stage, solder only the pin for the earth connection to the supply board to the screen at the top of the mother boards. Finally clean the boards with white spirit or alcohol and insulated their track side with plastic spray in the same manner as the busboards.

The mother boards can then be fitted in the enclosure. All connections to switches and potentiometers can then be made, as can those between the mother boards and busboards (at the right of the mother boards at the line section). Screened cable is not necessary as these connections are only a few centimetres long.

Next, make the connections between the supply board and mother boards.

The switching connections to the busboard may be made from flatcable terminated at both ends into a plug to mate with he corresponding sockets on the boards. It should be noted that socket K_1 should be fitted on the busboard 180° different from shown in Fig. 2.3. Thus, pin 1 should be located where pin 10 is shown.

Finally, make the connections between the MC/MD sockets and the associated inputs at the mother board, and those between the MC/MD amplifier output on the mother board and the busboard. These should be made from good-quality screened audio cable or flexible coaxial cable (TV type).

When all connections are made and checked, the mains may be switched on. Adjust P_1 and P_2 to obtain exactly ±18.5 V on the supply rails on the mother board.

Next, measure the direct voltage at pin 6 of IC_3; this should not be more negative than −14 V; if it is, lower the value of R_{15} until the correct level is obtained. This voltage depends to a large extent on what make of input transistors is used; normally, R_{15} need not be altered from its specified value. As a safety check, measure the direct voltage at pin 6 of IC_2; this should be not more than 5 mV, and preferably 0 V.

The preamplifier should easily meet the specifications given below: the prototypes exceeded the figures given in almost all cases; for instance, distortion measurements gave values of only about half the figures stated.

Technical specification

Input sensitivity

Phono:	MC (low)	0.1 mV into 47 kΩ
	MC (high)	0.2 mV into 47 kΩ
	MD (low)	2 mV into 47 kΩ
	MD (high)	4 mV into 47 kΩ
Tape, tuner, aux.		200 mV into 45 kΩ
CD		400 mV into 20 kΩ

Maximum input voltage at 1 kHz

Input - line out

Phono	MC (low)	1 mV
	MC (high)	2 mV
	MD (low)	20 mV
	MD (high)	40 mV
Tape, tuner, aux		2 V
CD		4 V

Input - tape out

Phono	MC (low)	6 mV
	MC (high)	12 mV
	MD (low)	120 mV
	MD (high)	240 mV

RIAA (IEC) correction

±0.2 dB over the frequency range of 20 Hz to 20 kHz. Standard input impedance: 47 kΩ; standard input capacitance: 50 pF. Values can be preset from 10 Ω to

47 kΩ and from 50 pF to 500 pF

Output (line out)

Nominal output voltage	1.2 V
Maximum output voltage	10 V
Output impedance	<100 Ω
Maximum output current	20 mA

Third harmonic distortion (at 1 kHz)

Output voltage		100 mV	1.2 V	10 V
Phono	MC (low)	<0.1%	<0.01%	<0.02%
	MC (high)	<0.05%	<0.01%	<0.02%
	MD (low)	<0.01%	<0.005%	<0.02%
	MD (high)	<0.01%	<0.005%	<0.02%
Tape, tuner, aux		<0.005%	<0.005%	<0.02%
CD		<0.005%	<0.005%	<0.02%

Over range 20 Hz to 20 kHz and output voltage of 1.2 V

Phono	MC	<0.02%
	MD	<0.01%
Tape, tuner, aux.		<0.008%
CD		<0.008%

Intermodulation distortion (60 Hz, 7 kHz, 4:1, SMPTE)

Tape, tuner, aux., CD	<0.003%

Signal-to-noise ratio (inputs short-circuited; output 1.2 V)

Phono	MC (low)	>70 dB
	MC (high)	>76 dB
	MD (low)	>86 dB
	MD (high)	>92 dB
Tape, tuner, aux		>105 dB
CD		>105 dB

Line amplifier (terminated into 47 kΩ)

Frequency range		10 Hz to 50 kHz (±0.1 dB)
		1.5 Hz to 500 kHz (–3 dB)
Phase shift		<±0.5° (15 Hz to 120 kHz)
Crosstalk (at 10 kHz)		
	line inputs (L↔R)	<–70 dB
	L/R to other inputs	<–80 dB
Slew rate		> 4 V μs^{-1}

Parts list

POWER SUPPLY BOARD

Resistors (all metal film):

R_1–R_4 = 1.8 Ω

R_5, R_{21}–R_{24} = 1 MΩ

R_6 = 2.7 kΩ

R_7 = 1.5 kΩ

R_8 = 220 Ω

R_9 = 120 Ω

R_{10}, R_{11}, R_{32}, R_{33} = 47 kΩ
R_{12}, R_{13} = 10 Ω
R_{14} = 680 Ω
R_{15} = 47 Ω
R_{16}–R_{20}, R_{25}–R_{29} = 10 kΩ
R_{30}, R_{31} = 1 kΩ
P_1, P_2 = 1 kΩ preset

Capacitors:
C_1–C_4 = 22 nF, 250 V, polypropylene
C_5, C_6 = 10 nF, 250 V, polypropylene
C_7, C_8 = 47 nF, 250 V, polypropylene
C_9, C_{10} = 4700 µF, 40 V, electrolytic
C_{11}, C_{12} = 100 nF
C_{13}, C_{14} = 4.7 µF, 25 V, electrolytic
C_{15}, C_{16} = 4700 µF, 25 V, electrolytic
C_{17} = 1000 µF, 40 V, electrolytic
C_{18} = 10 µF, 16 V, electrolytic
C_{19} = 100 µF, 16 V, electrolytic
C_{20}–C_{23}, C_{34} = 22 nF
C_{24}, C_{28} = 220 nF
C_{25}, C_{29}, C_{30}–C_{33} = 470 pF
C_{26} = 1 µF
C_{27} = 4.7 µF, 16 V, electrolytic

Semiconductors:
D_1–D_{10} = 1N4001
D_{11} = zener, 2.7 V, 400 mW
D_{12} = LED, red
D_{13}–D_{16} = 1N4148

Integrated circuits:
IC_1 = CD4001
IC_2 = CD4069
IC_3 = ULN2004
IC_4, IC_7 = 4093
IC_5 = 4098
IC_6 = 4063
IC_8 = LM317
IC_9 = LM337
IC_{10} = 7812

Miscellaneous:
S_1 = switch, 2 make contacts
S_2 = switch, rotary, 1-pole, 4-position
S_3 = switch, 1 make contact
F_1, F_2 = fuse, 800 mA, delayed action
2 off fuseholders
Mains transformer, 2×18 V, 0.83 A, e.g., ILP Type 11014*
3 off heat sinks, 6.8 K W^{-1}, e.g. Fischer SK59-37.5†
K_1, K_2 = 10-way PCB header

* Available from Jaytee Electronics Services, Telephone (01227) 375254
† Available from Dau (UK) Ltd; Telephone (01243) 553031

BUSBOARD (one channel)
Resistors (all 1% metal film):
R_{37}, R_{41}, R_{43} = 2.21 kΩ
R_{38}, R_{42}, R_{44} = 48.7 kΩ
R_{39} = 10.0 kΩ
R_{40} = 10.2 kΩ
R_{45} = 4.75 kΩ
R_{46} = 475 kΩ

Capacitors:
C_{33}–C_{37}, C_{39} = 100 nF, ceramic

Semiconductors:
D_1–D_6 = 1N4148

Relays:
Re_A–Re_F = 12 V sub-miniature relay, two-pole change-over for PCB mounting

Miscellaneous:
K_1 = 10-way PCB header
16 off screened phono sockets and mating plugs

MOTHER BOARD (one channel)
Resistors (all 1% metal film):
R_1 = 20 Ω
R_2 = 49.9 Ω
R_3 = 100 Ω
R_4, R_{50} = 1.0 kΩ
R_5 = 49.9 kΩ
R_6 = 150 Ω
R_7, R_8 = 1.5 kΩ
R_9–R_{11} = 392 Ω
R_{12} = 348 Ω
R_{13} = 3.48 kΩ
R_{14} = 3.16 kΩ
R_{15} = 22.1 kΩ
R_{16} = 1.21 kΩ
R_{17} = 16.5 Ω
R_{18}, R_{41}, R_{43} = 2.2 kΩ
R_{19} = 121 kΩ
R_{20} = 475 kΩ
R_{21}, R_{52} = 20 kΩ
R_{22} = 20 kΩ
R_{23} = 4.75 kΩ
R_{24} = 3.92 kΩ
R_{25}, R_{26}, R_{47} = 1 MΩ
R_{27} = 475 kΩ
R_{28} = 27.4 kΩ
R_{29} = 182 Ω

R_{30}, R_{33}, R_{34}, R_{36}, R_{54}–R_{57} = 10 Ω
R_{31}, R_{32} = 22 Ω
R_{35} = 6.8 kΩ
R_{48}, R_{49}, R_{51} = 10 kΩ
R_{53} = 100 kΩ
P_1 = 10 kΩ logarithmic potentiometer, e.g. Bourns Electronics
P_2 = 10 kΩ logarithmic stereo potentiometer, e.g. Alps RKGA-2 10 k AX2

Capacitors:
C_1 = 220 pF*
C_2, C_3 = 100 pF*
C_4 = 47 pF*
C_5, C_6, C_{12}, C_{14}, C_{40} = 10 µF†
C_7, C_{13}, C_{41} = 4.7 µF†
C_8 = 10 nF*
C_9, C_{11} = 330 nF*
C_{10} = 1 nF
C_{15} = 2.2 µF†
C_{16}, C_{17} = 470 nF†
C_{18} = 100 µF, 3 V, tantalum
C_{19}, C_{21}, C_{22}, C_{25}, C_{26}, C_{31}, C_{32}, C_{42}, C_{43}, C_{46}, C_{47} = 220 nF†
C_{20}, C_{29}, C_{30} = 100 µF, 25 V
C_{23}, C_{24}, C_{27}, C_{28}, C_{44}, C_{45}, C_{48}, C_{49} = 1000 µF, 25 V
C_{50}–C_{55} = 22 nF, ceramic
* polystyrene
† polypropylene

Semiconductors:
D_7 = 1N4148
D_8 = LED, red
T_1, T_2, T_3 = MAT-02
T_4 = 2N2291

Integrated circuits:
IC_1, IC_2, IC_4, IC_5 = OP-27
IC_3 = LF411

Miscellaneous:
S_1 = 2-pole DIP switch
S_2 = 8-pole DIP switch
S_3 = miniature SPST
Re_G = 12 V sub-miniature relay, 2-pole change-over, for PCB mounting, e.g. Siemens Type W11V23102-A006-A111*; Omron G2V-2; SDS DS2E-M-12
16 off silver- or gold-plated phono sockets for chassis mounting and mating plugs

* UK distributors: ElectroValue, telephone (01784) 33603 or (0161) 432 4945

Chapter 3

Preamplifier III

From many sources, it appears that not a few music lovers and audiophiles would like a preamplifier to have at least some of the following features: (a) a copying facility that is independent of the selector switch; (b) a facility to allow recording from one recorder to another; (c) a tone control with variable cut-off frequencies; (d) a headphone output with the facility that the main amplifier can be switched off when listening via the headphones only is required; (e) a dynamic pick-up input(a real surprise in these days of the compact disc). The preamplifier presented in this chapter meets all these requirements and still makes it possible for differently priced ICs to be used.

The source and record selector switches are housed on a busboard, together with the input and output sockets, while the remainder of the electronics is fitted on a motherboard. The block diagram in Fig. 3.1. shows that there are six inputs, including one for a

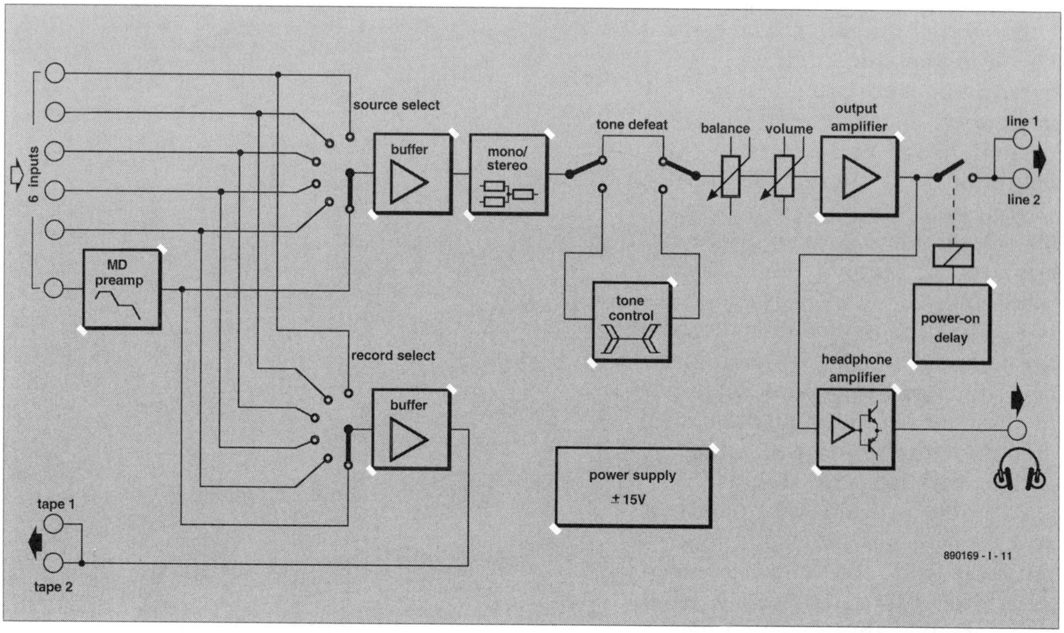

Fig. 3.1. Block diagram of Preamplifier III.

dynamic pick-up. If this is not needed, it may be transformed into a normal line input by a simple wire bridge. The CD input has provision for adding a board, such as a digital-to-analogue converter (DAC) at a later date.

The input signals are fed to two rotary switches: one for selecting the input to be processed and the other for selecting a signal to be recorded. In this way, it is, for instance, possible to record from a CD player and continue listening to the tuner. Standard rotary switches are used since these are much cheaper than the 12 relays that would otherwise be needed. The switches are located immediately adjacent to the inputs at the rear of the enclosure to avoid long signal paths.

Each switch is followed by a buffer, which in turn is followed by a stereo/mono selector which uses a summing op amp. This obviates the level differences that frequently occur in stereo/mono systems.

The tone control is a low-high design with two cut-off points at either end of the range. The control has been kept fairly limited to avoid overdriving subsequent stages; it is, however, more than adequate for normal usage and offers smooth operation. For those who do not want a tone control in any circumstances, it may be taken out of circuit by a simple 'tone defeat' switch.

The balance and volume controls are followed by the output amplifier, which provides the only amplification of the line signals. The op amp chosen for this stage is able to drive loads of ≥600 Ω.

A relay providing a delay at switch-on is provided at the output: this gives the unit a few seconds after the supply is switched on to stabilize during which period no signal is applied to the output.

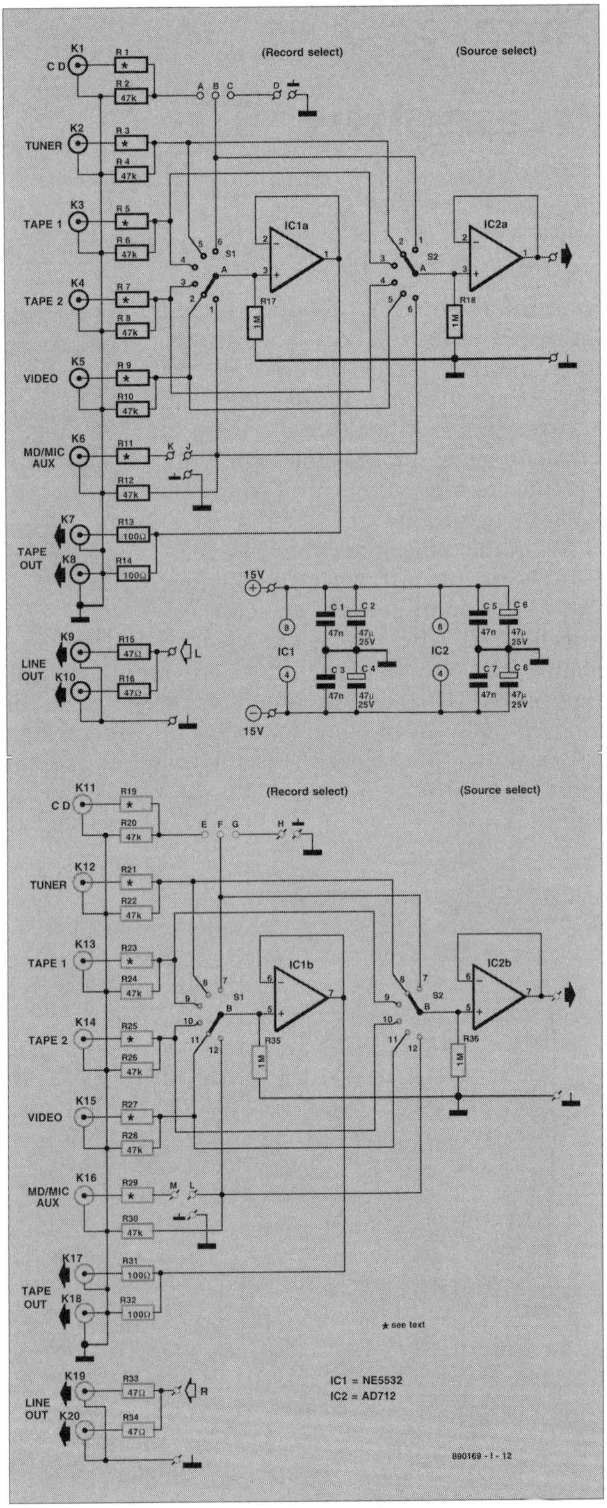

Fig. 3.2. Circuit diagram of the busboard.

The output amplifier also supplies part of the signal to a special headphone amplifier that consists of an op amp and discrete output stage. It provides enough power for driving low-resistance headphones.

Circuit description

The circuit diagrams in Fig. 3.2 (busboard) and 3.3 (motherboard) correspond closely to the drawings of the associated printed-circuit boards.

Figure. 3.2 shows the record and source selectors, S_1 and S_2 respectively, with associated buffers, IC_1 and IC_2, and all inputs and outputs, with the exception of that for the headphones. Each input is shunted by a potential divider, for instance, R_1-R_2 for the left-hand CD channel. These dividers largely determine the input impedance of 47 kΩ. It is advisable to use the dividers only if really necessary, since they may adversely affect the cross-talk between the left- and right-hand channels, as well as between the inputs. If dividers are not used, resistors R_1, R_3, R_5, and so on, should be replaced by wire bridges.

The CD input has some additional facilities. Normally, the analogue CD signals is applied directly to the CD input. A wire bridge between A and B (E and F) feeds the signal to the two rotary switches. Terminals C, D and ⊥ (G, H and ⊥) enable a digital input, connected to a separate digital-to-analogue amplifier, to be provided at a later stage.

The dynamic pick-up input also has additional facilities. Normally, the associate pre-amplifier on the motherboard is connected between K and J (left-hand channel) and between M and L (right-hand channel). If a dynamic pick-up input is not required, it may be converted to a microphone input or line input by linking J-K and M-L by wire bridges.

The op amps shown are not necessarily the least expensive types, but they are the best for the particular purpose as far as the prototype evidenced. This aspect will be reverted to later. For the moment, as an example, buffer IC_1 is a Type 5532, which is an inexpensive yet excellent double op amp, whereas buffer IC_2, a Type AD712, is a much dearer device. This type was found essential to obviate audible switching noises when S_2 is turned. These noises result from the change in total resistance at the input of the op amp: if, for instance, the CD input is used, R_{18} is shunted by R_2. If an op amp with transistor input, such as the NE5532, were used, the change in resistance would cause a corresponding change in the output of the op amp. An op amp with FET inputs, such as the AD712, hardly reacts to the change in resistance. If, notwithstanding this, a Type 5532 is used, R_{16} and R_{36} should be reduced to 220 kΩ. The input impedance will then drop to about 39 kΩ, and this means that the ratio of the potential divider will change slightly.

Apart from the inputs, the busboard also contains all the outputs: two tape outputs with an output impedance of 100 Ω, determined by R_{13}, R_{14}, R_{31} and R_{32}; and two line outputs, also with an output impedance of about 100 Ω, determined by R_{15}, R_{16}, R_{33} and R_{34}, and R_{55} and R_{80} on the motherboard.

Since the circuits of the left- and right-hand channels on the busboard in Fig. 3.3 are identical, they are described on the basis of that for the left-hand channel.

The amplifier for the dynamic pick-up is shown slightly away from the main circuit to emphasize its short connections to terminals J-K. The op amp chosen for this stage is a low-noise Type LT1028. If this is considered too expensive, a Type OP-27 or 5534 may be used. The RIAA (IEC) correction is provided by R_{81}–R_{83} and C_{58}–C_{61}. Network R_{12}-C_{62}-C_{63} forms a high-pass filter with a cut-off frequency of 20 Hz to conform to the RIAA requirement. If this frequency is considered too low, the values of C_{62} and C_{63} may be increased.

The input of the amplifier for the dynamic pick-up is direct-coupled to obviate any degradation of the signal by coupling capacitors. The input capacitance is determined primarily by C_{72}, whose value depends on what is needed by the pick-up. If this is not known, 47 pF may be assumed. Because of the direct coupling, the offset of the op amp depends on the internal resistance of the pick-up and P_6 is, therefore, provided to cancel the offset as appropriate.

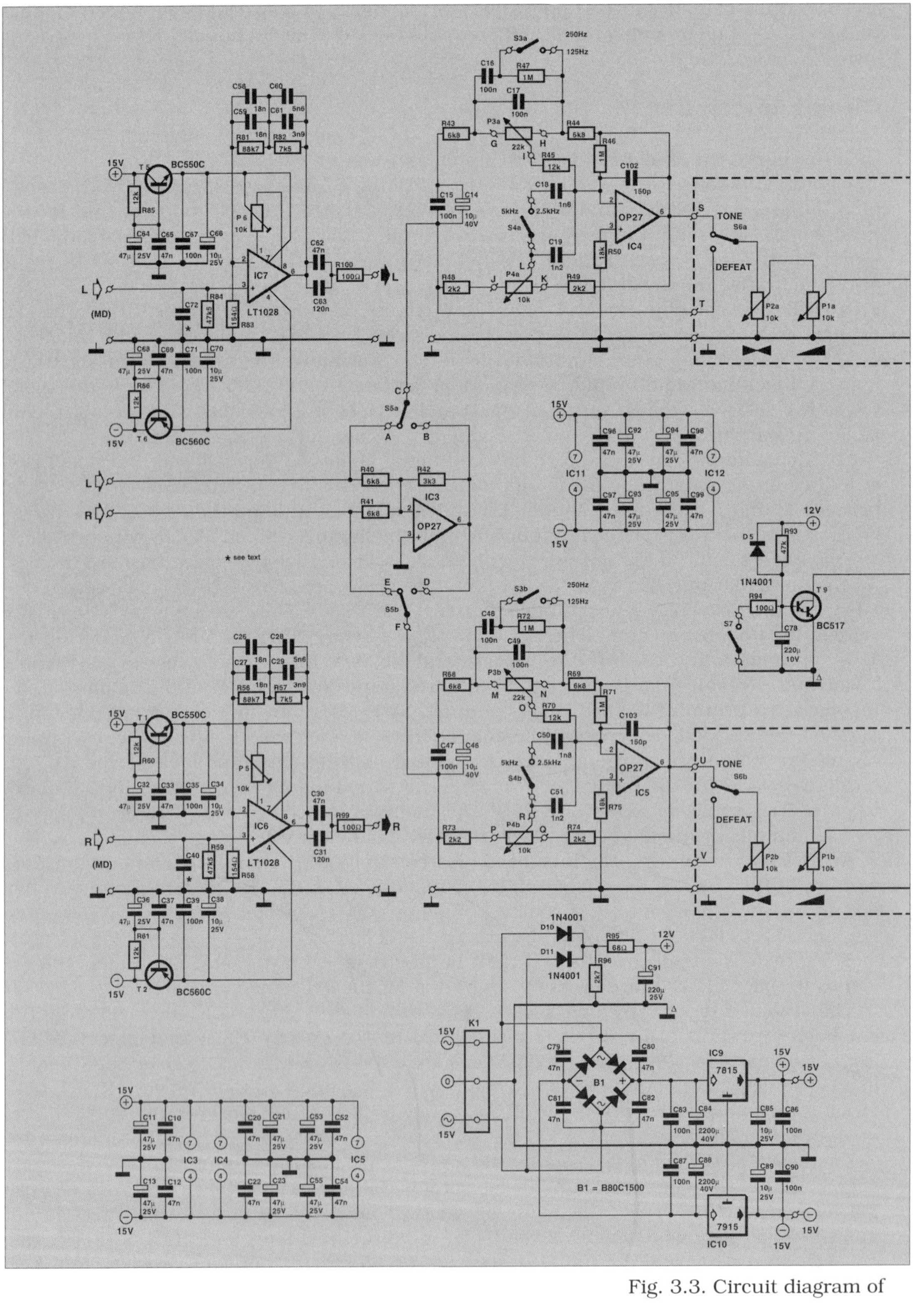

Fig. 3.3. Circuit diagram of

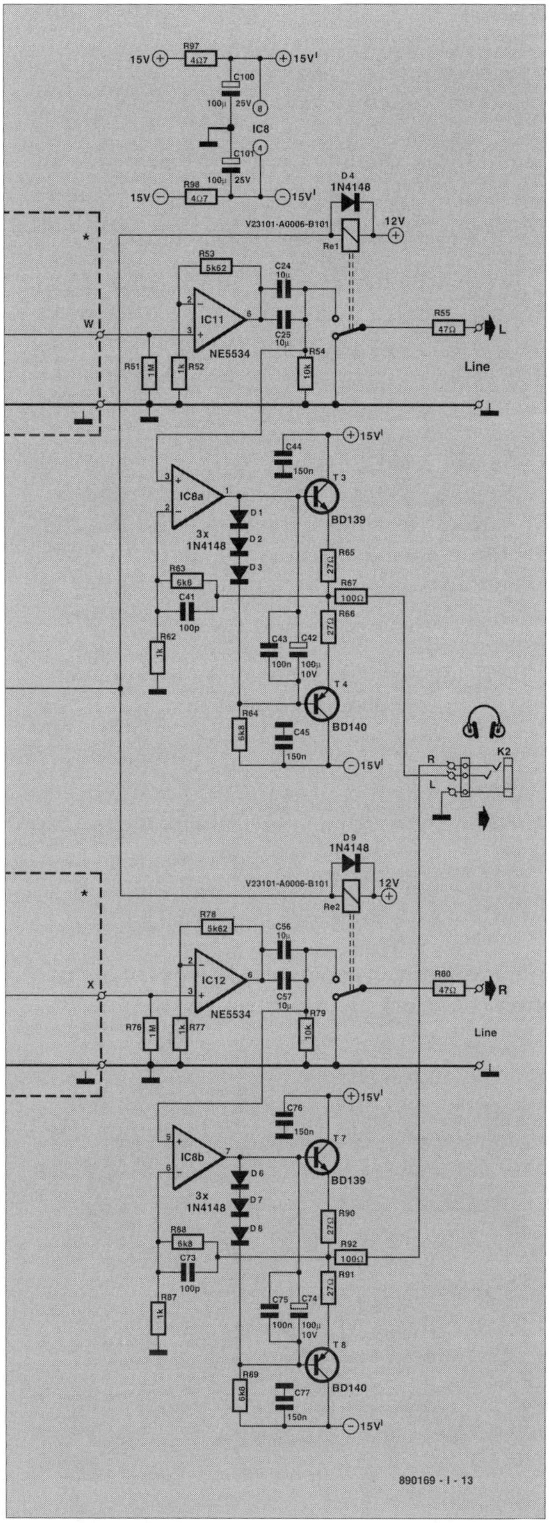

the motherboard.

The supply lines to this sensitive stage have been provided with additional filters (around T_5 and T_6) to even out any ripple on the ±15 V supply. In this context, it should be noted that the use of an LT1028 is justified only in conjunction with low-impedance (≤400 Ω) signal sources. If a standard high-impedance dynamic pick-up is used, an OP-27 will be perfectly all right, since the total noise then consists primarily of the thermal noise emanated by the pick-up.

The mono/stereo selector, S_5, is preceded by summing amplifier IC_3, which provides true addition of the left- and right-hand channels without any attenuation when stereo signals are processed as mono ones. Passive circuits always introduce some attenuation.

Although the design of the tone-control stage, IC_4, appears conventional, it has some unusual aspects. Normally, the connections to the ends of the carbon tracks of the treble potentiometers are via capacitors that determine the onset frequency of the control. If the onset frequency is to be altered, the value of both capacitors needs to be changed. In the present design, only one component per channel needs to be altered. Switches S_3 and S_4 enable two different onset frequencies to be selected. This arrangement makes the circuit slightly simpler as regards wiring and switching.

A 1 MΩ resistor, R_{45}, provides a feedback loop that prevents the output of the op amp from rising to unwanted heights if the wiper of the bass potentiometer occasionally loses contact with the carbon track. Capacitors C_{14} and C_{15} ensure that the offset of the input buffers is not amplified in the tone control stage. This parallel network of a bipolar and a film capacitor guarantees faithful signal processing even at high frequencies. The control range for various onset frequencies is shown in Fig. 3.4. If required, the entire tone control may be taken out of circuit by S_6.

It is recommended to use the very best quality potentiometers for the balance and volume controls, for instance, Alps types. These are not cheap, but they give consistently good performance over a long life. In some areas, a real balance control may be difficult or impossible to obtain (the correct type has half-silvered tracks which prevent any attenuation occurring in the centre position). If so, a standard linear stereo potentiometer may be used with a 2.2 kΩ resistor soldered between its wiper and the pole of S_{6a}; the attenuation in the centre position is then minimal.

The line amplifier, IC_{11}, is a Type NE5534, which has the advantage of being able to deliver more than 8 V r.m.s. into a 600 Ω load. Its amplification is set to ×6.6 to give a sensitivity at the line inputs of 150 mV r.m.s. for a nominal output of 1 V r.m.s. (if a 'real' balance control is used).

Immediately at the output of the op amp are two parallel-connected 10 μF capacitors, C_{24} and C_{25}, which isolate any offset in the preamplifier from the power amplifier used (remember that all op amps in the signal path are direct-coupled when the tone control is not in circuit). The layout of the printed-circuit board allows the use of polypropylene or metallized polypropylene capacitors.

The output contains a relay that provides a delay at switch-on to suppress any switching noise in the power amplifier. This relay may be switched off when listening via headphones only is required. The delay circuit is based on T_9. When the supply is switched on, capacitor C_{78} is charged slowly via R_{93}, so that it takes a while before the base-emitter junction voltage has reached the value at which the transistor begins to conduct. When the supply voltage is removed, C_{78} is discharged rapidly via D_5, which causes the relay to be deenergized almost instantly. The circuit has its own rectifier network, R_{95}-R_{96}-C_{91}-D_{10}-D_{11}, which makes rapid switch-off possible.

The headphone amplifier consists of a Type NE5532, IC_{8a}, and a discrete complementary output stage, T_3 and T_4. Three diodes, D_1–D_3, ensure proper class-A operation. the amplifier delivers sufficient power for driving low-impedance and insensitive headphones (with the exception of electro-static ones that normally require more power). The 100 Ω resistor in the output line limits the maximum current which can rise to 100 mA with an output impedance of 8 Ω. The amplification of the stage is set to a value at which clipping of the line output voltage of 1 V just does not occur.

The supply lines to output stages IC_8, T_3, T_4, T_7 and T_8 have their own decoupling, provided by R_{97}, R_{98}, C_{100} and C_{101}, to prevent any feedback to the other op amps in the preamplifier when the output current is high.

The power supply for the preamplifier is simple and conventional: a bridge rectifier, B_1, decoupled by capacitors C_{79}–C_{82}, buffer capacitors C_{84} and C_{89}, and two regulators, IC_9

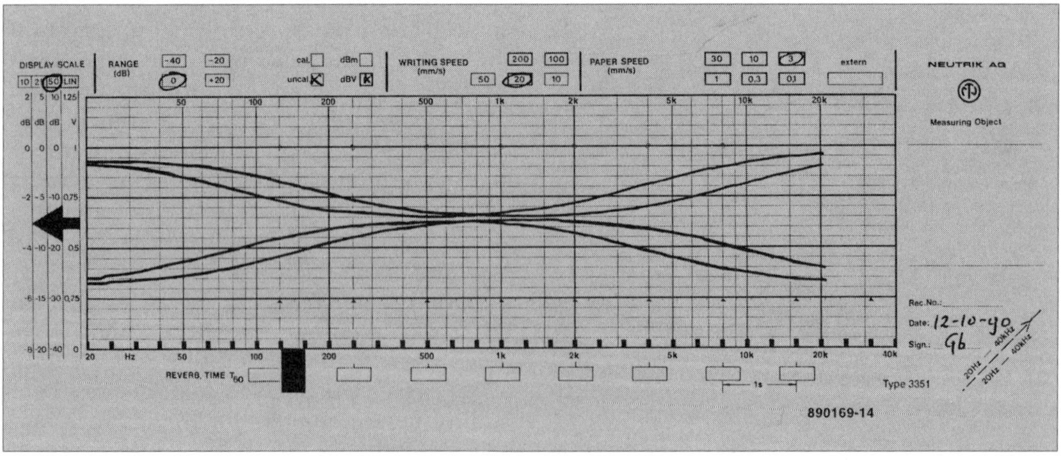

Fig. 3.3. Tone control characteristics. For purists, there is a tone-defeat switch.

and IC_{10}.

Choice of components

The quality of a preamplifier depends to a large extent on the op amps used and on the types of capacitor in the signal paths. In the present design, there is only one capacitor in the signal path: at the output. The motherboard has provision for two capacitors in parallel: C_{24} and C_{25}. These should preferably be metallized polypropylene types, but since at the values specified these are fairly expensive, standard polypropylene types may be used.

The capacitors at the output of the dynamic pick-up section should be metallized polyproplylene: their value keeps the cost down. There are some capacitors in the tone control circuit, which is unavoidable. Since this circuit degrades the performance to some extent in any case, the quality of these capacitors is of not much consequence.

There is a wide choice of suitable op amps. The quality of the preamplifier, however, is not necessarily in direct proportion to the price of the op amps chosen. It is really a matter of choosing the op amps on their merit and to individual requirements.

IC_1. In this position, a low-noise type with small offset voltage is preferred. The Signetics Type 5532 is a good and economical choice. A slightly faster version is the 5535, while the SM2132 is even faster. National Semiconductor's LM833 is, as far as its specification is concerned, almost identical to the 5532. More expensive, but very fast, is Analog Devices' D712 (which has the added advantage of FET inputs). Then there is the OP-270 (a double OP-27) or the OP-249, which has a high slew rate and FET inputs. In the Burr-Brown range, there are the OPA2107 and OPA2604, both with FET inputs. An inexpensive but still good choice is the TL072, which has FET inputs but a rather higher noise figure than the other types.

IC_2. The same considerations as for IC_1 apply, but in this position it is even more important to have a type with a low offset voltage, because all deviations between input and output are amplified. In view of the switching clicks and level changes that occur when a different input channel is selected, op amps with FET inputs are preferred. Both conditions are met by the AD712, OP-249, OPA2107 and OPA2604. The least expensive choice is the TL072 (but remember its noise figure).

IC_3. In this position, a low-noise op amp is essential: the OP-27, 5534A and SSM2134 are suitable. The last two types are not unit-gain stable, however, which makes it necessary for a 22 pF capacitor to be soldered

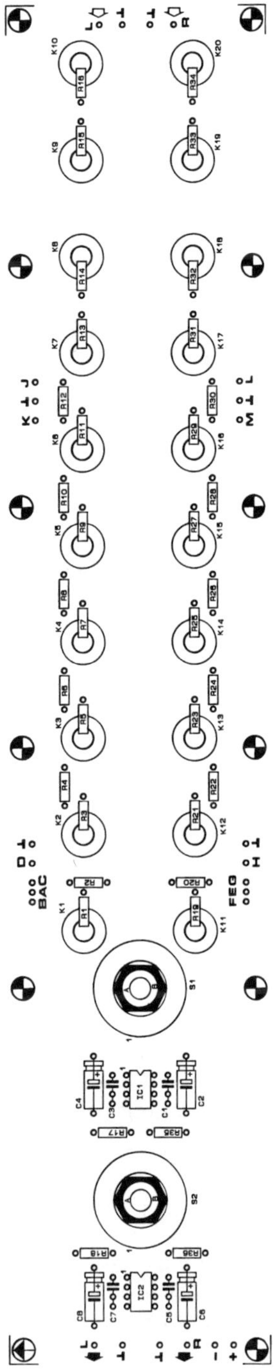

Fig. 3.5. Printed-circuit busboard (component layout) reduced to 70.7% (=A3→A4). For track layout and overlay, see page 225.

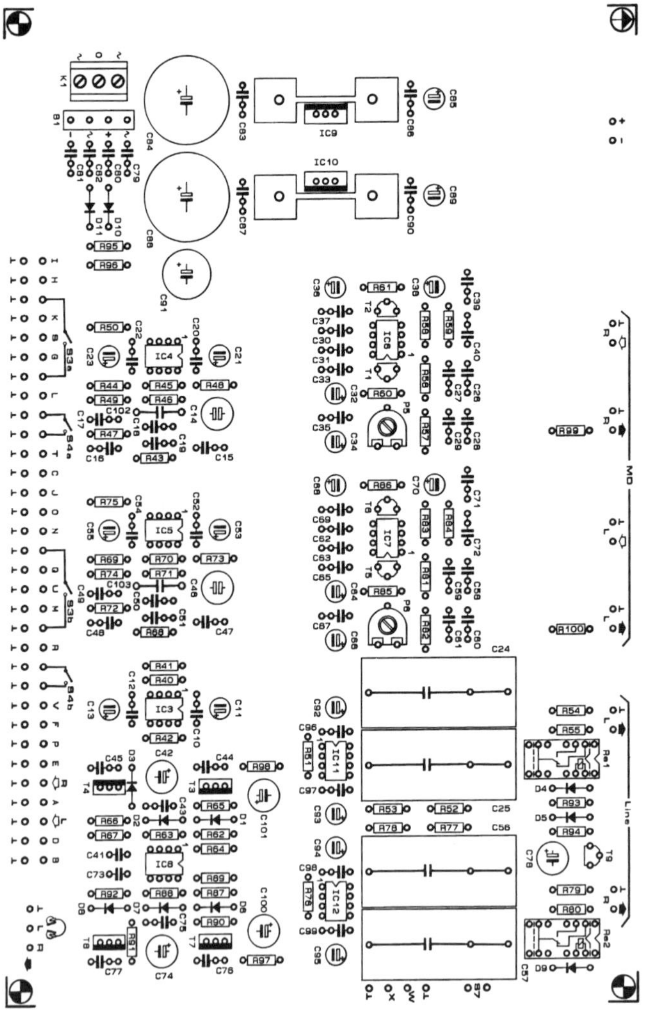

Fig. 3.6. Printed-circuit motherboard (component layout) reduced to 70.7% (=A3→A4). The track layout and component overlay are given on pages 227 and 229.

between pins 5 and 8 at the track side of the motherboard. FET inputs are not necessary. Again, a TL072 may be used where economy is important.

IC_4 and **IC_5**. The same considerations as for IC_3 apply. Remember the 22 pF capacitor if a 5534A or SSM2134 is used.

IC_6 and **IC_7**. Because very low noise in the dynamic pick-up circuit is essential, a Linear Technology Type LT1028 would be ideally suitable in these positions, were it not for a tendency to become unstable (for reasons that are not clear) and, therefore, the same manufacturers' Type LT1115 is preferred. This type has virtually the same specification as the 1028, but does not become unstable. A less expensive choice is the OP-27, while the 5534A and SSM2134 would also be suitable. These last two types require, apart from the 22 pF capacitor mentioned earlier, a change of value to 100 kΩ for presets P_5 and P_6.

IC_8. In this, the headphone amplifier, position, very low noise is again a prime requirement, and this is met by the Signetics Type 5532 or 5535, an SSM2132, an LM833 or an OP-27. If the preamplifier is required to deliver regularly fairly large output voltages, that is, larger than the nominal 1 V r.m.s., a type with FET inputs should be used in this position. The inputs of the 5532, for instance, are provided with protection diodes that may cause small direct voltages at the line output if the amplifier is overdriven.

IC_{11} and **IC_{12}**. The main reason for using a 5534 in these positions is its facility to provide sufficient current to low impedances (≤600 Ω), but the SSM2134 is also suitable. Most op amps with FET inputs do not perform well with such low impedances. If the load on the preamplifier is never likely to be below 2 kΩ, an AD711 or TL071 may also be used.

Construction

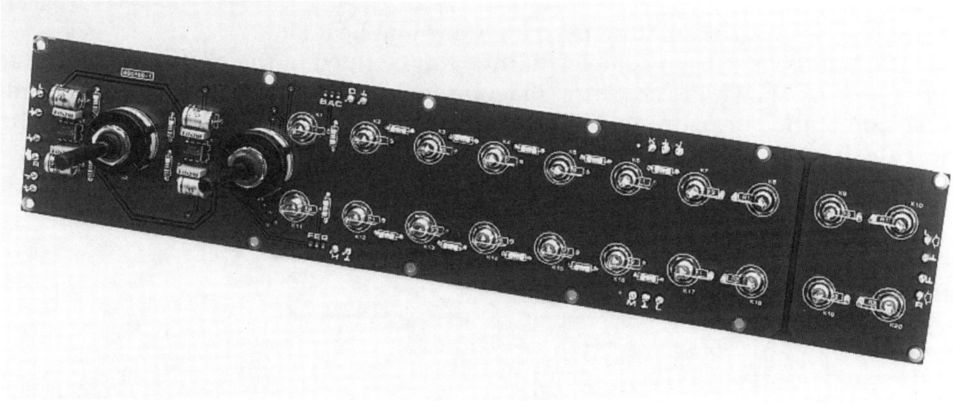

Fig. 3.7. Finished busboard.

It is better not to use IC sockets: each extra point of contact, and this applies particularly to such sockets, degrades the quality of any audio circuit. Soldering the ICs directly to the boards is preferred, although it is admitted hat sockets enable the quick and trouble-free exchanging of ICs to compare their performance.

Populating the boards should not present undue difficulties. Fit soldering pins at all input and output points: they will make the wiring later on that much easier. On the motherboard, solder a wire bridge between J and K and between L and M if the dynamic pick-up amplifier is not likely to be used. Wire bridges are also required between A and B

Fig.3.8. Finished motherboard.

and between E and F.

Note that both voltage regulators need an individual heat sink.

When the boards have been completed, they can be fitted in the enclosure once this has been prepared. As shown in the wiring diagram in Fig. 3.9, the motherboards fits into he left-hand side of the enclosure to enable the extension spindles of the rotary switches to just clear the heat sinks of the voltage regulators. At the right-hand side, there is space for

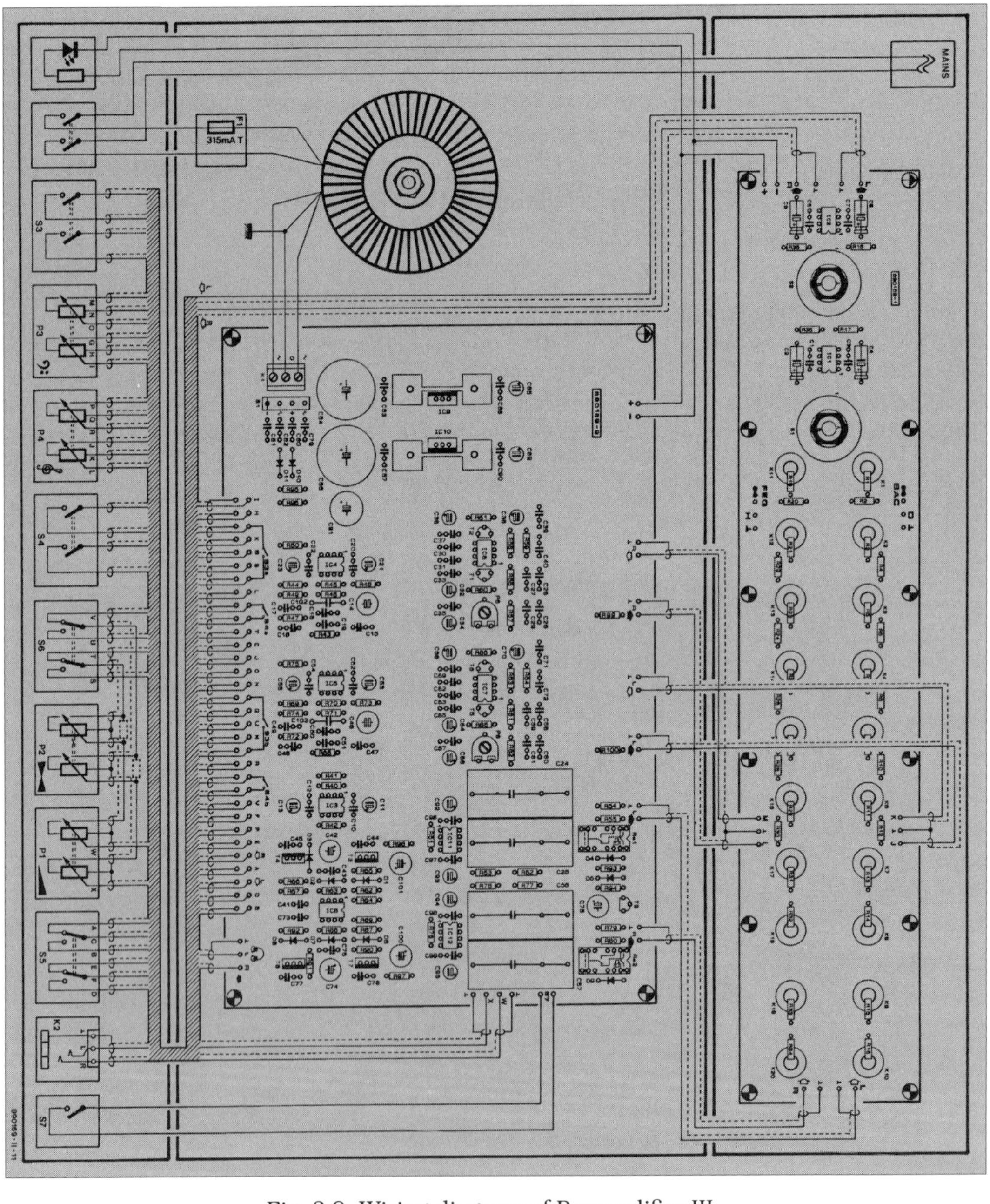

Fig. 3.9. Wiring diagram of Preamplifier III.

the mains transformer.

The layout of the front panel as suggested in Fig. 3.10 determines the location of the busboard at the back of the enclosure. Depending on the dimensions of the case, a mains input plug with integral fuse or a mains cable entry with strain relief may be used. In the latter case,a fuse holder must be fitted within the enclosure, preferably near the mains transformer.

Photocopies of the front panel drawing in Fig. 3.10 (scale 3.4) and of the busboard in Fig. 3.7 are convenient templates for drilling the needed holes in the front and rear panels respectively.

When the enclosure is ready, the boards and the various controls can be fitted. If Bourn potentiometers are used, make sure that these have a 6.3 mm spindle and not the usual 6 mm one.

Finally, the inter-wiring can be carried out. Take good care to insulate all mains-voltage carrying parts. All signal wires should be screened, but their diameter should not exceed 5 mm to prevent difficulties in soldering them to the motherboard where they are very close together.

The wiring between busboard and motherboard should be carried out first, followed by that between the boards and the controls on the front panel, and finally the power lines.

Although the screen of each and every signal line is connected to a separate soldering pin, only that at the volume and balance controls is actually connected to the earth line: that of all other screened cables is left unconnected. The enclosure earth is connected to the mains earth and this is the only real earth connection of the preamplifier.

The output socket of the headphone amplifier must be well insulated from the enclosure to avoid earth loops.

Technical data

Sensitivity	
dynamic pick-up	2.4 V r.m.s.
line inputs	150 mV r.m.s.
Input impedance	47 kΩ
Output voltage	
nominal	1 V r.m.s.
maximum	9.5 V r.m.s.
Output impedance	<100 Ω
Bandwidth (load = 10 kΩ)	
dynamic pick-up	20 Hz to 20 kHz ±3 dB
line inputs	5 Hz to 1 MHz ±0.5 dB
Signal-to-noise ratio (inputs short-circuited)	
dynamic pick-up	>80 dB (linear)
line inputs	>100 dB (linear)
Channel separation (Z_{source} = <600 Ω)	
1 kHz	>100 dB
20 kHz	>70 dB

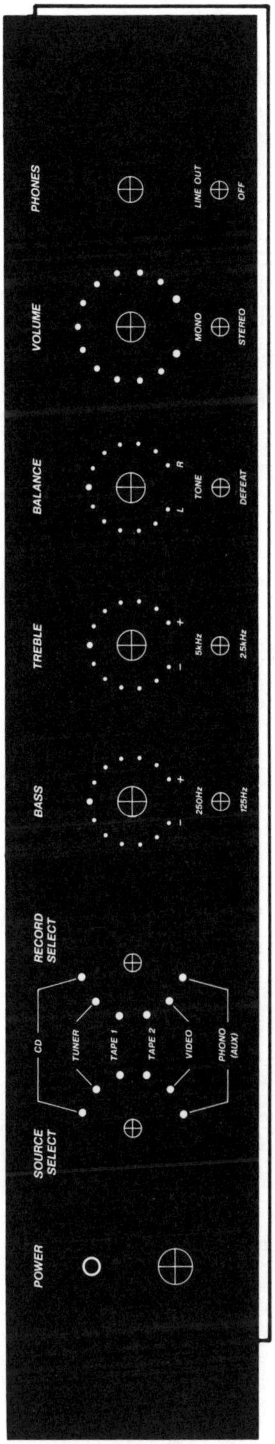

Fig. 3.10. Suggested front panel layout (reduced to 75%).

Cross-talk

1 kHz	>100 dB
20 kHz	>80 dB

Harmonic distortion (line in, U_0 = 1 V)

20 Hz to 20 kHz	<0.003%

Intermodulation distortion (line in, U_0 = 1 V)

250 Hz/4 kHz, 4:1)	<0.005%

Parts list

R_1, R_3, R_5, R_7, R_9, R_{11}, R_{19}, R_{21}, R_{23}, R_{25}, R_{27}, R_{29} = see text
R_2, R_4, R_6, R_8, R_{10}, R_{12}, R_{20}, R_{22}, R_{24}, R_{26}, R_{28}, R_{30} = 47 kΩ
R_{13}, R_{14}, R_{31}, R_{32} = 100 Ω
R_{15}, R_{16}, R_{33}, R_{34} = 47 Ω
R_{17}, R_{18}, R_{35}, R_{36} = 1 MΩ
R_{40}, R_{41}, R_{43}, R_{44}, R_{63}, R_{64}, R_{68}, R_{69}, R_{88}, R_{89} = 6.8 kΩ
R_{42} = 3.3 kΩ
R_{45}, R_{60}, R_{61}, R_{70}, R_{85}, R_{86} = 12 kΩ
R_{46}, R_{47}, R_{51}, R_{71}, R_{72}, R_{76} = 1 MΩ
R_{48}, R_{49}, R_{73}, R_{74} = 2.2 kΩ
R_{50}, R_{75} = 18 kΩ
R_{52}, R_{77} = 1 kΩ, 1%
R_{53}, R_{78} = 5.62 kΩ, 1%
R_{54}, R_{79} = 10 kΩ
R_{55}, R_{80} = 47 Ω
R_{56}, R_{81} = 88.7 kΩ, 1%
R_{57}, R_{82} = 7.5 kΩ, 1
R_{58}, R_{83} = 154 Ω, 1%
R_{59}, R_{84} = 47.5 kΩ, 1%
R_{62}, R_{87} = 1 kΩ
R_{65}, R_{66}, R_{90}, R_{91} = 27 Ω
R_{67}, R_{92}, R_{94}, R_{99}, R_{100} = 100 Ω
R_{93} = 47 kΩ
R_{95} = 68 Ω, ½ W
R_{96} = 2.7 kΩ
R_{97}, R_{98} = 4.7 Ω

Potentiometers:
P_1 = 10 kΩ, logarithmic stereo potentiometer
P_2 = 10 kΩ balance control
P_3 = 22 kΩ (25 kΩ) potentiometer
P_4 = 10 kΩ linear potentiometer
P_5, P_6 = 10 kΩ preset

Capacitors:
C_1, C_3, C_5, C_7 = 47 nF, ceramic
C_2, C_4, C_6, C_8 = 47 μF, 25 V
C_{10}, C_{12}, C_{20}, C_{22}, C_{33}, C_{37}, C_{52}, C_{54}, C_{65}, C_{69}, C_{79}, C_{80}–C_{82}, C_{96}–C_{99} = 47 nF, ceramic
C_{11}, C_{13}, C_{21}, C_{23}, C_{32}, C_{36}, C_{53}, C_{55}, C_{64}, C_{68}, C_{92}–C_{95} = 47 μF, 25 V, radial
C_{14}, C_{46} = 10 μF, 40 V, bipolar, radial
C_{15}–C_{17}, C_{35}, C_{39}, C_{43}, C_{47}–C_{49}, C_{67}, C_{71}, C_{75}, C_{83}, C_{86}, C_{87}, C_{90} = 100 nF
C_{18}, C_{50} = 1.8 nF
C_{19}, C_{51} = 1.2 nF

C_{24}, C_{25}, C_{56}, C_{57} = 10 μF, metallized polypropylene (see text)
C_{26}, C_{27}, C_{58}, C_{59} = 18 nF, 1%, polypropylene
C_{28}, C_{60} = 5.6 nF, 1%, polypropylene
C_{29}, C_{61} = 3.9 nF, 1%, polypropylene
C_{30}, C_{62} = 47 nF, 1%, polypropylene
C_{31}, C_{63} = 120 nF, 1%, polypropylene
C_{34}, C_{38}, C_{66}, C_{70}, C_{85}, C_{89} = 10 μF, 25 V, radial
C_{40}, C_{72} = 47 pF, polystyrene
C_{41}, C_{73} = 100 pF, polystyrene
C_{42}, C_{74} = 100 μF, 10 V, radial
C_{44}, C_{45}, C_{76}, C_{77} = 150 nF
C_{78}, C_{91} = 220 μF, 25 V, radial
C_{84}, C_{88} = 220 μF, 40 V, radial
C_{100}, C_{101} = 100 μF, 25 V, radial
C_{102}, C_{103} = 150 pF, polystyrene

Semiconductors:
B_1 = B80C1500
D_1–D_4, D_6–D_9 = 1N4148
D_5, D_{10}, D_{11} = 1N4001
T_1, T_5 = BC550C
T_2, T_6 = BC560C
T_3, T_7 = BD139
T_4, T_8 = BD140
T_9 = BC517

Integrated circuits:
IC_1, IC_8 = NE5532
IC_2 = AD172 (see text)
IC_3, IC_4, IC_5 = OP-27
IC_6, IC_7 = LT1028CN8
IC_9 = 7815
IC_{10} = 7915
IC_{11}, IC_{12} = NE5534

Miscellaneous:
S_1, S_2 = 2-pole, 6-position rotary switch (PCB mounting)
S_3, S_4 = DPST miniature toggle switch
S_5, S_6 = DPCO miniature toggle switch
S_7 = SPST miniature toggle switch
K_1–K_{20} = 3.5 mm audio socket (chassis fitting)
Re_1, Re_2 = 12 V relay, 1 change-over contact
1 off 6 mm stereo audio socket
2 off heat sink for IC_9 and IC_{10}
1 off 3-way terminal block (PCB fitting)
1 off DPST mains press button switch
1 off fuse holder with315 mA slow action fuse
1 off strain relief sleeve
1 off mains transformer, secondary 2×15 V, 1 A

Part 2

Power amplifiers

Chapter 4

60 W power amplifier

In spite of their relatively modest TO220 case, the International Rectifier HEXFETS used in the present 60-watt output amplifier can cope with fairly large voltages and currents. The amplifier is absolutely symmetrical from input to output. Its mechanical design is such that it can be accommodated on a fairly small printed-circuit board, including the electrolytic capacitors of the power supply. The design objective was to arrive at a not too complex unit that nevertheless offered excellent performance, was fairly simple to build and could be reproduced relatively easily. The result is a straightforward amplifier without any unnecessary gimmicks.

A bipolar transistor may be considered a current-amplifying device that enables a (relatively) large current to be controlled by a much smaller one. A field-effect transistor (FET) behaves differently: it is a sort of variable resistance whose conduction is controlled by a voltage. It follows that the drives of these devices are quite different: an important consideration in the design of an output amplifier. A bipolar transistor needs a base current before it can function, whereas a FET can be driven almost without any energy. All it needs is a control voltage; the current it draws is negligible. When power FETs first came on to the market, many designers thought that they would simplify the design of output amplifiers beyond belief. That quickly proved to be not so, however, because power FETs have a fairly large capacitance between the gate and the drain/source channel (sometimes of the order of a few nanofarads). This means that at high audio frequencies the driver stages need to deliver fairly large transfer currents to keep the bandwidth sufficiently large.

It may well be asked what advantage(s) a FET offers. In a bipolar power transistor, it is difficult to combine high voltage, large current, and wide bandwidth, because its operation must remain within the Safe Operating Area—SOA. It is not enough to just look at the peak voltage and current in the relevant data sheet. By virtue of modern production techniques, FETs can be fabricated that can handle high voltages (100 V and more) and, in spite of their modest dimensions, large currents. It is, therefore, much simpler to design an output amplifier with reasonable power output with power FETs than with power transistors. Of course, there are other requirements as well, such as slew rate and matching of complementary semiconductors, and ...

Circuit description

A symmetrical design has the advantage that it minimizes problems with distortion, particularly that associated with even harmonics. Therefore, the input stages consist of two differential amplifiers, T_1-T_2 and T_3-T_4. To keep the cost down, these do not use expensive dual devices, but discrete transistors. Performance is excellent, particularly if the transistors are matched.

A differential amplifier is one of the best means of combining two electrical signals: here, the input signal and the feedback signal. The amplification of the stage is determined mainly by the ratio of the collector and emitter resistances (in the case of T_1-T_2 these are R_9, R_{10}, R_{11} and R_{12}). These form a sort of local feedback: limiting the amplification reduces the distortion.

Two *RC* networks (R_3-C_3 and R_4-C_4) limit the bandwidth of the differential amplifiers

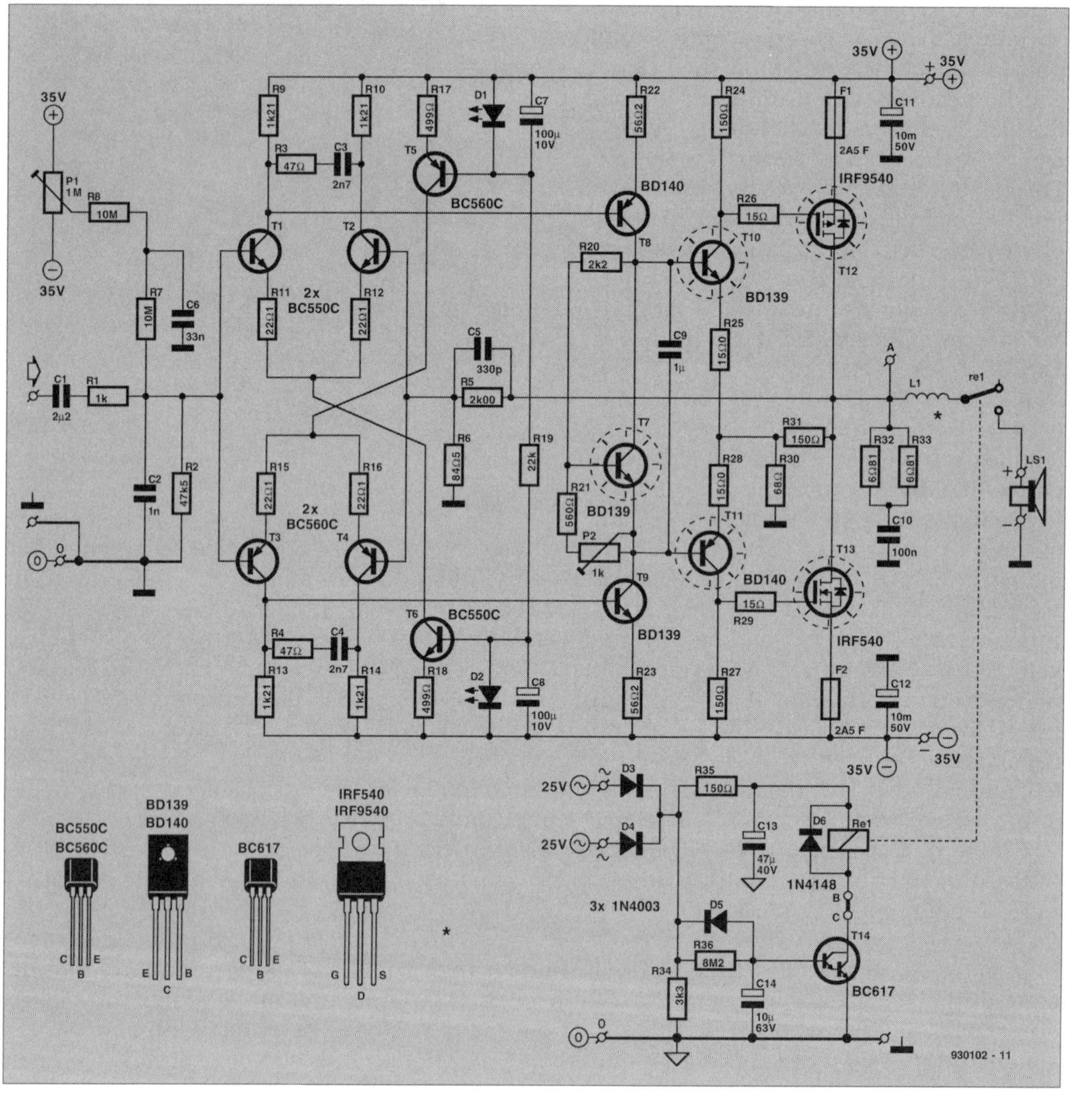

Fig. 4.1. Circuit diagram of the 60 W power amplifier.

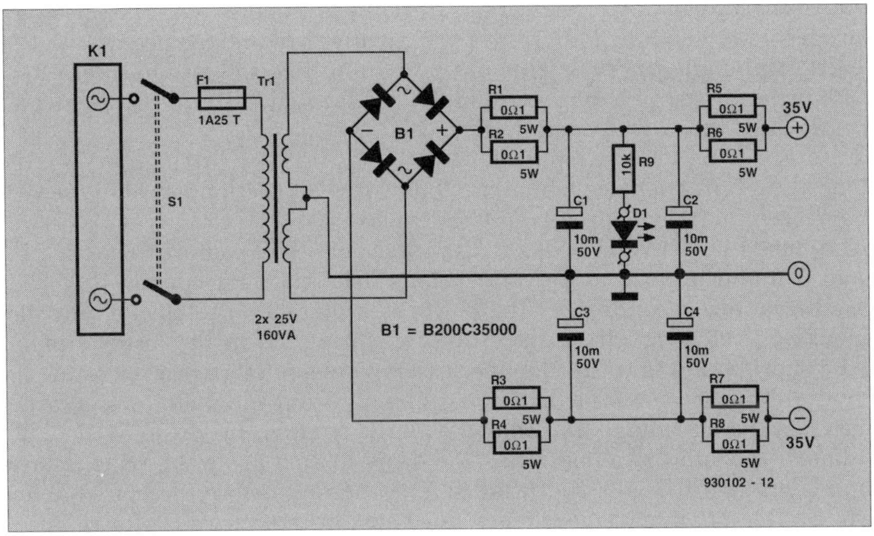

Fig. 4.2. Circuit diagram of the power supply.

and these determine, to a degree, the open-loop bandwidth of the entire amplifier (which is 6.5 kHz).

The d.c. operating point of the differential amplifiers is provided by two current sources. Transistor T_6, in conjunction with R_{18} and D_2, provides a constant current of about 2 mA for T_1-T_2. Transistor T_5, with R_{17} and D_1, provides a similar current for T_3-T_4. The combination of a transistor and an LED creates a current source that is largely independent of temperature, since the temperature coefficients of the LED and the transistor are virtually the same. It is, however, necessary that these two components are thermally coupled (or nearly so) and they are, therefore, located side by side on the printed-circuit board.

In the input stage, C_1 is followed by a low-pass section, R_1-C_2, which limits the bandwidth of the input to a value that the amplifier can handle. Resistor R_2 is the base resistor of T_1 and T_3. So far, this is all pretty normal. Network P_1-R_7-R_8 is somewhat out of the ordinary, however. It forms an offset control to adjust the direct voltage at the output of the amplifier to zero. Such a control is normally found **after** the input stage. The advantage of putting it before that stage is that the inputs of the differential amplifiers are exactly at earth potential, which means that the noise contribution of their base resistors is negligible.

The signals at the collectors of T_1 and T_3 are fed to pre-drivers T_8 and T_9. Between these transistors is a 'variable zener' formed by T_7 which, in conjunction with P_2, serves to set the quiescent current of the output FETs.

The output of the pre-drivers is applied to T_{10} and T_{11}, which drive HEXFETs T_{12} and T_{13}. This power section has local feedback (R_{30}-R_{31}).

The design of T_{10}–T_{13} is a kind of compound output stage, since the drain of the power FETs is connected to the output terminal. Note that T_{12} is a p-channel FET and T_{13} an n-channel type. Therefore, the stage provides current amplification as well as voltage amplification. The voltage amplification is limited to ×3 by the local feedback resistors (R_{30}-R_{31}). Here again, this feedback serves to reduce the distortion. The overall feedback of the amplifier is provided by R_5-R-C_5.

Fuses are provided in the source lines of the HEXFETs. Power FETs have an inherent current limitation by virtue of their positive temperature coefficient: when the device gets hot, its drain-source resistance rises and this reduces the current through it. The fuses and this property provide adequate protection against brief short-circuits. Note that the HEXFETs used can handle peak currents of up to 75 A. Electrolytic capacitors C_{11} and C_{12} (10,000 μF each and part of the power supply) are located close to the FETs, so that the heavy currents have only a short path to follow.

At the output is a Boucherot network, R_{32}-R_{33}-C_{10}, that ensures an adequate load on the amplifier at high frequencies, since the impedance of the loudspeaker, because of its inductive character, is fairly high at high frequencies.

Inductor L_1 limits any current peaks that may arise with capacitive loads.

The signal is finally applied to the loudspeaker, LS_1, via relay contact Re_1. The relay is not energized for a few seconds after the power is switched on to obviate any plops from the loudspeaker. Such plops are caused by brief variations in the direct supply voltage arising in the short period that the amplifier needs to reach its correct operating level.

The supply voltage for the relay is derived directly from the mains transformer via D_3 and D_4. This has the advantage that the relay is deactuated, by virtue of the low value of C_{13}, immediately the supply voltage fails. The delay in energizing the relay is provided by T_{14} in conjunction with R_{36} and C_{14}. It takes a few second before the potential across C_{14} has risen to a value at which T_{14} switches on. This darlington transistor requires a base voltage of not less than 1.2 V before it can conduct.

The power supply—see Fig. 4.2—is traditional, apart from the resistors, R_5–R_8 in the

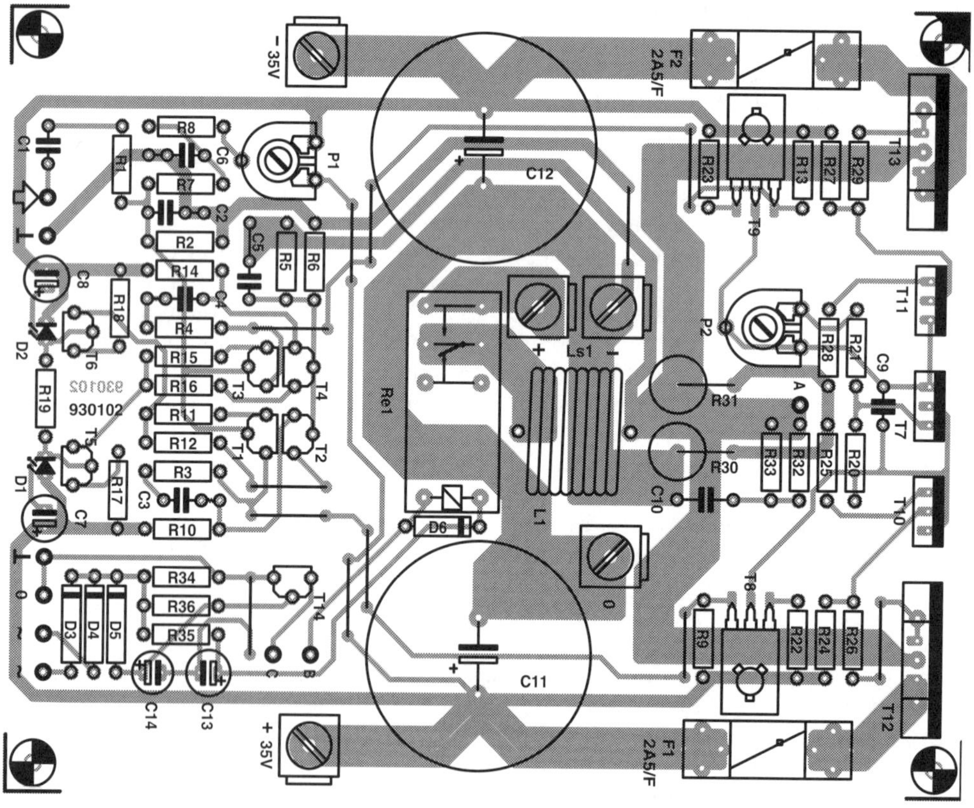

Fig. 4.3. Printed-circuit board (component layout) for the 60 W power amplifier. The track layout is given on page 231.

power lines. These limit, to some degree, the very large peak charging currents to electrolytic capacitors C_{11} and C_{12}. Moreover, together with these capacitors, they form a filter that prevents most spurious voltages from reaching the amplifier. Measurements on the prototype showed that this was particularly evident at frequencies below 500 Hz.

Construction

The design of the printed-circuit board for the amplifier (Fig. 4.3) takes good account of the large currents that flow in the amplifier. This has given rise to a couple of tracks being paralleled instead of combined, so that the effect of currents in the power section on the input stages is minimal.

Populating the board is straightforward. Although not strictly necessary, it is advisable to match the transistors used in the differential amplifiers. This may be done conveniently on an h_{fe} tester by measuring the amplification at a collector current of about 1 mA. If such a tester is not available, use a base resistor that results in a collector current of about 1 mA measured with a multimeter. With the same resistor, test a number of other transistors and note the collector currents. Mount the selected pairs on the board and pack them closely together with a 5 mm wide copper ring (made from a piece of 12 mm copper water pipe) as shown in Fig. 4.4.

Inductor L_1 consists of six turns, inner diameter 16 mm ($^5/_8$ in), of insulated copper wire 1.5 mm ($^1/_{16}$ in) thick.

The large transistors are located on one side of the board, so that they can be fixed directly to the heat sink. They must be isolated with the aid of ceramic washers.

The two sizes indicated on the board for T_{12} and T_{13} may be ignored: they are a precaution for possible different types of transistor at a later stage.

Connections from the power supply and to the loudspeaker are by means of terminal blocks that can be screwed on to the board.

Mount the two amplifier boards, mains transformers and electrolytic capacitors in a suitable enclosure. The wiring diagram for one channel is given in Fig. 4.5.

It is advisable to measure the supply voltages before they are connected to the amplifiers. Also, turn P_2 to maximum (wiper towards R_{33}) before connecting the power supply to the amplifiers. Set input presets P_1 to the centre of their travel. A few seconds after the supply has been switched on, the relay should come on. Connect a multimeter (1 V direct voltage range) and adjust P_1 until the meter reads zero (both channels!).

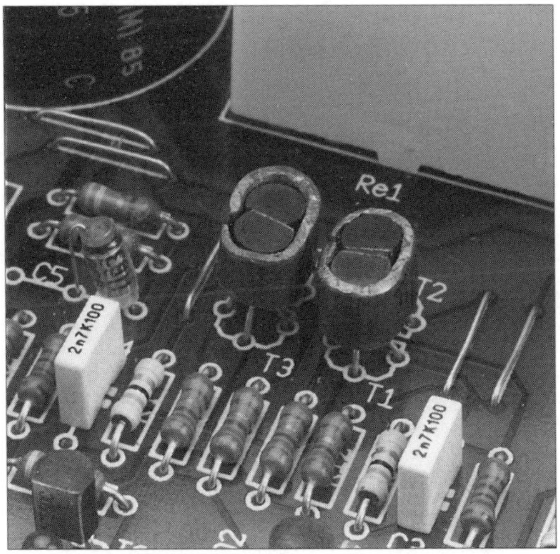

Fig. 4.4. The differential amplifiers should be clamped together with a DIY copper ring.

Switch the supply off again and insert a multimeter (1 A d.c. range) in one of the supply lines; do not substitute it for one of the fuses, since that would affect the operating point of the relevant power FET. Switch the supply on again, wait 5–10 minutes (when the current has stabilized) and adjust P_2 for a meter reading of 330 mA. After about half an hour, the current will remain steady at about 230 mA. The quiescent current through the output transistors is then around 200 mA. Switch off the supply, remove the meter from the supply line and repeat the above procedure with the other

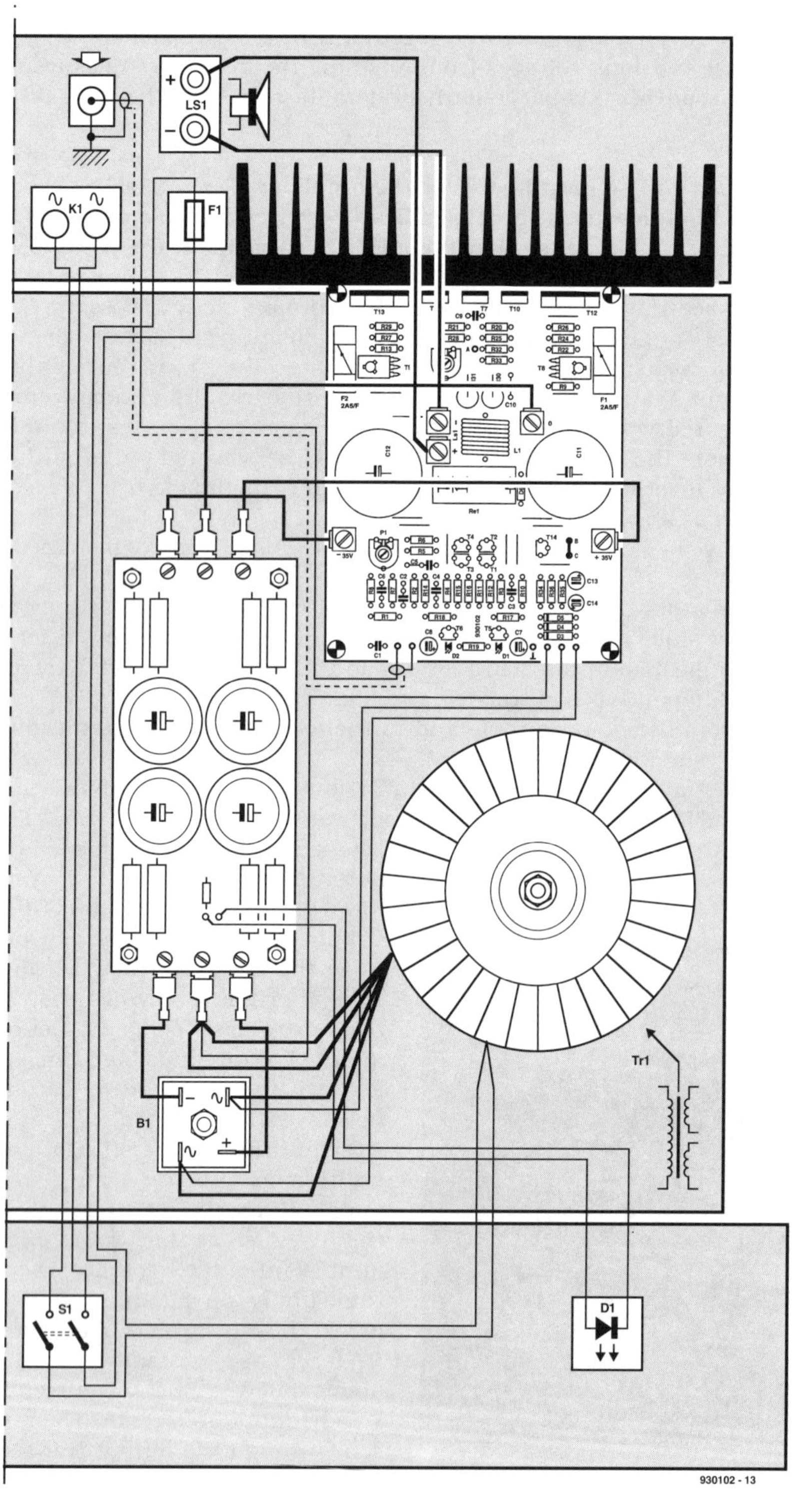

Fig. 4.5. Wiring diagram for the 60 W power amplifier (one channel only).

channel.

Finally, recheck the direct voltages at the outputs of the amplifiers and, if necessary, readjust P_1 slightly.

The loudspeakers must be 4-ohm or 8-ohm types, whose impedance must not drop below 3 Ω. It is not permissible to connect two 4-ohm units in parallel to the amplifier, because that would give problems when large drive signals are applied to the FETs.

Parts list (one channel)

Resistors:

R_1 = 1 kΩ
R_2 = 47.5 kΩ, 1%
R_3, R_4 = 47 Ω
R_5 = 2.0 kΩ, 1%
R_6 = 84.5 Ω, 1%
R_7, R_8 = 10 MΩ
R_9, R_{10}, R_{13}, R_{14} = 1.21 kΩ, 1%
R_{11}, R_{12}, R_{15}, R_{16} = 22.1 Ω, 1%
R_{17}, R_{18} = 499 Ω, 1%
R_{19} = 22 kΩ
R_{20} = 2.2 kΩ
R_{21} = 560 Ω
R_{22}, R_{23} = 56.2 Ω, 1%
R_{24}, R_{27} = 150 Ω, 1%
R_{25}, R_{28} = 15.0 Ω, 1%
R_{26}, R_{29} = 15 Ω
R_{30} = 68 Ω, 5 W
R_{31} = 150 Ω, 5 W
R_{32}, R_{33} = 6.81 Ω, 0.6 W, 1%
R_{34} = 3.3 kΩ
R_{35} = 150 Ω
R_{36} = 8.2 MΩ
P_1 = 1 MΩ preset
P_2 = 1 kΩ preset

Capacitors:

C_1 = 2.2 µF, 50 V, polypropylene
C_2 = 1 nF
C_3, C_4 = 2.7 nF
C_5 = 330 pF, polystyrene, axial
C_6 = 33 nF
C_7, C_8 = 100 µF, 10 V, radial
C_9 = 1 µF
C_{10} = 100 nF
C_{11}, C_{12} = 10,000 µF, 50 V, radial, for PCB mounting
C_{13} = 47 µF, 40 V, radial
C_{14} = 10 µF, 63 V, radial

Inductors:

L_1 = air-core, 0.1 mH (see text)

Semiconductors:

D_1, D_2 = 3 mm LED, red (1.6 V drop at 3 mA)

Fig. 4.6. Top view of the completed amplifier with enclosure removed.

D_3–D_5 = 1N4003
D_6 = 1N4148
T_1, T_2, T_6 = BC550C
T_3–T_5 = BC560C
T_7, T_9, T_{10} = BD139
T_8, T_{11} = BD140
T_{12} = IRF9540
T_{13} = IRF540
T_{14} = BC617

Miscellaneous;
Re_1 = relay, 24 V, 1 make contact (e.g., Siemens V23056-A0105-A101*)
F_1, F_2 = fuse, 2.5 A, fast, with holder for PCB mounting
Ceramic washers (5) for T_7, T_{10}–T_{13}
Terminal block (5) (see text)
Heat sink, 0.6 K W^{-1} (e.g., Fischer SK85**)

Power supply:
Mains transformer, 2 × 25 V, 160 VA
Mains on-off switch with indicator
Fuse 1.25 A, slow with holder
Bridge rectifier Type B200C35000
Electrolytic capacitor (4), 10,000 μF, 50 V
Resistor (8) 0.1 Ω, 5 W

* ElectroValue, 3 Central Trading Estate, Staines, TW18 4UX, ☎ (0784) 442 253. Private customers welcome.

** Dau (UK) Ltd, 7075 Barnham Road, Barnham, West Sussex PO22 0ES. ☎ (0243) 553 031. Trade only, but information as to your nearest dealer will be given by telephone.

The HEXFET structure

As implied, the HEXFET structure involves an hexagonal device geometry. At the core is a radically new hexagonal, cellular structure as illustrated. It is this hexagonal geometry, along with advanced MOS processing, that gives the HEXFET an on-state resistance, R_{DSon}, one-third of that possible with the best previous MOSFET technology, in a given die size.

A planar, non-V-groove structure, the HEXFET conducts current vertically. For high packaging density, it uses a silicon-gate structure. The density of the hexagonal source cells on the top surface of the silicon die is over half a million cells per square inch. Electrons flow from a source cell through the channel which is around the periphery of that cell and then into the drain body. The bottom surface of the drain body is in electrical and thermal contact with the holder.

The efficient hexagonal source pattern, the silicon gate, and advanced MOS processing techniques combine to produce the HEXFET's unique performance.

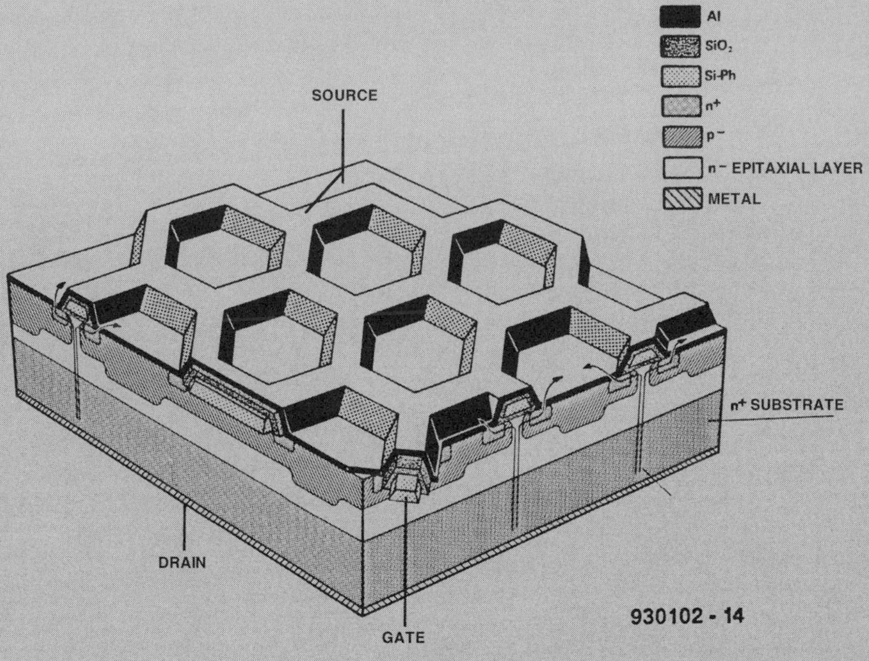

Chapter 5

90 W power amplifier

The amplifier discussed in Chapter 4 has one small drawback: it delivers 'only' 60 W into 8 Ω (or 120 W into 4 Ω). Otherwise, it is a first class amplifier that provides excellent music reproduction, which is evidenced not so much by measurement as by audition. It has a quality not unlike that of a valve amplifier. To meet the requirements of readers who want rather more power, it has been upgraded to provide around 90 W into 8 Ω (about 160 W into 4 Ω). By a stroke of good fortune, a pair of IGBTs (Insulated Gate Bipolar Transistors) proved ideal replacements for the HEXFETs used in the 60 W design. Apart from the figures for power output, the technical specification remains virtually the same (see page 64).

Although IGBTs are quite different from HEXFETs, the board for the 60 W design can be used without any modification, whence the duplicated holes for the output transistors on the printed-circuit board In fact, the circuit has hardly changed. The most noticeable alteration is the replacement of the fuses in the source lines of the power FETs by emitter resistors for the IGBTs. The only other changes are in the value of two resistors in the compensating circuit of the input stage, of one in the quiescent-current circuit, and of one resistor and two capacitors in the protection circuit. This means that anyone who has built the 60 W amplifier can quickly modify it to the upgraded version.

One item needs to be replaced, however: the mains transformer. This is obvious, since more power can not be obtained from the same supply voltage/current. The mains transformer needs to provide 2×30 V at 3.75 A, which will result in a direct voltage of ±43 V.

Circuit description

The circuit diagram of the 90 W amplifier is given in Fig. 5.1. Changed with respect to the 60 W version are T_{12}, T_{13}, RF_1, RF_2, R_3, R_4, R_{21}, R_{35}, C_{13} and C_{14}. Also, to improve performance at high frequencies, a damping resistor has been added to, or rather in, inductor L_1. Finally, to improve the noise figure, the impedance of input filter R_1-C_2 has been lowered.

A symmetrical design has the advantage of minimizing problems with distortion, particularly that associated with even harmonics. Therefore, the input stages consist of two differential amplifiers, T_1-T_2 and T_3-T_4. To keep the cost down, these do not use expensive dual devices, but discrete transistors. Performance is excellent, particularly if the transistors are matched.

A differential amplifier is one of the best means of combining two electrical signals: here, the input signal and the feedback signal. The amplification of the stage is determined mainly by the ratio of the collector and emitter resistances (in the case of T_1-T_2 these are R_9, R_{10}, R_{11} and R_{12}). These provide a form of local feedback: limiting the amplification reduces the distortion.

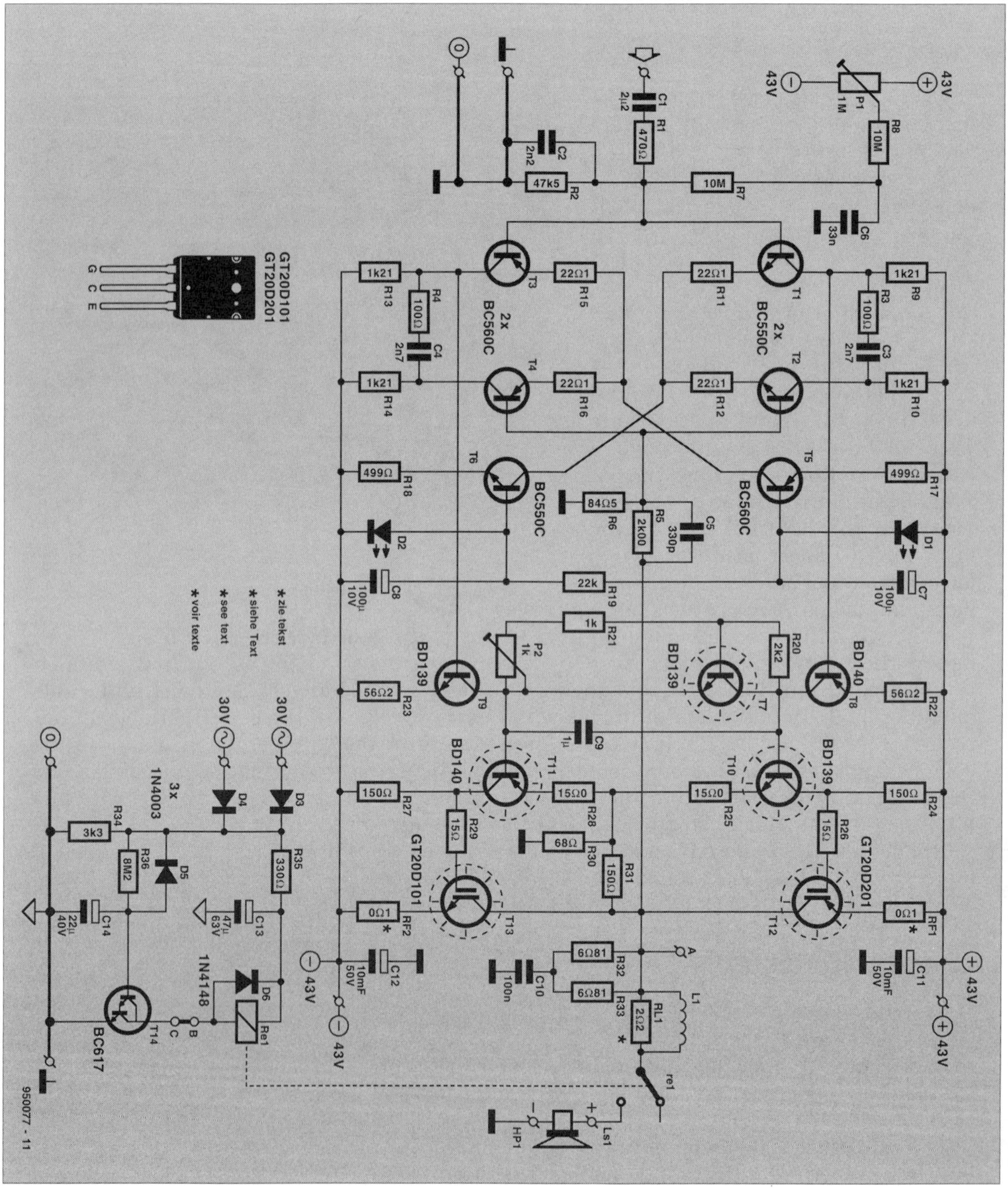

Fig. 5.1. Circuit diagram of the 90 W power amplifier.

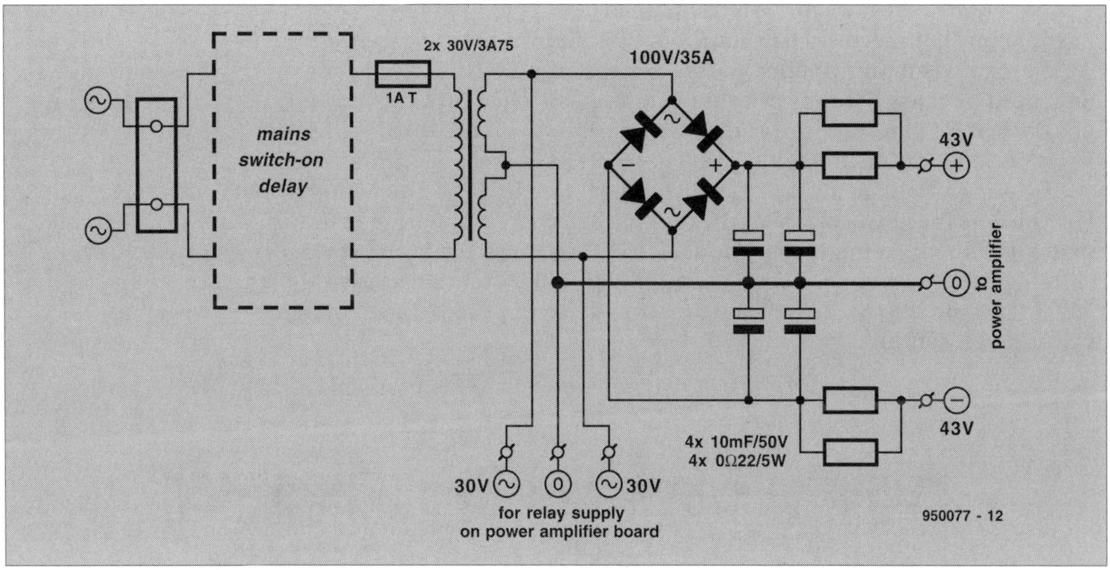

Fig. 5.2. Circuit diagram of the power supply for one mono 90 W power amplifier.

Two *RC* networks (R_3-C_3 and R_4-C_4) limit the bandwidth of the differential amplifiers and these determine, to a degree, the open-loop bandwidth of the entire amplifier.

The d.c. operating point of the differential amplifiers is provided by two current sources. Transistor T_6, in conjunction with R_{18} and D_2, provides a constant current of about 2 mA for T_1-T_2. Transistor T_5, with R_{17} and D_1, provides a similar current for T_3-T_4. The combination of a transistor and an LED creates a current source that is largely independent of temperature, since the temperature coefficients of the LED and the transistor are virtually the same. It is, however, necessary that these two components are thermally coupled (or nearly so) and they are, therefore, located side by side on the printed-circuit board.

In the input stage, C_1 is followed by a low-pass section, R_1-C_2, which limits the bandwidth of the input to a value that the amplifier can handle. Resistor R_2 is the base resistor of T_1 and T_3. So far, this is all pretty normal. Network P_1-R_7-R_8 is somewhat out of the ordinary, however. It forms an offset control to adjust the direct voltage at the output of the amplifier to zero. Such a control is normally found after the input stage. The advantage of putting it before that stage is that the inputs of the differential amplifiers are exactly at earth potential, which means that the noise contribution of their base resistors is negligible.

The signals at the collectors of T_1 and T_3 are fed to pre-drivers T_8 and T_9. Between these transistors is a 'variable zener' formed by T_7 which, in conjunction with P_2, serves to set the quiescent current of the output transistors.

The output of the pre-drivers is applied to T_{10} and T_{11}, which drive IGBTs T_{12} and T_{13}. This power section has local feedback (R_{30}-R_{31}).

The design of T_{10}–T_{13} is a kind of compound output stage, since the collector of the power transistors is connected to the output terminal. The voltage amplification is limited to ×3 by the local feedback resistors (R_{30}-R_{31}). Here again, this feedback serves to reduce the distortion. The overall feedback of the amplifier is provided by R_5-R-C_5.

Electrolytic capacitors C_{11} and C_{12} (10,000 μF each and part of the power supply) are located close to the IGBTs, so that the heavy currents have only a short path to follow.

At the output is a Boucherot network, R_{32}-R_{33}-C_{10}, that ensures an adequate load on the amplifier at high frequencies, since the impedance of the loudspeaker, because of its inductive character, is fairly high at high frequencies.

Inductor L_1 limits any current peaks that may arise with capacitive loads.

The signal is finally applied to the loudspeaker, LS_1, via relay contact Re_1. The relay is not energized for a few seconds after the power is switched on to obviate any plops from the loudspeaker. Such plops are caused by brief variations in the direct supply voltage arising in the short period that the amplifier needs to reach its correct operating level.

The supply voltage for the relay is derived directly from the mains transformer via D_3 and D_4. This has the advantage that the relay is deactuated, by virtue of the low value of C_{13}, immediately the supply voltage fails. The delay in energizing the relay is provided by T_{14} in conjunction with R_{36} and C_{14}. It takes a few seconds before the potential across C_{14} has risen to a value at which T_{14} switches on. This darlington transistor requires a base voltage of not less than 1.2 V before it can conduct.

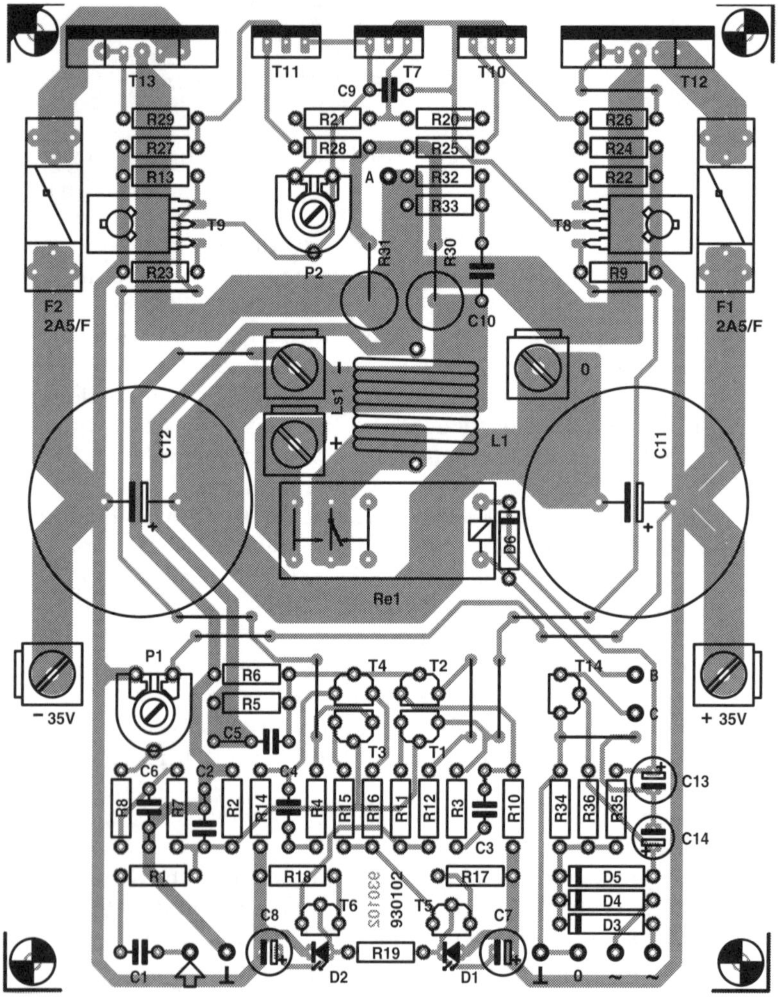

Fig. 5.3. Component layout of the printed-circuit board for the 90 W power amplifier. The track lauout is on page 233.

The power supply—see Fig. 5.2—is traditional, apart from the resistors, R_5–R_8, in the power lines. These limit, to some degree, the very large peak charging currents drawn by electrolytic capacitors C_{11} and C_{12}. Moreover, together with these capacitors, they form a filter that prevents most spurious voltages from reaching the amplifier. Measurements on the prototype showed that this was particularly evident at frequencies below 500 Hz.

Construction

The design of the printed-circuit board for the amplifier (Fig. 5.3) takes good account of the large currents that flow in the amplifier. Therefore, a couple of tracks have been paralleled instead of combined, so that the effect of currents in the power section on the input stages is minimal.

Populating the board is straightforward. Although not strictly necessary, it is advisable to match the transistors used in the differential amplifiers. This may be done conveniently on an h_{fe} tester by measuring the amplification at a collector current of about 1 mA. If such a tester is not available, use a base resistor that results in a collector current of about 1 mA measured with a multimeter. With the same resistor, test a number of other transistors and note the collector currents. Mount the selected pairs on the board and pack them closely together with a 5 mm wide copper ring (made from a piece of 12 mm copper water pipe—for an example, see Fig. 4.4).

Inductor L_1 consists of six turns, inner diameter 16 mm ($^5/_8$ in), of insulated copper wire 1.5 mm ($^1/_{16}$ in) thick. Mount RL_1 inside the coil.

The large transistors are located on one side of the board, so that they can be fixed directly to the heat sink. They must be insulated from the heat sink with the aid of ceramic washers.

The two sizes indicated on the board for T_{12} and T_{13} were explained earlier.

Connections from the power supply and to the loudspeaker are by means of terminal blocks that can be screwed on to the board.

Mount the two amplifier boards, mains transformers and electrolytic capacitors in a suitable enclosure. The wiring diagram for one channel is given in Fig. 5.5.

It is advisable to measure the supply voltages before they are applied to the amplifiers. Also, turn P_2 to maximum (wiper towards R_{33}) before connecting the power supply to the amplifiers. Set input presets P_1 to the centre of their travel. A few seconds after the supply has been switched on, the relay should come on. Connect a multimeter (1 V direct voltage range) and adjust P_1 until the meter reads zero (both channels!).

Switch the supply off again and connect a multimeter (100 mV d.c. range) across RF_1 or RF_2. Switch on the supply and adjust P_2 for a meter reading of about 10 mV: this corresponds to a quiescent current of 100 mA through T_{12} and T_{13}. After about half an hour, the current will have stabilized at about 200 mA (meter reading of about 20 mV). Readjust P_2 slightly if required. Note that owing to the positive temperature coefficient of IGBTs, the quiescent current does not increase but drops with rising dissipation.

Finally, recheck the direct voltages at the outputs of the amplifiers and, if necessary, readjust P_1 slightly.

The loudspeakers must be 4-ohm or 8-ohm types, whose impedance must not drop below 3 Ω. It is not permissible to connect two 4-ohm units in parallel to the amplifier, because that would give problems when large drive signals are applied to the IGBTs.

Parts list (one channel)

Resistors:
R_1 = 470 Ω
R_2 = 47.5 kΩ, 1%
R_3, R_4 = 100 Ω
R_5 = 2.0 kΩ, 1%

Fig. 5.4. Good thermal coupling between the transistors and heat sink ensures along life of the devices.

R_6 = 84.5 Ω, 1%
R_7, R_8 = 10 MΩ
R_9, R_{10}, R_{13}, R_{14} = 1.21 kΩ, 1%
R_{11}, R_{12}, R_{15}, R_{16} = 22.1 Ω, 1%
R_{17}, R_{18} = 499 Ω, 1%
R_{19} = 22 kΩ
R_{20} = 2.2 kΩ
R_{21} = 1 kΩ
R_{22}, R_{23} = 56.2 Ω, 1%
R_{24}, R_{27} = 150 Ω, 1%
R_{25}, R_{28} = 15.0 Ω, 1%
R_{26}, R_{29} = 15 Ω
R_{30} = 68 Ω, 5 W
R_{31} = 150 Ω, 5 W
R_{32}, R_{33} = 6.81 Ω, 0.6 W, 1%
R_{34} = 3.3 kΩ
R_{35} = 330 Ω
R_{36} = 8.2 MΩ
RF_1, RF_2 = 0.1 Ω, 5 W
RL_1 = 2.2 Ω, 5 W (fit inside L_1)
P_1 = 1 MΩ preset

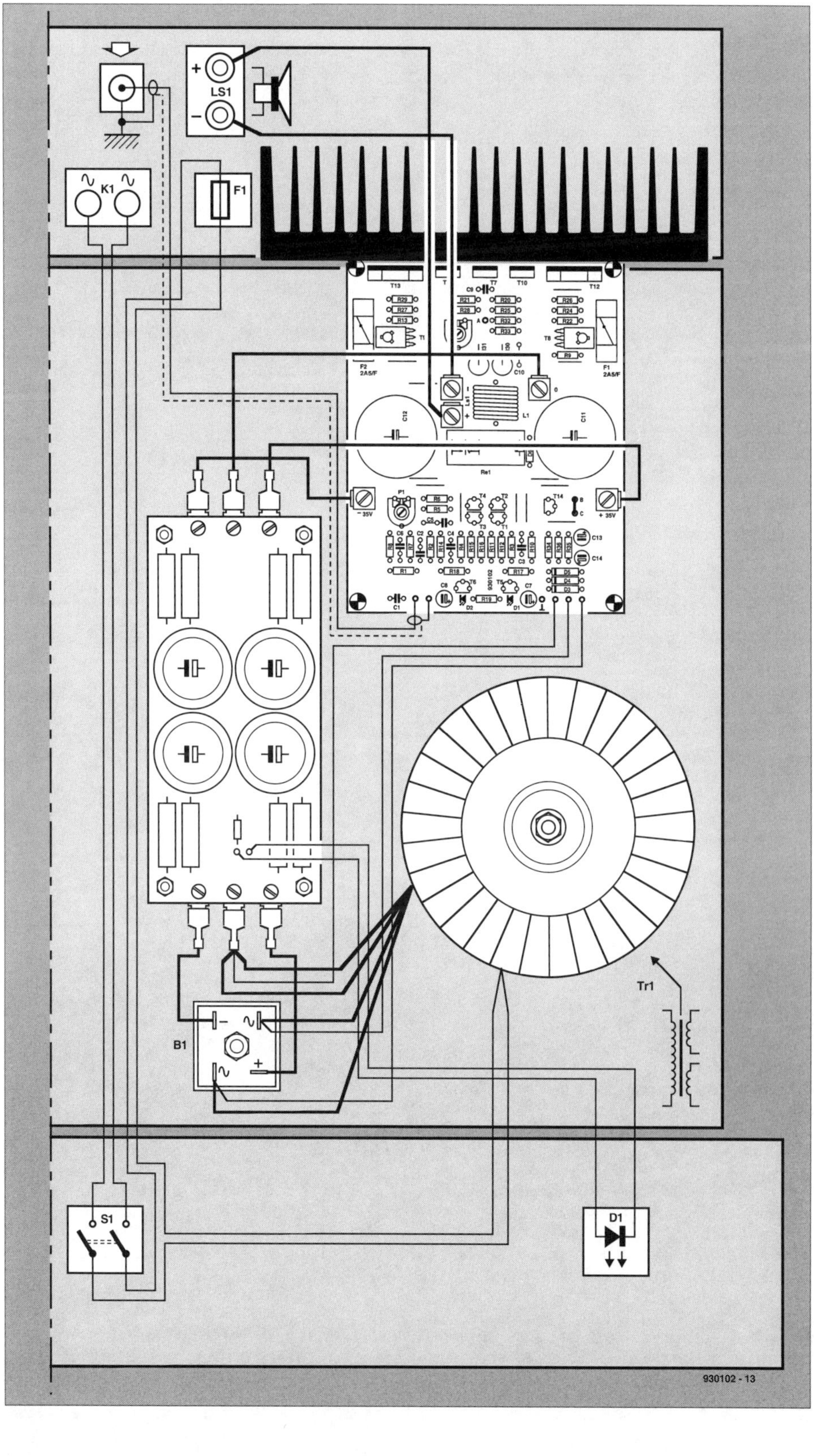

Fig. 5.5. Wiring diagram of the 90 W power amplifier.

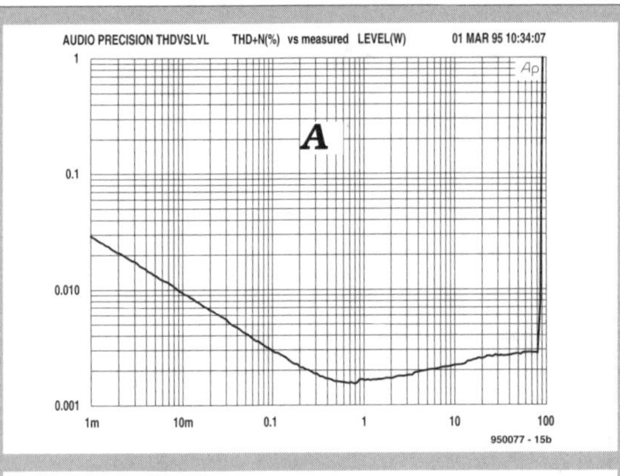

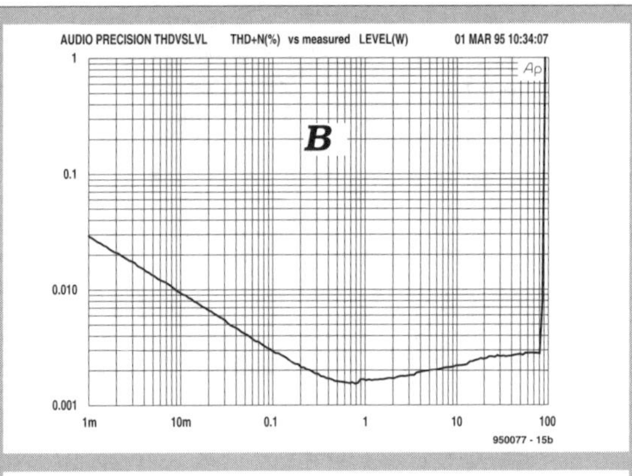

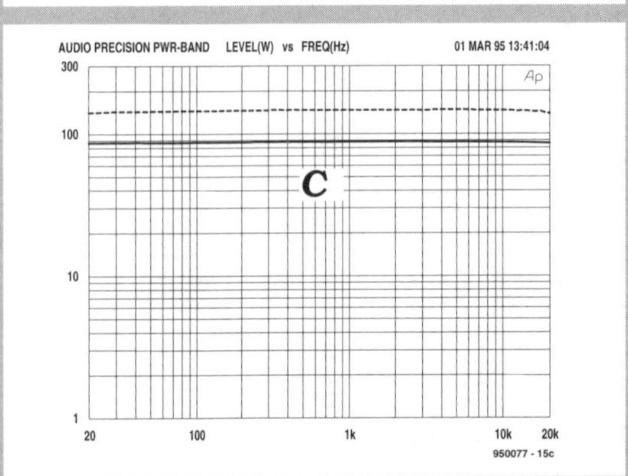

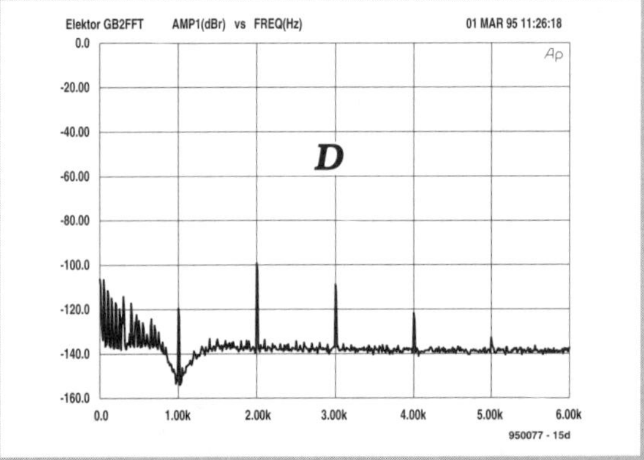

Brief technical data

Input sensitivity	1.1 V r.m.s.
Input impedance	47.7 kΩ
Power output (1 kHz, 0.1% THD)	88 W into 8 Ω
	146 W into 4 Ω
Music power (1 kHz burst, 5 cycles on, 5 cycles off)	94 W into 8 Ω
	167 W into 4 Ω
Power bandwidth (40 W into 8 Ω)	1.5 Hz – 115 kHz
Slew rate	>35 V μs^{-1}
Signal-to-noise ratio (1 W into 8 Ω)	105 dB (A-weighted)
	101 dB (linear 22 Hz – 22 kHz)
Harmonic distortion (1 W into 8 Ω)	0.002% (1 kHz)
(80 W into 8 Ω)	0.003% (1 kHz)
	<0.05% (20 Hz – 20 kHz)
Intermodulation distortion (50 Hz:7 kHz; 4:1)	0.002% (1 W into 8 Ω)
	0.003% (40 W into 8 Ω)
Dynamic intermodulation distortion (rectangular 3.15 kHz + sine wave 15 kHz)	0.0025% (1 W into 8 Ω)
	0.002% (80 W into 8 Ω)
Damping factor (at 8 Ω)	>600 (1 kHz)
	>400 (20 kHz)

Measurements for the characteristics shown were made with an Audio Precision analyser.

A shows the total harmonic distortion (THD+N) from 20 Hz to 20 kHz. The solid curve refers to 1 W into 8 Ω and the dashed one to 75 W into 8 Ω.

B shows the distortion at 1 kHz as a function of drive (bandwidth 22 Hz – 22 kHz; load 8 Ω). The sharp bend at the end of the curve is the clipping point.

C shows the maximum power output when the distortion is 0.1%. It shows that the power is independent of frequency, whether the load is 8 Ω (solid curve) or 4 Ω (dashed curve).

D shows a Fourier analysis of a 1 kHz signal (1 W into 8 Ω) with the fundamental suppressed. The 2nd, 3rd and 4th harmonics can be seen, but they are attenuated, respectively, by 100 dB, 110 dB and 120 dB with respect to the fundamental frequency.

P_2 = 1 kΩ preset

Capacitors:
C_1 = 2.2 μF, 50 V, polypropylene
C_2 = 2.2 nF
C_3, C_4 = 2.7 nF
C_5 = 330 pF, polystyrene, axial
C_6 = 33 nF
C_7, C_8 = 100 μF, 10 V, radial
C_9 = 1 μF, polypropylene, pitch 5 mm
C_{10} = 100 nF
C_{11}, C_{12} = 10,000 μF, 50 V, radial, for PCB mounting
C_{13} = 47 μF, 63 V, radial
C_{14} = 22 μF, 40 V, radial

Inductors:
L_1 = air-core, 0.1 mH (see text)

Semiconductors:
D_1, D_2 = 3 mm LED, red (1.6 V drop at 3 mA)
D_3–D_5 = 1N4003
D_6 = 1N4148
T_1, T_2, T_6 = BC550C
T_3–T_5 = BC560C
T_7, T_9, T_{10} = BD139
T_8, T_{11} = BD140
T_{12} = GT20D201
T_{13} = GT20D101
T_{14} = BC617

Miscellaneous;
Re_1 = relay, 24 V, 1 make contact (e.g., Siemens V23056-A0105-A101*)
F_1, F_2 = fuse, 2.5 A, fast, with holder for PCB mounting
Ceramic washers (5) for T_7, T_{10}, T_{11}
Mica washers (2) for T_{12}, T_{13}
Terminal block (5) (see text)
Heat sink, 0.6 K W^{-1} (e.g., Fischer SK85**)

Power supply:
Mains transformer, 2 × 30 V, 375 VA
Mains on-off switch with indicator
Fuse 1 A, slow with holder
Bridge rectifier Type B200C35000
Electrolytic capacitor (4), 10,000 μF, 50 V
4 off Resistor 0.22 Ω, 5 W

* ElectroValue, 3 Central Trading Estate, Staines, TW18 4UX, ☎ (01784) 442 253. Private customers welcome.

** Dau (UK) Ltd, 7075 Barnham Road, Barnham, West Sussex PO22 0ES. Tel. (01243) 553 031. Trade only, but information as to your nearest dealer will be given by telephone.

Chapter 6

100 W power amplifier

The simplified circuit diagram in Fig. 6.1. clearly shows the symmetrical design of the amplifier. The input stage is formed by differential amplifiers T_1, T_3, whose gain is limited to about 40 dB (×100). The stage is coupled to differential amplifiers T_2, T_4, whose gain is around 10 dB (×3). These amplifiers are linked via optoisolators to drivers T_{12}, T_{13}, which form the link between the input and output stages. The zener symbol between the drivers indicates a variable zener transistor, which enables the accurate setting of the quiescent current through the output transistors. The output stage consists of transistors T_{16} and T_{17}, each of which drives two parallel-connected power transistors: T_{18}, T_{19} and T_{20}, T_{21} respectively.

TECHNICAL DATA

Input sensitivity	1 V r.m.s.
Input impedance	46.5 kΩ
Output power (0.1% THD)	100 W into 8 Ω
	175 W into 4 Ω
Music power (500 Hz burst: 5 periods on, 5 periods off)	105 W into 8 Ω
	185 W into 4 Ω
Power bandwidth (50 W into 8 Ω)	1.5 Hz – 220 kHz
Slew rate	>50 V μs^{-1}
Signal-to-noise ratio (1 W into 8 Ω)	>102 dB (A weighted)
Harmonic distortion (B = 10 Hz – 80 kHz)	
at 1 W into 8 Ω	<0.0025% (1 kHz)
at 90 W into 8 Ω	<0.0015% (1 kHz)
	<0.015% (20 Hz – 20 kHz)
Intermodulation distortion (50 Hz : 7 kHz; 4 : 1)	<0.002% (1 W in to 8 Ω)
	<0.003% (50 W into 8 Ω)
Dynamic IM distortion (block 3.15 kHz with 15 kHz sine wave)	<0.0025% (1 W into 8 Ω)
	<0.0015% (100 W into 8 Ω)
Damping factor (8 Ω output)	>1000 (1 kHz)
	>290 (20 kHz)

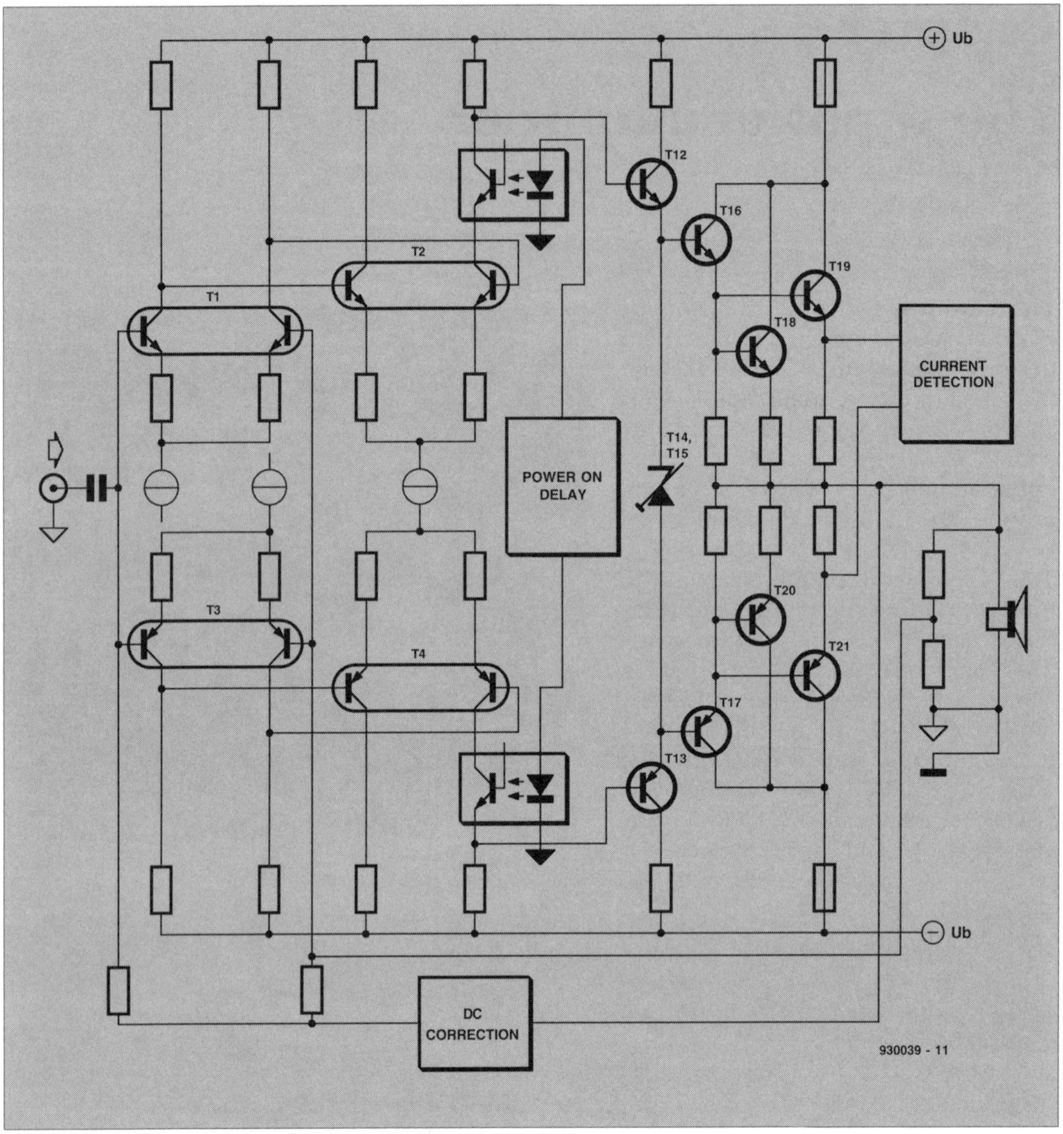

Fig. 6.1. Simplified circuit diagram of the amplifier (one channel).

The optoisolators provide suppression of the on/off switching clicks, which, owing to the absence of an output relay, is essential. The POWER ON DELAY (which is part of a protection circuit on a separate PCB) ensures that the LEDs in the optoisolators light up only gradually after switch-on. Consequently, the optotransistors in the collector circuits of T_{12} and T_{13} come into conduction slowly, resulting in the power supply to the output transistors building up gradually. This arrangement effectively prevents annoying clicks at switch-on. When the mains is switched off, the LEDs go out rapidly, resulting in an abrupt removal of power from the output transistors. Consequently, the output stage gets no time to produce irritating switch-off phenomena.

The protection circuit uses no relay and relies on the current through the emitter resistors of the output transistors and the output voltage for correct operation. If, for whatever reason, the current rises unduly or a direct voltage appears at the output, two actions take place: the optoisolators are cut off, so that the power to the output stages is removed instantly, and the supply lines are shorted to earth with the aid of triacs, A somewhat drastic, but very effective and reliable means of safeguarding the amplifier from compression and distortion at large drive voltages.

Not shown in Fig. 6.1 is a thermal protection circuit, which, again via the optoisolators, removes the power from the output transistors if the temperature of these devices rises unduly.

Amplifier circuit

The circuit diagram of the amplifier (single channel) in Fig. 6.2 is, in large parts, similar to that of the 'Output amplifier for ribbon loudspeakers' (2). The differences lie mainly in the much higher supply voltage (2 × 42 V instead of 2 × 15 V) and the much lower peak output current. Other, smaller, differences are a number of altered component values, new types of transistor in various positions, lower value fuses, fewer parallel-connected emitter resistors for the output transistors and an additional capacitor, C_{fb}, in the feedback loop. Also, the separate sense lines to the loudspeaker(s) are no longer there, and the connections between IC_3 and the supply lines have been broken (D_9 and D_{10} are omitted), because the supply voltage for the DC correction is now provided by an auxiliary supply.

At the input, there is a combination of a high-pass section, C_1-R_1, and a low-pass section, C_2-R_2. The first filter prevents any direct voltage entering the amplifier, while the second restricts the upper limit of the input bandwidth to about 300 kHz.

The signal is applied to differential amplifiers T_1 and T_3. Since the maximum permissible supply voltage of the excellent dual transistors MAT02 and MAT03 is too low, these types had to be replaced by a 2N2914 and a BFX36 respectively. Frequency compensation of the input stage is provided by networks R_5-C_3 and R_{10}-C_4. The open-loop bandwidth of the amplifier is around 14 kHz. The gain of the amplifiers is determined by the ratio of their collector and emitter resistors.

Current sources T_5 and T_6 guarantee stable operation of T_1 and T_3. This is augmented by the arrangement of holding the current through reference LEDs D_1 and D_2 stable with the aid of a third current source: T_9.

The next stage consists of dual transistors T_2 and T_4, which are controlled by current sources T_7 and T_8. Again, the current through LEDs D_3 and D_4 is held stable by current sources T_{10} and T_{11}.

Diodes D_1–D_4 must be of the same type with a forward voltage of 1.55–1.65 V to ensure that the operating points of the various transistors remain unchanged.

The optoisolators for the power-on delay are inserted in the collector leads of T_{2b} and T_{4b}. Their LEDs are controlled by the external protection circuit via B1 and B2.

The collectors of predrivers T_{12} and T_{13} are connected via T_{14} and T_{15}. The latter transistors form a zener transistor, the voltage across which, and thus the quiescent current through the output transistors, is set as appropriate with P_1.

Each of the two halves of the output stage, a super emitter follower, consists of a driver, T_{16} and T_{17}, and two power transistors, T_{18}, T_{19} and T_{20}, T_{21} respectively. The power transistor pairs are connected in parallel, but have their own emitter resistors. Transistor T_{22} monitors the voltage drop across the emitter resistors via potential divider R_{44}-R_{45}. If the total peak output current exceeds 18 A, T_{22} switches on, which causes the external protection circuit to be actuated.

A 6.3 A fuse has been incorporated in the collector line of each of the power transistors. The + and – taps in these lines are connected to the triacs (+ and –) in the protection circuit—see Fig. 6.2. If the fuses blow, D_5 and D_6 light to indicate this. In practice this will happen hardly ever, because the primary fuse will almost certainly blow first.

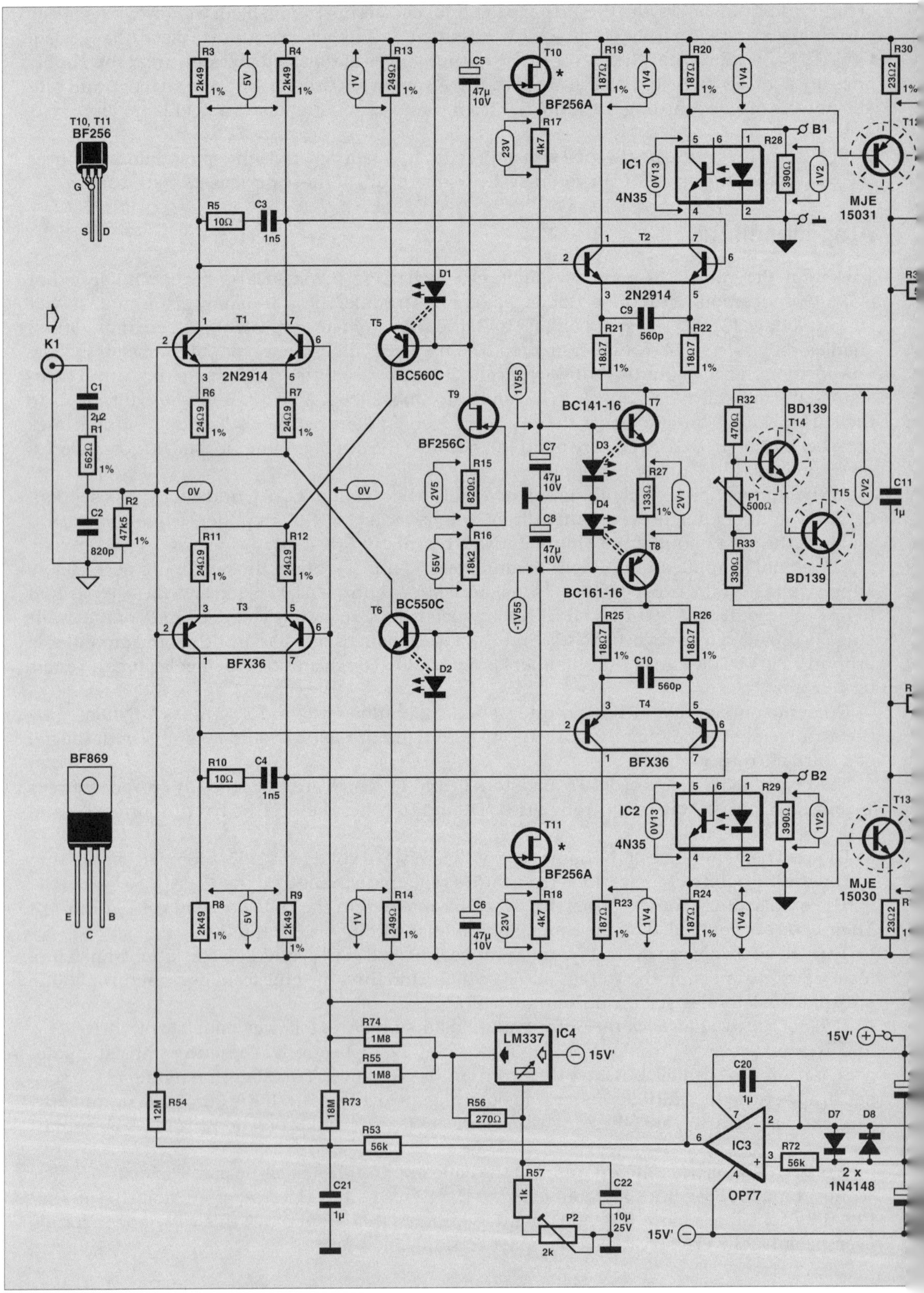
T10, T11
BF256
G
S
D
K1
C1
2μ2
R1
562Ω
1%
C2
820p
R2
47k5
1%
BF869
E
C
B
R3
2k49
1%
5V
R4
2k49
1%
1V
R13
249Ω
1%
C5
47μ
10V
T10
BF256A
R17
4k7
23V
R19
187Ω
1%
1V4
R20
187Ω
1%
1V4
R30
23Ω2
IC1
4N35
0V13
R28
390Ω
1V2
B1
MJE
15031
R5
10Ω
C3
1n5
T2
2N2914
D1
T1
2N2914
T5
BC560C
0V
C9
560p
R21
18Ω7
R22
18Ω7
R6
24Ω9
R7
24Ω9
T9
BF256C
1V55
BC141-16
T7
D3
R15
820Ω
2V5
C7
47μ
10V
R27
133Ω
2V1
R32
470Ω
BD139
T14
P1
500Ω
T15
2V2
C11
1μ
R16
18k2
55V
D4
C8
47μ
10V
T8
BC161-16
R33
330Ω
BD139
R11
24Ω9
R12
24Ω9
-1V55
T3
BFX36
T6
BC550C
D2
R25
18Ω7
R26
18Ω7
C10
560p
T4
BFX36
R10
10Ω
C4
1n5
IC2
4N35
0V13
R29
390Ω
1V2
B2
T13
T11
BF256A
MJE
15030
R8
2k49
R9
2k49
R14
249Ω
C6
47μ
10V
R18
4k7
R23
187Ω
R24
187Ω
R74
1M8
R55
1M8
LM337
IC4
15V'
15V'
C20
1μ
R54
12M
R73
18M
R56
270Ω
R53
56k
IC3
OP77
R72
56k
D7
D8
2 x
1N4148
C21
1μ
R57
1k
C22
10μ
25V
P2
2k
15V'

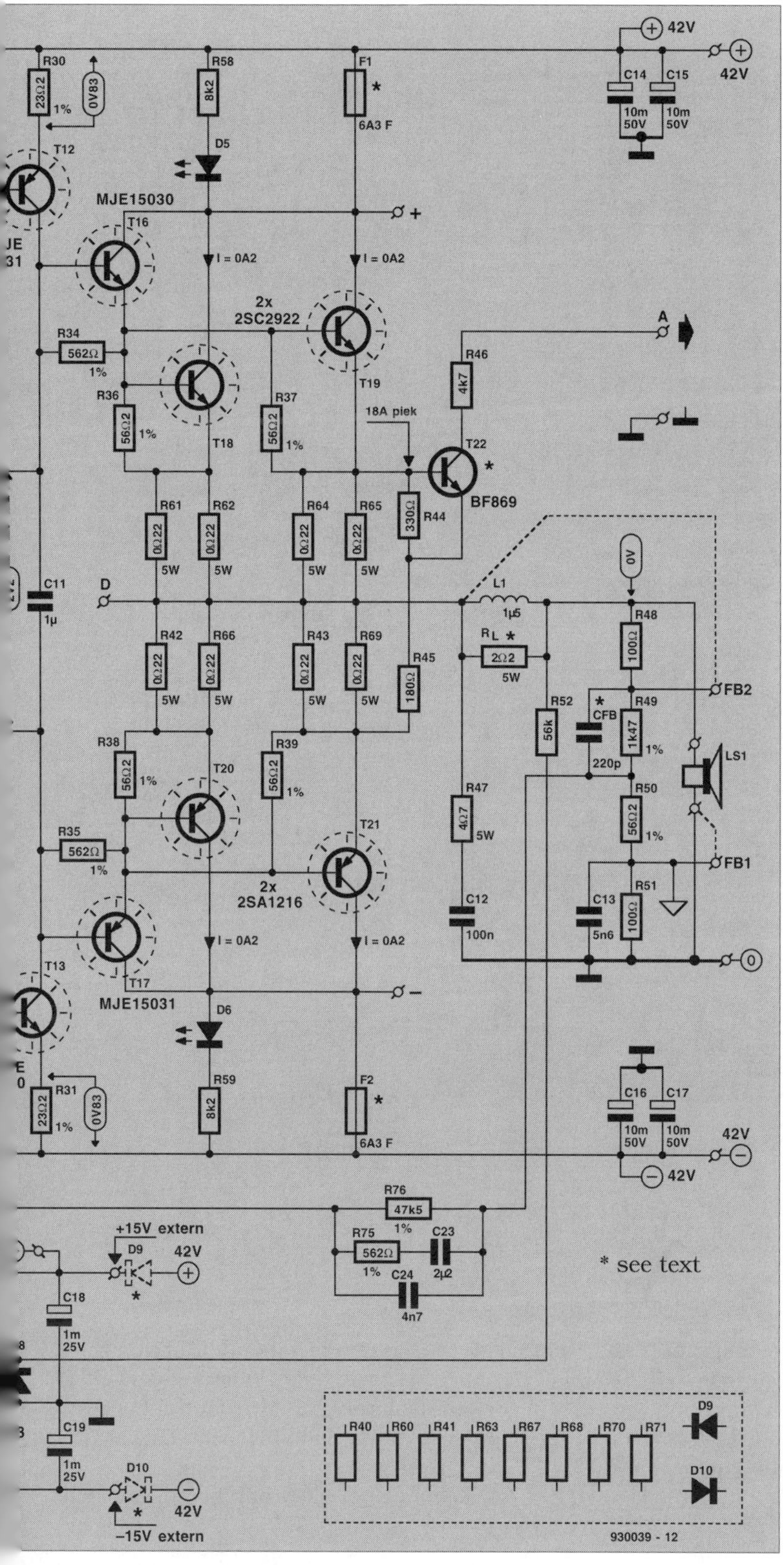

Boucherot network R_{47}-C_{12} ensures that the amplifier remains loaded adequately at high frequencies.

Output power is supplied to the loudspeaker(s) via inductor L_1, which limits the rise time of the signal when the load is capacitive, and potential divider R_{48}-R_{51}. The ratio $R_{49}:R_{50}$ determines the feedback factor. The feedback voltage at the junction of these resistors is applied to the bases of T_{1b} and T_{3b}. Capacitor C_{fb} accelerates the action of the feedback circuit. Network R_{75}-R_{76}-C_{23}-C_{24} serves to keep the base impedances at the left-hand and right-hand halves of the differential amplifier equal to ensure optimum common-mode suppression.

An additional stabilized voltage is produced with the aid of IC_4 to limit the offset voltage at the output to a minimum. The voltage is used to compensate the base currents of the input stage via R_{55} and R_{74} to such a degree that the direct voltage at the basis is virtually nil. Integrator IC_3 removes any residual direct voltage (caused by, for instance, temperature

Fig. 6.2. Circuit diagram of the 100 W power amplifier. The double dashed lines indicate that thermal coupling is required.

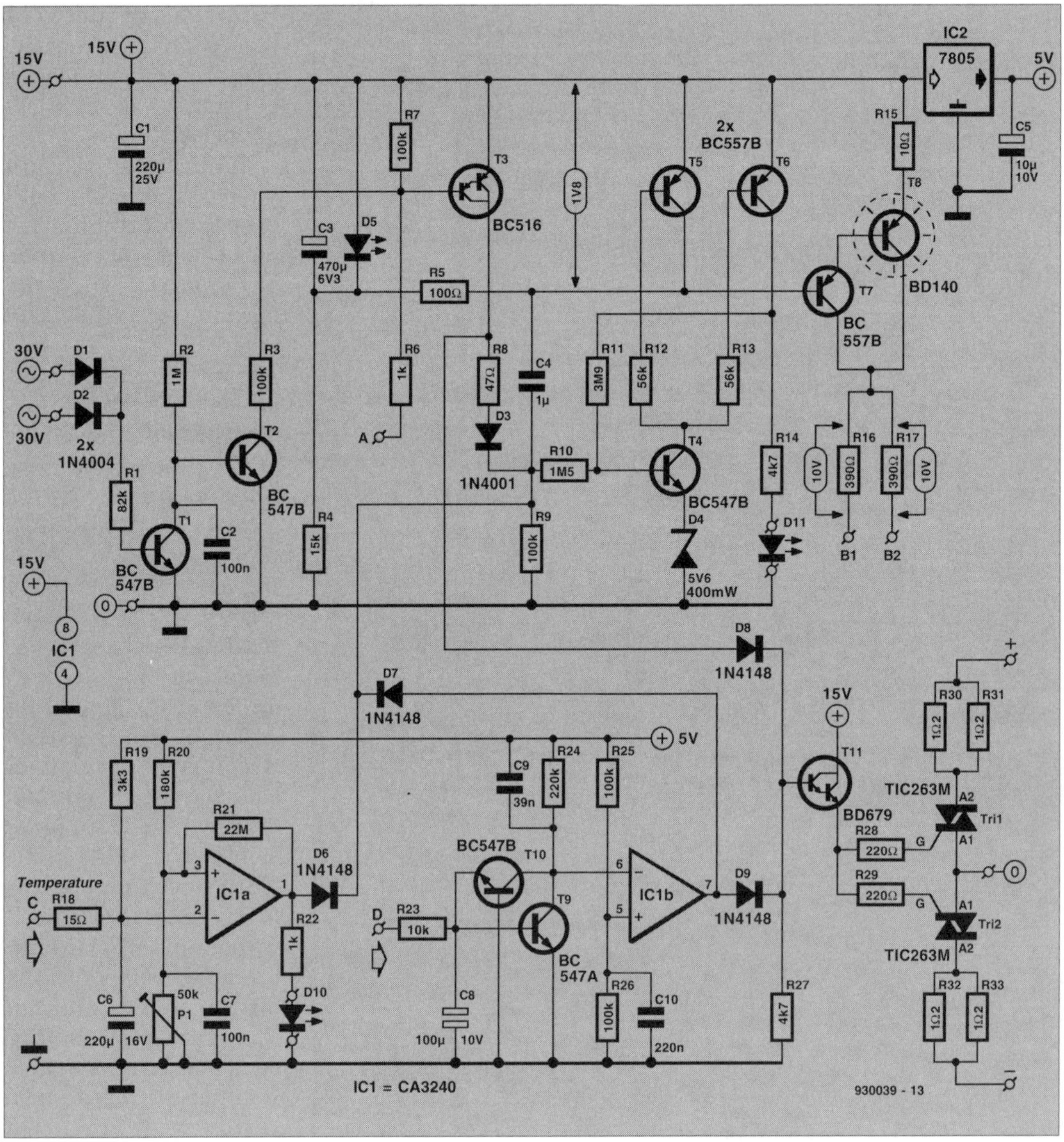

Fig. 6.3. The protection circuit does not use relays.

changes). Its output is used to adjust the base voltage of T_1 and T_3 via R_{53}-C_{21}, R_{54} and R_{73}.

Protection circuit

The protection circuit, whose diagram is shown in Fig. 6.3., serves a number of functions.
Switch-on click. When the power is switched on, C_3 is charged slowly via R_4. After a short while, darlington T_7-T_8 comes on, whereupon the LEDs in the optoisolators begin to conduct gradually. When the potential across C_3 has risen to 1.7–1.8 V, the diodes light brightly. This situation remains because D_5 holds the base voltage of T_7 at a fixed level.
Switch-off click. As long as the mains is switched on, T_1 conducts every half period (since it

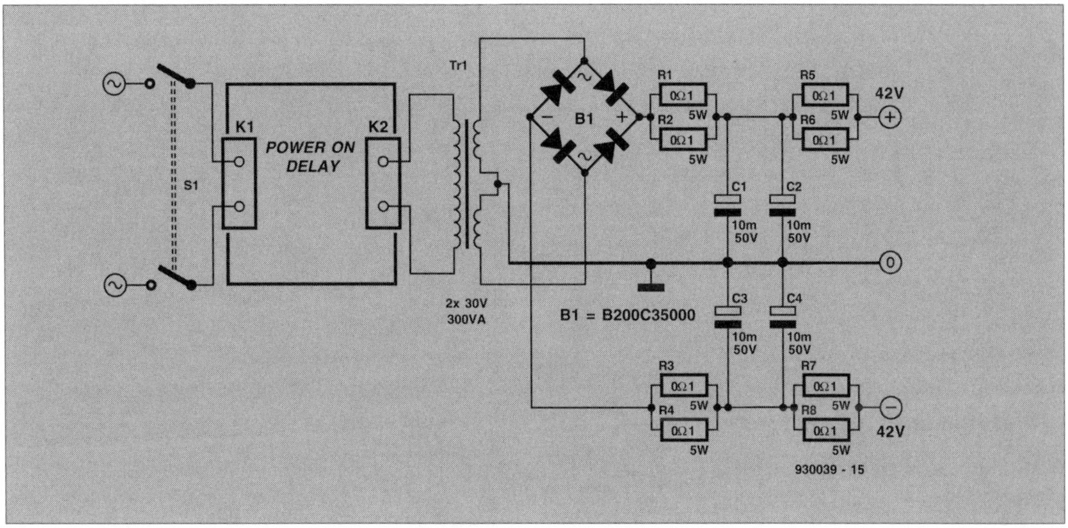

Fig. 6.4. Circuit diagram of the power supply.

is connected directly to the secondary of the mains transformer via D_1 and D_2), whereas T_2 and T_3 are off. When the mains is switched off, T_2 and T_3 are switched on within half a period. Schmitt trigger T_4-T_6 changes state, whereupon T_5 conducts and T_7 and T_8 are cut off, so that the optoisolators are deactuated. This state is indicated by the lighting of D_{11}.

Over-current. Current monitor T_{22} in the amplifier is linked to terminal A via the auxiliary circuit (Fig. 6.5). If the output current of the amplifier rises unduly, T_3 switches on, whereupon the optoisolators are deactuated as described earlier. At the same time, triacs Tri_1 and Tri_2 are switched on by T_{11} via D_8 and short-circuit the supply lines. If this drastic action is not acceptable, D_8 may be omitted.

Temperature. Transistor T_{12} functions as a temperature sensor diode which is mounted on the heat sink of T_{14}–T_{21}. The voltage across the transistor is compared by IC_{1a} with a preset (P_1) reference voltage. If the sensor voltage drops (temperature rises) below the reference potential, IC_{1a} changes state and the optoisolators are deactuated via D_6. Once the temperature has dropped below a safe value, the amplifier is switched on again.

Direct voltage. If there is a direct voltage greater than ±0.6 V at the output of the amplifier (point D), either T_9 or T_{10} is switched on via low-pass section R_{23}-C_8. In either case, the negative input of IC_{1b} is pulled to earth and its output changes from low to high. This results in the triacs being switched on via D_9 and T_{11}. At the same time, D_7 causes the Schmitt trigger to change state, resulting in the optoisolators being deactuated. This function may be disabled by omitting D_7 and D_9.

Power supply

The power supply for the amplifier (see Fig. 6.4) is a traditional design. A double-pole mains switch and switch-on delay circuit are followed by the usual combination of transformer, bridge rectifier and electrolytic smoothing, buffer and reservoir capacitors. The series resistors limit the peaks in the charging currents of the electrolytic capacitors and, together with these capacitors, form an effective mains filter.

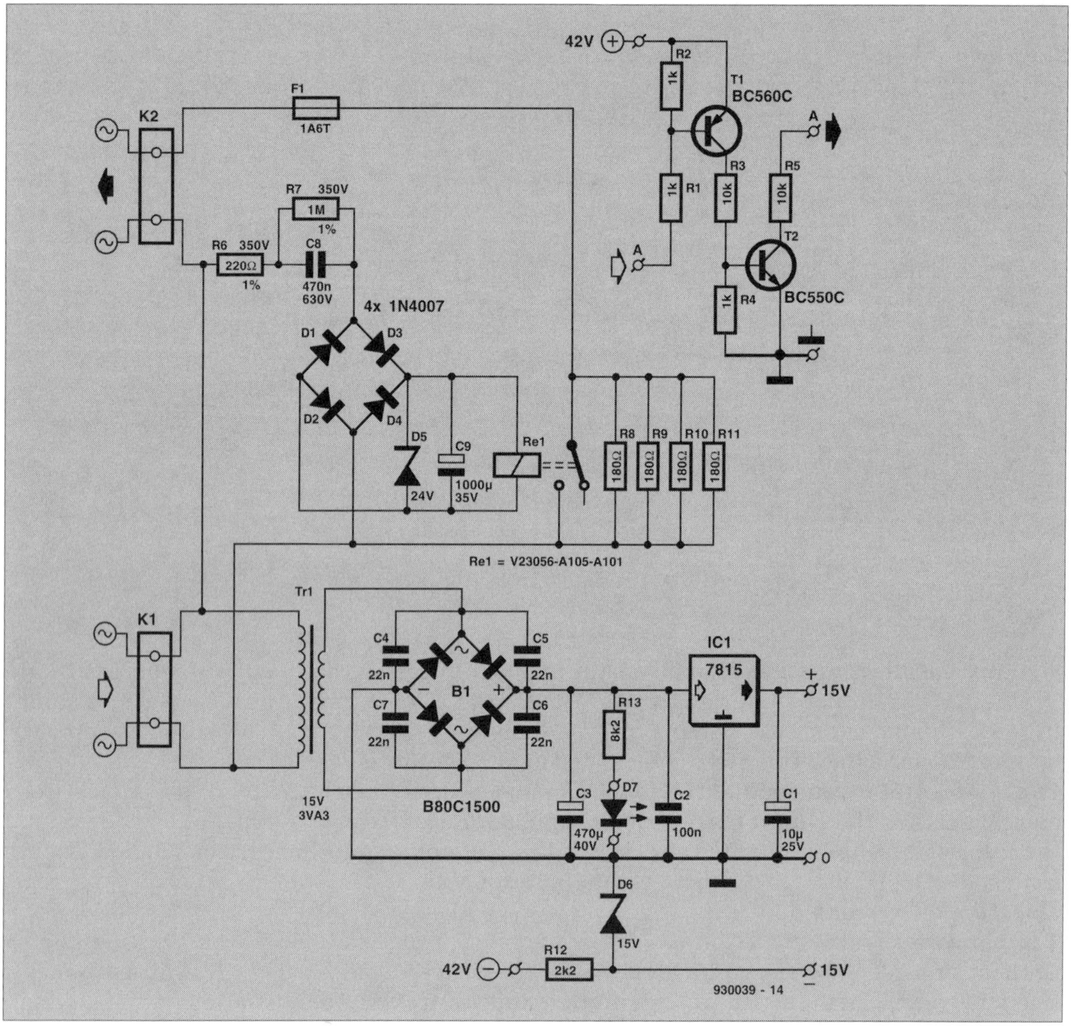

Fig. 6.5. Diagram of the auxiliary circuit.

Auxiliary circuit

The auxiliary circuit (see Fig. 6.5) consists of a ±15 V supply and a mains on delay.

The ±15 V supply is provided by Tr_1, B_1 and various capacitors. The +15 V line is stabilized by a Type 7815 regulator, but the −15 V line by a simple zener diode, D_6, since that line is further stabilized in the amplifier by a Type LM337 regulator (IC_4). Mains on indication is provided by R_{13} and D_7.

The section based on T_1 and T_2 acts as a monitor for the current protection. It is inserted between amplifier output A and terminal A on the protection board. The circuit transforms the potential measured by current sensor T_{22} into a much lower voltage (relative to earth) suitable for the protection circuit.

The design of the mains on delay circuit between K_1 and K_2 is straightforward. As soon as the mains at K_1 is switched on, the amplifier is powered immediately via K_2, but the current to it is limited severely by R_8–R_{11}. This means that the electrolytic capacitors in the main power supply are charged relatively slowly and that the mains fuse does not have to cope with a sub-

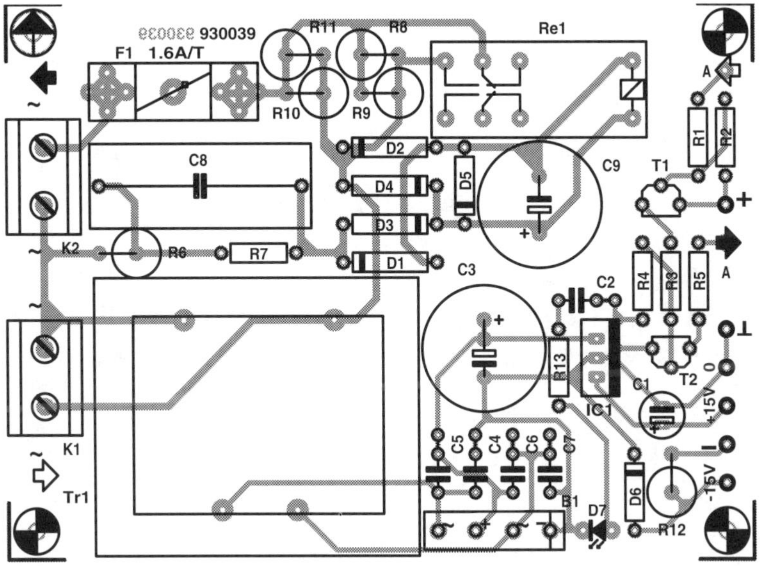

Fig. 6.6. Component layout of the PCB for the auxiliary circuit. For track layout, see page 235.

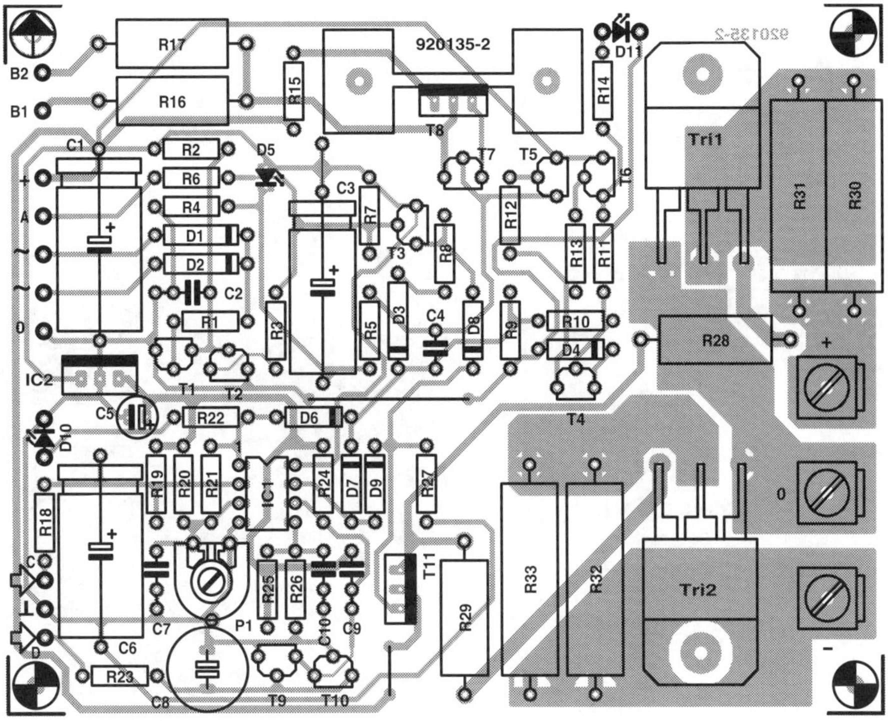

Fig. 6.7. Component layout of the PCB for the amplifier. The track layout is given on page 235.

stantial surge. Within a short while, C_9 is charged via R_6, C_8 and D_1–D_4 to such an extent that the potential across it is sufficient to energize relay Re_1. Parallel-connected resistors R_8–R_{11} are then short-circuited by the relay contact and full power is applied to the amplifier.

Diode D_5 limits the voltage across C_9, while R_7 ensures that 'a.c. resistor' C_8 is discharged

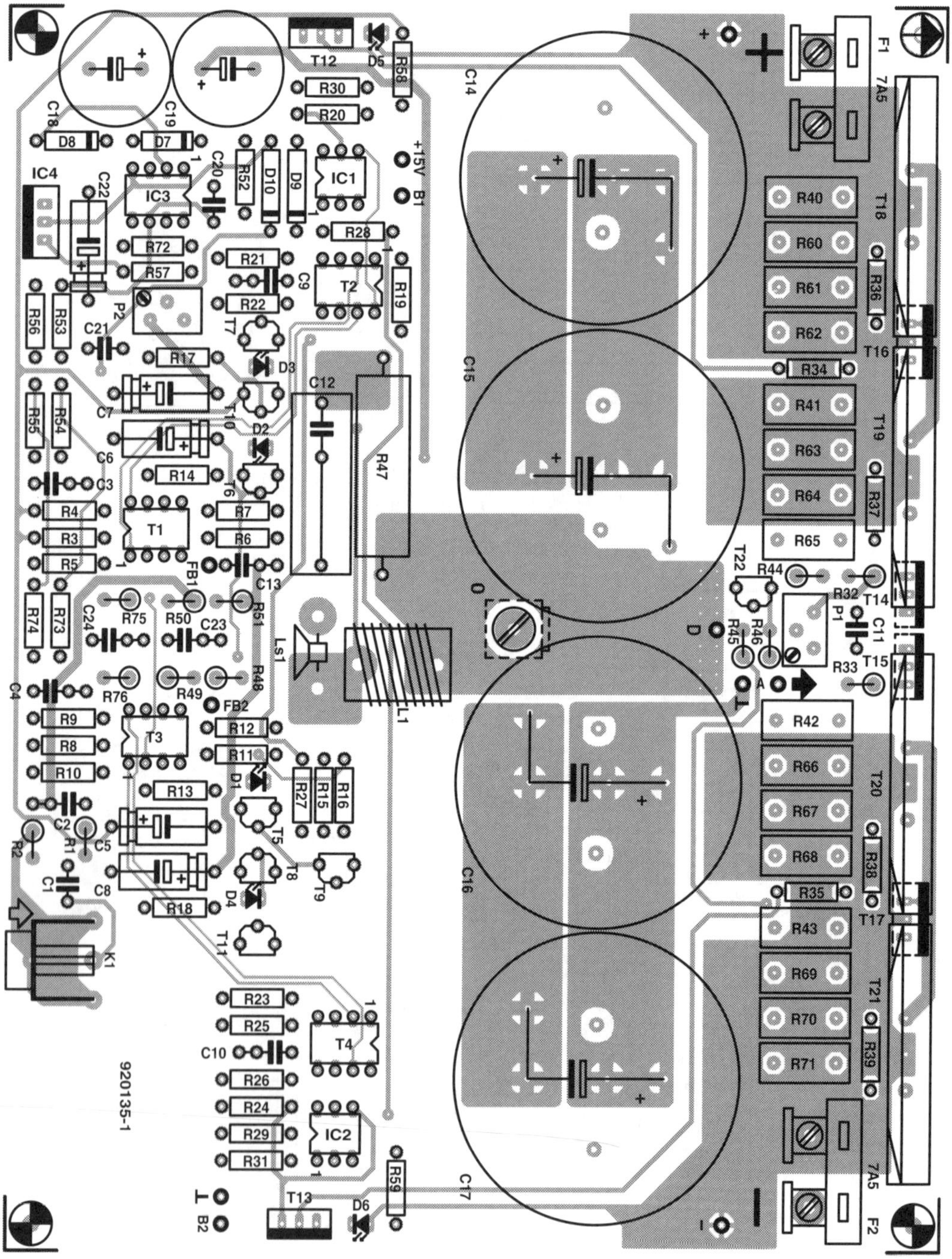

Fig. 6.8. Component layout of the PCB for the protection circuit.
The track layout and component overlay are given on pages 237 and 239.

rapidly when the supply is switched off.

Construction

The printed-circuit board design for the auxiliary circuit is shown in Fig. 6.6, and the component layout of the PCB for the protection circuit and the amplifier in Fig. 6.8 and Fig. 6.7 respectively. There is no PCB design for the power supply of Fig. 6.4, so this must be constructed on a prototyping board.

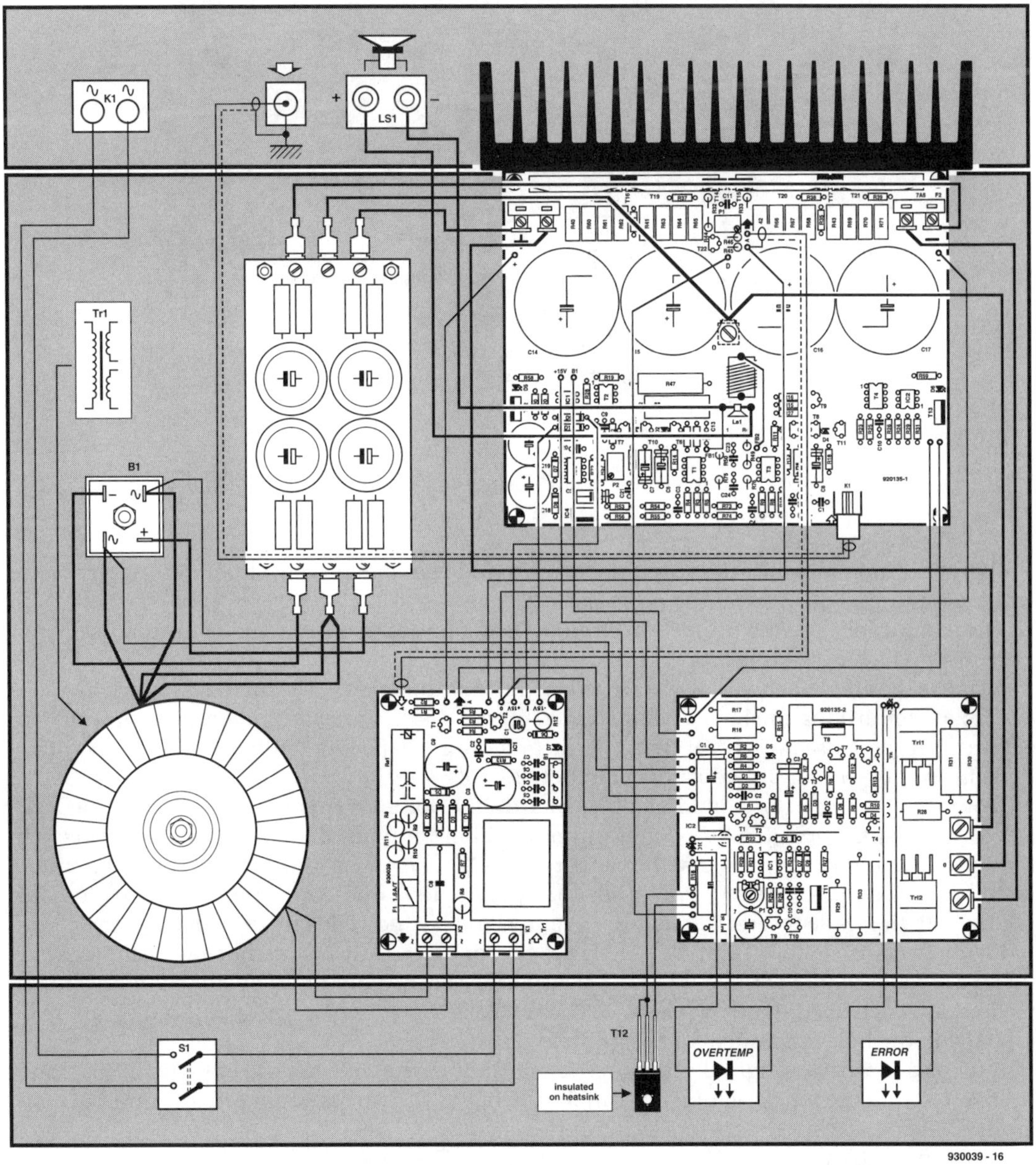

Fig. 6.9. Wiring diagram for the 100 W amplifier (one channel).

Fig. 6.10. Completed single-channel (mono) amplifier module.

Building the auxiliary circuit is straightforward and consists merely of populating the board with reference to Fig. 6.6 and the parts list.

The same applies to the protection circuit. Do not forget the heat sink for T_8 and use heavy duty terminals for the connections to the triacs.

As far as the amplifier board is concerned, a full description was given in Ref. 2, which includes a template for drilling the heat sink (repeated on page 81). The following description will concentrate on the more important details only and on the aspects resulting from the modifications in the present amplifier over that in Ref. 2.

Mount most of the components on the board, but, for convenience's sake, leave the large electrolytic capacitors till a little later. Although different types of dual transistor are used in the T_1–T_4 positions, they are also housed in a TO-78 enclosure and they are pin-compatible with the earlier types. The types of transistor used in the T_7 and T_8 positions are also pin-compatible with the earlier types, but they are slightly larger, which adds to the density on the board.

Field-effect transistors T_{10} and T_{11} need to be modified slightly before they can be fitted. Since their gate must not be connected to the board, the relevant pin must be linked to the source pin and as much of it cut off as possible. The result is a FET with only two pins which are to be inserted in the holes intended for the drain and source.

Since the board was originally intended for a BC550C, the collector pin of T_{22} (a BF869) must be inserted into the hole originally meant for the base, and the base pin into the original commector hole. Take care not to short-circuit these pins.

The same sort of thing needs to be done with T_{12} and T_{13}. Since the types now used are housed in a TO-220 case, they must be mounted on the board the wrong way around, that is, with the

Fig. 6.11. Mono amplifier module seen from above with cover of enclosure removed.

metal plane at the inside.

Resistor R_L must be mounted centrally inside air-cored inductor L_1. The coil consists of 10 turns of 1.5 mm diameter enamlled copper wire wound on a 15 mm diameter tube. Make small loops at the ends of the wire so that the coil can be mounted on the board with small bolts and nuts.

There are no holes provided for C_{fb}: this capacitor should be soldered at the track side of the board directly across R_{49}.

Diodes D_9 and D_{10} must no**t** be used in the present amplifier. Fit PCB pins in the holes intended for these diodes and link them to the ±15 V supply on the auxiliary circuit board.

Since sense lines to the loudspeaker(s) are no longer used, terminal FB2 must be linked to the junction L_1-R_{47}, and FB1 to the earth terminal of the loudspeaker (clamp together with the loudspeaker cable in the spring-loaded terminal).

If car-type fuses of 6.3 A are unobtainable, use a glass fuse mounted in a traditional holder soldered on to the screw type terminals.

Use flat car-type connectors for terminating the supply lines and also for the output lines (these come at the underside of the board).

The board is intended to form a (lateral) 'T' with the heat sink. Screw the power transistors

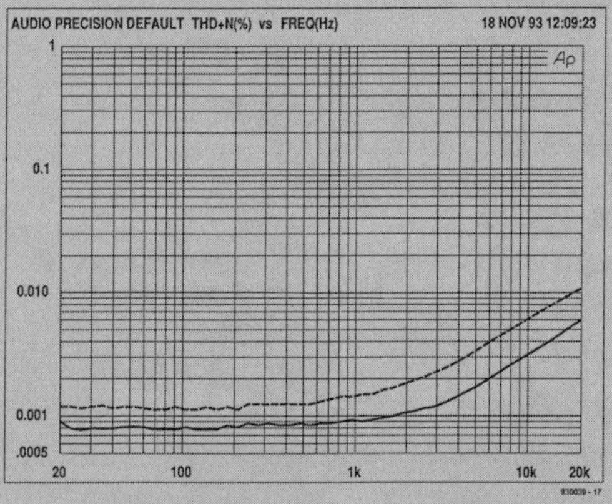

A

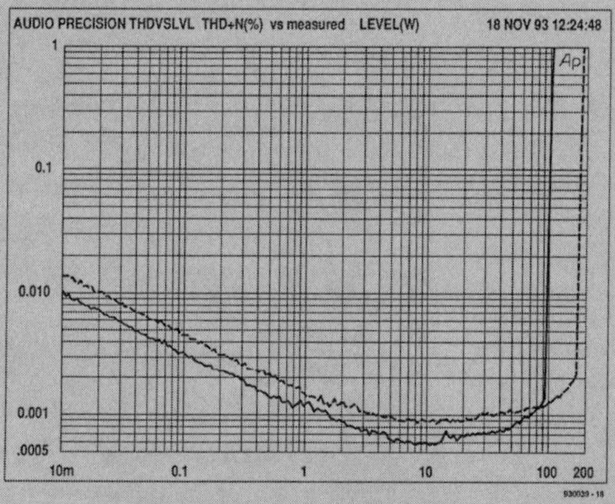

B

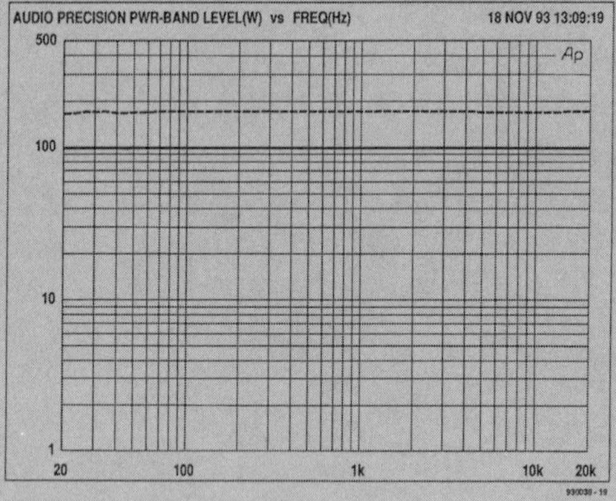

C

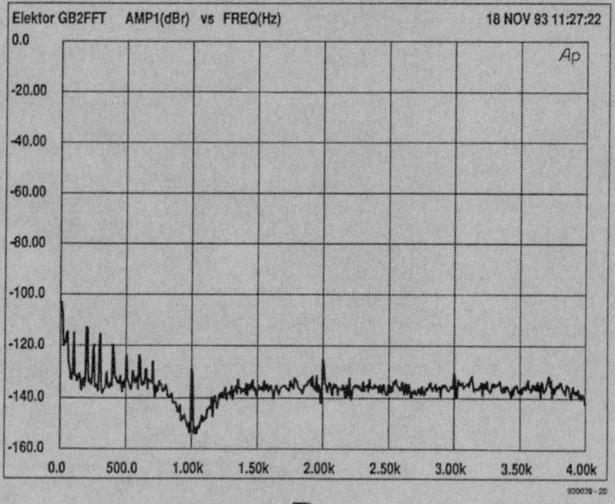

D

The results of the measurements on the prototype are, without exception, very good. Noise and residual distortion are far removed from the human limits of observation. Pulse response and damping factor are almost ideal, although that is a strong word to use.

Listening tests confirm the measured values. Even the most demanding recordings from our record library were reproduced without any discernible hitch.

The four characteristics shown above, obtained with an Audio Precision Analyser, illustrate these statements.

Characteristic A shows the total harmonic distortion plus noise (THD + N) over the frequency range 20 Hz to 20 kHz at an output power of 50 W.

Characteristic B illustrates the distortion at 1 kHz as a function of the drive level over the frequency range 22 Hz to 22 kHz. It is clear that between 10 W and 20 W the limits of the analyser were reached. The sharp bend at the right-hand side of the curve is the clipping point.

Characteristic C shows the maximum power at a distortion of 0.1%; it will be seen that within the audio range the power is completely independent of frequency.

Characteristic D gives the Fourier analysis of a 1 kHz signal at an output of 1 W into 8 Ω with the fundamental frequency suppressed. The attenuation of the 2nd harmonic is about 125 dB, while the 3rd harmonic is lost in the noise.

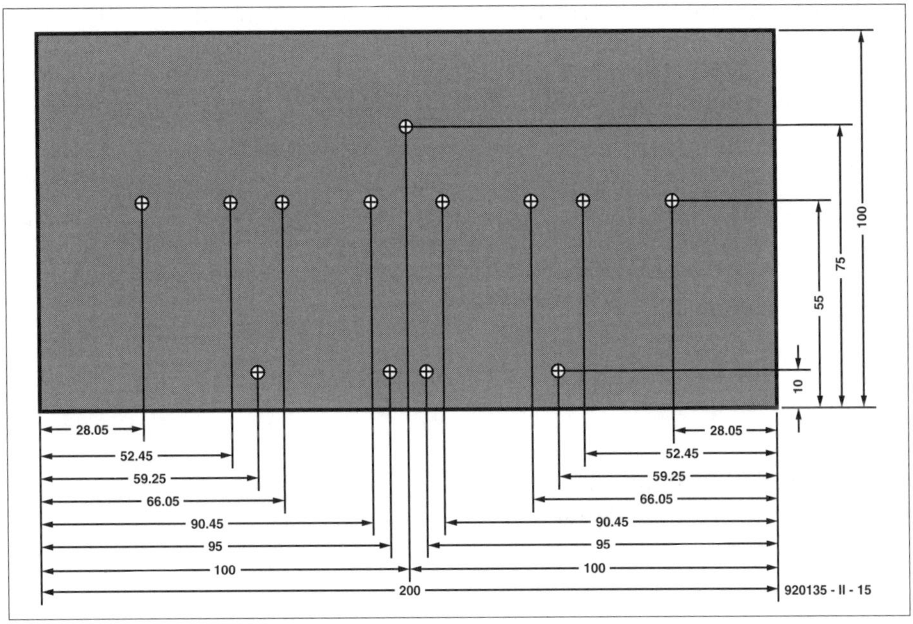

Fig. 6.12. Template for drilling the heat sink.

and T_{14}–T_{17}, mounted on the amplifier board, to the heat sink: use insulating washers in all cases. Do not yet fit T_{12} to the upper half of the heat sink: see under 'Alignment'.

A completed single-channel amplifier module is shown in Fig. 6.10. Its wiring diagram is given in Fig. 6.9. Use heavy-duty wire (≥4 mm^2 cross-section) for the power lines, output lines to the loudspeaker(s) and the lines between the amplifier board and the triacs on the protection board. Use screened cable for the input lines. Such cable is also recommended to link points A on the three boards.

It proved impossible to obtain an enclosure suitable for housing two amplifier modules to form an integrated stereo amplifier; the prototype stereo amplifier therefore consists of two individual mono amplifier modules.

Alignment

Set P_1 to maximum resistance: check this with an ohmmeter. Switch on the mains, and measure the voltage at the various points shown in Fig. 6.2 and Fig. 6.3. If these are all as specified, measure the voltage across one of the emitter resistors of T_{18}–T_{21} with a digital voltmeter and adjust P_1 until the meter reads 22 mV (which corresponds to a quiescent current of 200 mA per transistor).

Next, again with a digital voltmeter, measure the direct voltage at the base of T_1 or T_3; adjust P_2 to make the reading exactly zero.

Then, measure the output voltage of IC_4, which must be between 0 V and –11 V. If the voltage is more negative, reduce the values of R_{55} and R_{74} by about 10% (after switching off the amplifier!).

Next, check the output voltage of IC_3 (pin 1). After the amplifier has been switched on for a little while, this voltage should remain within ±10 V. If this is not the case, reduce R_{73} to the next lower E12 value.

Finally, connect T_{12} via a short length of cable to terminals C and earth on the protection board. Place the transistor in a dish of warm water (about 60 °C), making sure that the device's terminals are not short-circuited by the water. Adjust P_1 on the protection board until D_{10} on the same board just lights. Then, fit T_{12} to the heat sink. Bear in mind that the temperature of the heat sink even in normal operation becomes at least 20 °C higher than ambient.

References:

1. 'Medium power a.f. amplifier', *Elektor Electronics*, October/November 1990
2. 'Output amplifier for ribbon loudspeakers', *Elektor Electronics*, November/December 1992.

PARTS LIST (One channel)

AUXILIARY CIRCUIT

Resistors:

R_1; R_2; R_4 = 1 kΩ
R_3; R_5 = 10 kΩ
R_6 = 220 Ω 1 W; e.g., PR01 (Philips)
R_7 = 1 MΩ, 350 V; e.g., SFR25H (Philips)
R_8–R_{11} = 180 Ω, 5 W
R_{12} = 2.2 kΩ, 1 W; e.g., PR01
R_{13} = 8.2 kΩ

Capacitors:

C_1 = 10 µF, 25 V radial
C_2 = 100 nF ceramic
C_3 = 470 µF, 40 V, radial
C_4–C_7 = 22 nF, ceramic
C_8 = 470 nF, 630 V
C_9 = 1000 µF, 35 V, radial

Semiconductors:

B_1 = B80C1500
D_1–D_4 = 1N4007
D_5 = 24 V, 1.4 W
D_6 = 15 V, 0.4 W
D_7 = LED green, low current
T_1 = BC560C
T_2 = BC550C

Integrated circuits:

IC_1 = 7815

Miscellaneous:

K_1; K_2 = 2-way PCB terminal block, pitch 7.5 mm
Tr_1 = 15 V, 3.3 VA, e.g., VTR3115 (Monacor/Monarch)
Re_1 = V23056-A105-A101 (Siemens)
F_1 = 1.6 A, slow, with PCB holder

PROTECTION BOARD

Resistors:

R_1 = 82 kΩ
R_2 = 1 MΩ
R_3; R_7; R_9; R_{25}; R_{26} = 100 kΩ

R_4 = 15 kΩ
R_5 = 100 Ω
R_6; R_{22} = 1 kΩ
R_8 = 47 Ω
R_{10} = 1.5 MΩ
R_{11} = 3.9 MΩ
R_{12}; R_{13} = 56 kΩ
R_{14}; R_{27} = 4.7 kΩ
R_{15} = 10 Ω
R_{16}; R_{17} = 390 Ω, 1 W
R_{18} = 15 Ω
R_{19} = 3.3 kΩ
R_{20} = 180 kΩ
R_{21} = 22 MΩ
R_{23} = 10 kΩ
R_{24} = 220 kΩ
R_{28}; R_{29} = 220 Ω, 1 W
R_{30}–R_{33} = 1.2 Ω, 5 W
P_1 = 50 kΩ preset H

Capacitors:
C_1 = 220 µF, 25 V
C_2; C_7 = 100 nF
C_3 = 470 µF, 6.3 V
C_4 = 1 µF
C_5 = 10 µF, 10 V, radial
C_6 = 220 µF, 16 V
C_8 = 100 µF, 10 V, radial, bipolar
C_9 = 39 nF
C_{10} = 220 nF

Semiconductors:
D_1,D_2 = 1N4004
D_3 = 1N4001
D_4 = 5.6V, 0.4 W
D_5 = LED green
D_6–D_9 = 1N4148 (fit w. connectors on PCB pins)
D_{10} = LED orange
D_{11} = LED, red, high efficiency
T_1; T_2; T_4; T_{10} = BC547B
T_3 = BC516
T_5; T_6; T_7 = BC557B
T_8 = BD140 + heatsink
T_9 = BC547A
T_{11} = BD679

Integrated circuits:
IC_1 = CA3240
IC_2 = 7805

Miscellaneous:
Tri_1;Tri_2 = TIC263M
3 off screw-mount spade terminal for PCB mounting

AMPLIFIER BOARD

Resistors:

R_1; R_{34}; R_{35}; R_{75} = 562 Ω, 1%
R_2; R_{76} = 47.5 kΩ, 1%
R_3; R_4; R_8; R_9 = 2.49 kΩ, 1%
R_5; R_{10} = 10 Ω
R_6; R_7; R_{11}; R_{12} = 24.9 Ω, 1%
R_{13}; R_{14} = 249 Ω, 1%
R_{15} = 820 Ω
R_{16} = 18.2 kΩ, 1%
R_{17}; R_{18} = 4.7 kΩ
R_{19}; R_{20}; R_{23}; R_{24} = 187 Ω, 1%
R_{21}; R_{22}; R_{25}; R_{26} = 18.7 Ω, 1%
R_{27} = 133 Ω, 1%
R_{28}; R_{29} = 390 Ω
R_{30}; R_{31} = 23.2 Ω, 1%
R_{32} = 470 Ω
R_{33}; R_{44} = 330 Ω
R_{36}–R_{39}; R_{50} = 56.2 Ω, 1%
R_{42}; R_{43}; R_{61}; R_{62}; R_{64}; R_{65}; R_{66}; R_{69} = 0.22 Ω, 5 W, low-inductance
R_{40}; R_{41}; R_{60}; R_{63}; R_{67}; R_{68}; R_{70};R_{71} = not used
R_{45} = 180 Ω
R_{46} = 4.7 kΩ
R_{47} = 4.7 Ω, 5 W
R_{48}; R_{51} = 100 Ω
R_{49} = 1.47 kΩ, 1%
R_{52}; R_{53}; R_{72} = 56 kΩ
R_{54} = 12 MΩ
R_{55}; R_{74} = 1.8 MΩ
R_{56} = 270 Ω
R_{57} = 1 kΩ
R_{58}; R_{59} = 8.2 kΩ
R_{73} = 18 MΩ
R_L (in L_1) = 2.2 Ω, 5 W
P_1 = 500 Ω multiturn, vertical preset (Bourns 3296Y)
P_2 = 2 kΩ multiturn, vertical preset (Bourns 3296Y)

Capacitors:

C_1; C_{23} = 2.2 µF, 50 V, polypropyleneT
C_2 = 820 pF, polystyrene
C_3, C_4 = 1.5 nF
C_5–C_8 = 47 µF, 10 V
C_9; C_{10} = 560 pF, polystyrene
C_{11}; C_{20}; C_{21} = 1 µF
C_{12} = 100 nF
C_{13} = 5.6 nF
C_{14}–C_{17} = 10 000 µF, 50 V, radial, for PCB mounting
C_{18}; C_{19} = 1000 µF, 25 V, radial
C_{22} = 10 µF, 25 V
C_{24} = 4.7 nF
C_{fb} = 220 pF, polystyrene (mount at track side)

Semiconductors:

D_1–D_4 = LED, red (V_F = 1.6 V)
D_5 = LED, red (high efficiency)

D_6 = LED, green (high efficiency)
D_7; D_8 = 1N4148
D_9; D_{10} = not used
T_1; T_2 = 2N2914
T_3; T_4 = BFX36
T_5 = BC560C
T_6 = BC550C
T_7 = BC141-16
T_8 = BC161-16
T_9 = BF256C
T_{10}; T_{11} = BF256A (interconnect gate and source; cut off gate)
T_{12}; T_{17} = MJE15031
T_{13}; T_{16} = MJE15030
T_{14}; T_{15} = BD139
T_{18}; T_{19} = 2SC2922 (Toshiba)
T_{20}; T_{21} = 2SA1216 (Toshiba)
T_{22} = BF869 (interchange base and collector)

Integrated circuits:
IC_1;IC_2 = 4N35
IC_3 = OP77
IC_4 = LM337

Miscellaneous:
L_1 = 10 turns (15 mm dia.) of 1.5 mm dia. enamelled copper wire
K_1 = phono socket for PCB mounting (or two PCB pins).
F_1;F_2 = 6.3 A (with fuseholder).
5 off spade terminal, screw type.
1 off heat sink max. 0.5 K W^{-1}.
2 off heat sink for T_{12} and T_{13}

POWER SUPPLY & ENCLOSURE
Tr_1 = ring core transformer, sec. 2×30V, 300 VA, e.g., 71017 (Amplimo or ILP)
B_1 = B200C35000 (200 V/35 A bridge)
R_1–R_8 = 0.1 Ω, 5 W
C_1–C_4 = 10 000µF, 50 V
1 off mains socket
1 off mains switch
2 off phono socket (gold-plated)
2 off loudspeaker terminal or banana socket
Enclosure : e.g., ESM Type ET38/13 (300 mm deep) (Maplin)

Chapter 7

200 watt power amplifier

The amplifier has only a moderately high open-loop gain, but very fast ring-emitter power transistors are used to obtain a very great open-loop bandwidth and power bandwidth. Another property is the low transient distortion. Lead-compensation networks, which increase the amplification from a given frequency upwards, optimize the phase behaviour and bandwidth. Lag compensation is used to improve the slew rate. A block diagram of the amplifier is shown in Fig. 7.1.

Design considerations

The design of a power amplifier can be based on high loop amplification coupled with severe negative feedback or on low loop amplification and consequent less severe negative feedback. Most designs are of the first kind because this offers a simple way of keeping the harmonic distortion to a minimum. However, when the input signal is large and has a frequency that lies above the open-loop bandwidth of the amplifier, it may happen that some stages are driven into saturation. This can result in strong (well audible) intermodulation bursts. Since the largest voltage changes of an audio signal occur at and near the zero crossings, this is where saturation tends to occur. The resulting distortion sounds very much like cross-over distortion.

These problems can be obviated by limiting the open loop amplification and the feedback factor. This results in a much greater open-loop bandwidth, which reduces the risk of the frequency of the input signal being outside the bandwidth. It has a drawback, however, in that, because of the reduced feedback, the harmonic distortion is larger. This does not matter all that much because the human ear is not nearly as sensitive to total harmonic distortion (THD) as to cross-over distortion and transient intermodulation distortion (TIM). For instance, an amplifier with 0.3% THD and 0.003% TIM will in practice sound much more agreeable than one with 0.003% THD and 0.3% TIM.

Apart from low open loop amplification, the various stages of a good amplifier must have a large bandwidth to ensure that the total open-loop bandwidth – if at all possible – exceeds the audio frequency range. This can be arranged with the aid of lead compensating networks which increase the stage amplification of frequencies above a certain point. These networks also optimize the phase behaviour of the stage.

Another important measure is the frequency or lag compensation which limits the open loop bandwidth. This compensation determines the slew rate of the amplifier and must be located as close to the input as possible so as to be able to limit the input signal before this is being amplified.

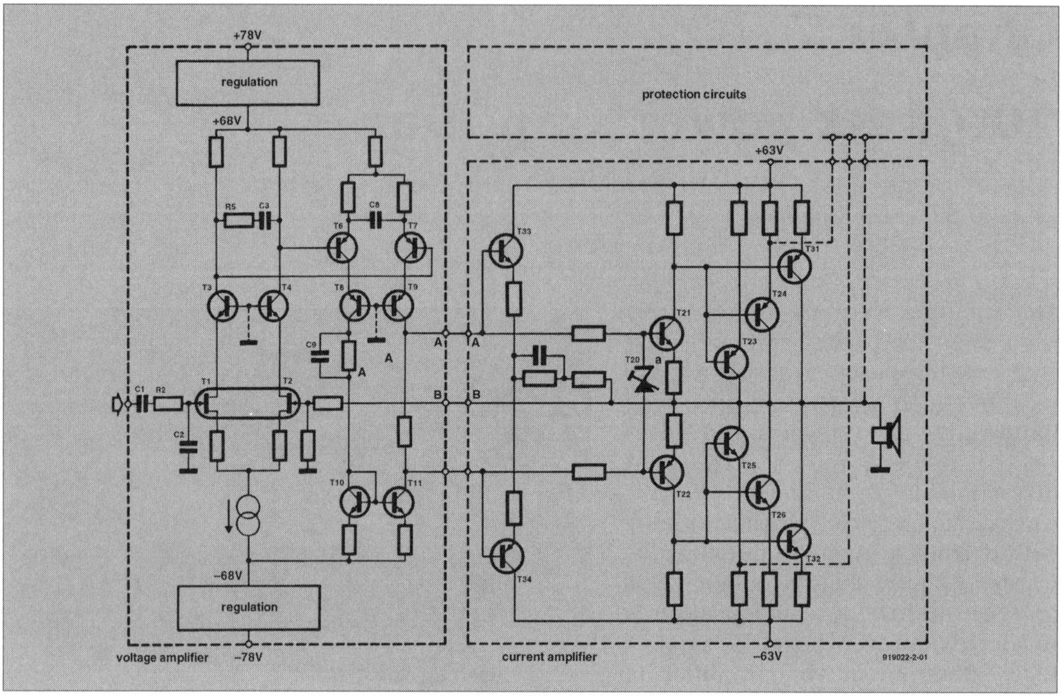

Fig. 7.1. Block diagram of the 200 W power amplifier.

Finally a note about feedback. Often, a different feedback factor is used for alternating and direct voltages by means of a capacitor in the feedback loop. This does not affect the stability of the amplifier, but it can create some other problems (often, because of the large capacitance required, an electrolytic capacitor is used with all the adverse consequences of this). In a good design and with good temperature stability of the various stages, this is not necessary, however: the feedback factors for alternating voltages and direct voltages can be identical.

Practical considerations

The design considerations help to draw up a splendid amplifier on paper, but in practice, there will be some dificulties. To keep within a given cost and using not too esoteric components, a compromise between THD and TIM figures was found necessary. This resulted in an open loop amplification of just over 2000 and an open loop bandwidth of 10 kHz. The amplification proved sufficient to achieve acceptable THD figures, but, with the transistors used, the bandwidth could not be extended to 20 kHz, even with extensive lead compensation networks. Moreover, good phase behaviour also meant that the bandwidth had to be restricted to about 10 kHz. Note that this is still is a good bandwidth if it is borne in mind that an amplifier with high open loop amplification (10^5–10^6) normally has an open loop bandwidth of some tens of hertzs.

The best location for the bandwidth determing components, i.e., the lag compensation, was found to be between the branches of the first differential amplifier.

Stage coupling is direct in all cases to ensure identical a.c. and d.c. amplification. To make this possible, the input stage uses a dual field-effect transistor (FET): not a cheap solution, but one that ensures good stability. The amplification of such a stage is not high, but in the present design this does not matter.

Although the prototypes have a slew rate approaching 100 V μs^{-1}, in the practical amplifier this has been restricted to 50 V μs^{-1}.

The design

The foregoing considerations led to the design shown in simplified form in the block diagram of Fig. 7.1. The circuit consists of two distinct parts: a current amplifier and a voltage amplifier. The input of the voltage amplifier is a differential amplifier, formed by a dual FET, in whose drain lines a cascode network is placed. This enables the drain-source potential of the FETs to be kept at an acceptable level and also largely eliminates the drain-gate capacitance of the FETs, which results in a large bandwidth. The diferential amplifier is followed by another, which is however formed by bipolar transistors, and a current mirror, T_{10}-T_{11}. Network R_5-C_3 is a frequency compensating network, while C_8 and C_9 provide lead compensation.

The input of the current amplifier consists of a 'variable zener' transistor, T_{20}, which sets the quiescent current, and a symmetrical two-part output stage: a driver and two parallel-connected output transistors. Note that the output transistors are not arranged as emitter followers, as is usual, but provide the output at their collectors. This arrangement, called a compound stage, is a kind of darlington, in which high internal feedback results in low distortion and a low output impedance.

The voltage amplifier is supplied from a stabilized supply line, the level of which is 4 V higher than that to the current amplifier, so that the potential drop across output transistors remains small even at maximum, drive levels.

Circuit description

The input of the voltage amplifier is formed by a dual FET in a cascode arrangement, which ensures great bandwidth and good linearity. The drain-source voltage is held at an acceptable level by dividing the supply voltage over the FETs and bipolar transistors.

Fig. 7.2. The completed voltage amplifier board in situ.

Each part of the circuit contained within dashed lines—see Fig. 7.3.—is built on a separate board. Thus, there are four boards: one for the voltage amplifier, one for the current amplifier, one for the protection circuits, and one for the power supply.

The only capacitor in the signal path is C_1 at the input. From there the signal is applied to the differential amplifier via low-pass filter R_2-C_2, which has a cutoff frequency of about 200 kHz. This limits the input bandwidth and thus the slew rate before any amplification takes place. Dual FET T_1-T_2 forms the differential input amplifier; the feedback is applied to the gate of T_2. Transistors T_3 and T_4 and the dual FET form a cascode circuit which holds the direct voltage at about 20 V via R_{12}-R_{13}-R_{14}-D_1-D_2. Resistors R_9 and R_{10} limit the amplification to about ×3.5. So as to keep the bandwidth as large as feasible, the value of the collector resistors, R_7 and R_8/P_1, has been kept as low as practicable. Small variations in the d.c. setting can be negated with preset P_1. Frequency (lag) compensation is provided by R_5-C_3; the capacitor determines the open-loop cutoff point, while the resistor improves the phase behaviour which enhances the stability at this large bandwidth. The d.c. setting for the input differential amplifier is provided by constant current source T_5.

The second differential amplifier, T_6-T_7, together with T_8 and T_9, forms a cascode circuit to achieve the maximum possible bandwidth. The output signal of T_8 appears at B via current mirror T_{10}-T_{11}. Thus, the difference in direct voltage between the signals applied to the current amplifier from A and B is determined by the quiescent current setting of T_{20}. Lead compensation capacitors C_8 and C_9 ensure the maximum possible bandwidth of this stage.

The current amplifier consists of drivers T_{21} and T_{22}, and output transistors T_{23}-T_{24} and T_{25}-T_{26}, in the earlier mentioned compound arrangement. Because of resistors R_{55} and R_{62} there is also some voltage amplification.

Heavy-duty diodes D_9 and D_{10} protect the output transistors against any high-voltage peaks that may emanate from the loud-

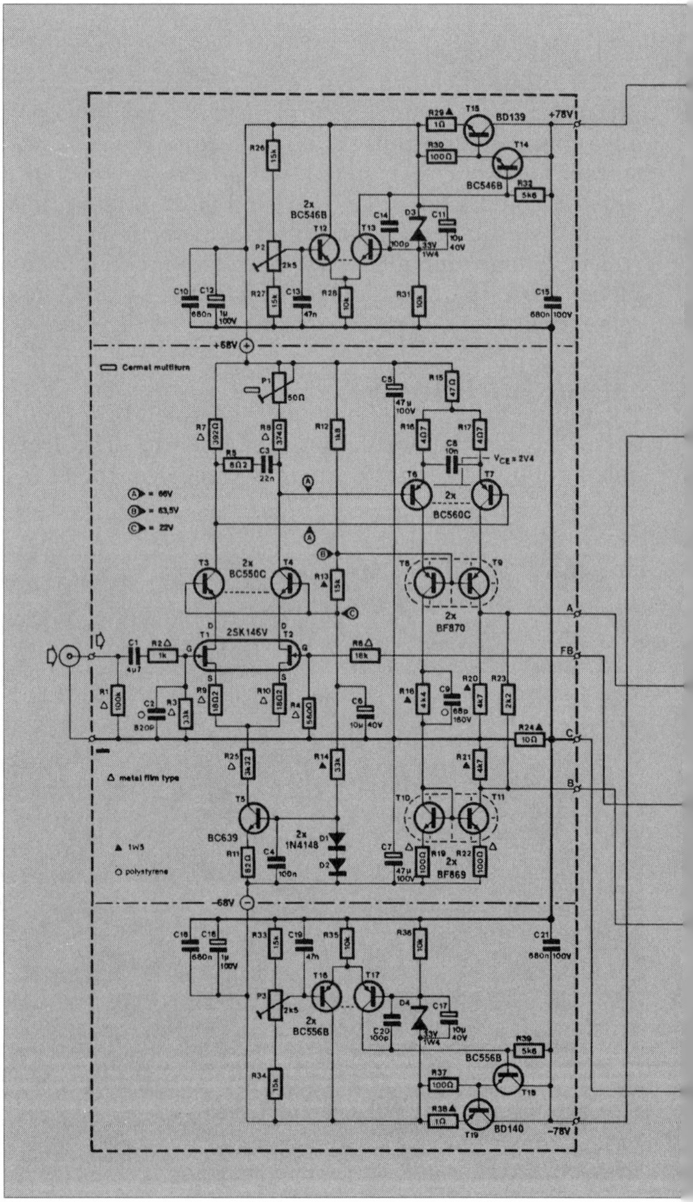

Fig. 7.3. Circuit diagram

speaker system.

Variable zener T_{20} sets the voltage drop across T_{21}, R_{50}, R_{55} and T_{22}, and thus that across R_{49} and R_{56}, which determines the quiescent current through the output transistors. The zener is fitted on the heat sink for the drivers and output transistors for good thermal coupling to ensure that the quiescent current remains level even at rising temperatures. Since the quiescent current is 100 mA per transistor, the output amplifier can easily process small signals in its Class-A operating area.

Boucherot network R_{64}-C_{28} at the output of the amplifier ensures that the amplifier remains properly loaded even at high frequencies. Inductor L_1 limits the output current

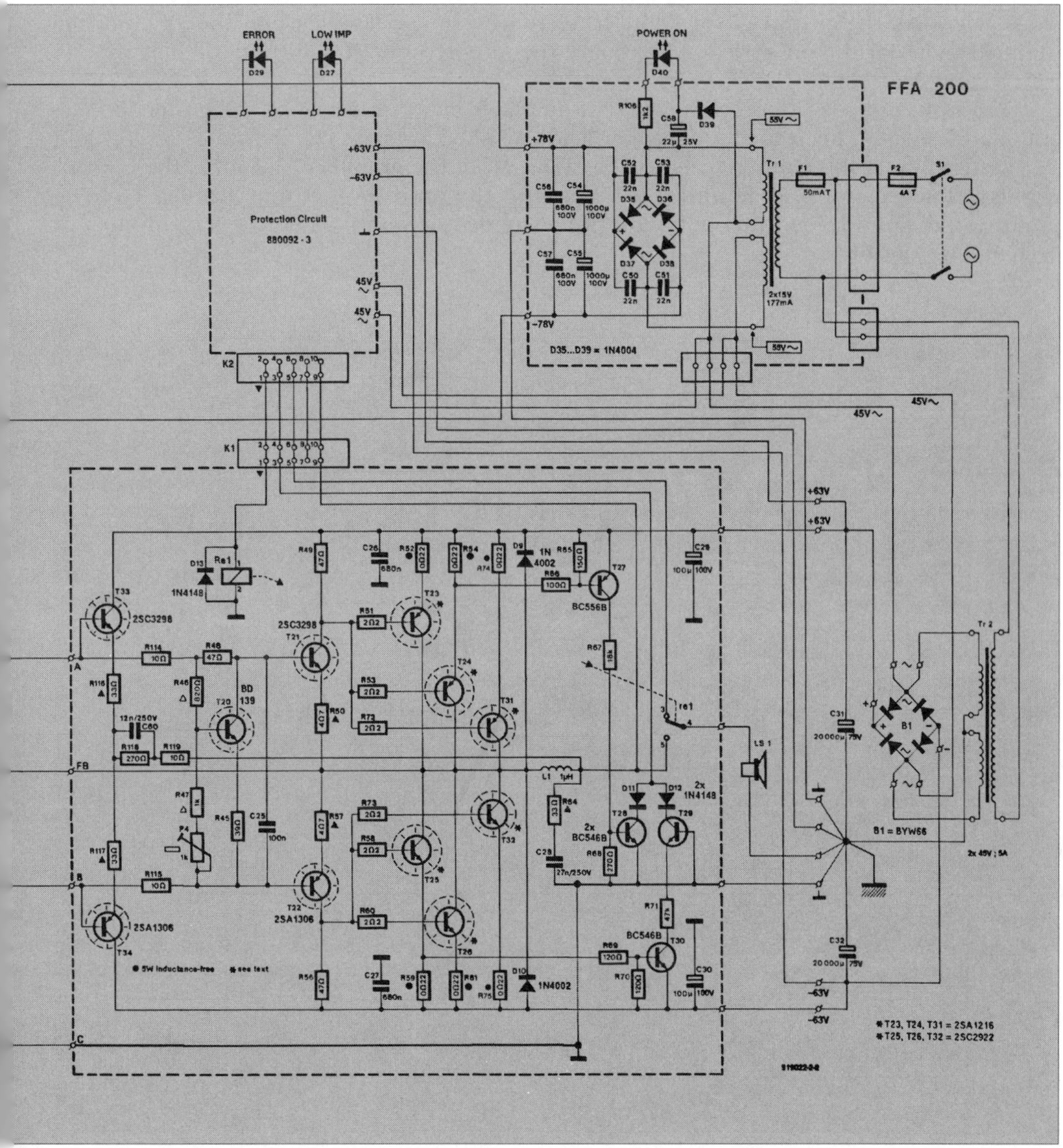

of the 200 W power amplifier.

when the load becomes too capacitive.

Part of the output signal is tsken from the input of L_1 and fed back to T_2 via R_6. The value of resistors R_4 and R_6 in the feedback loop is such that the voltage amplification is ×32. The input sensitivity of the amplifier is then 1.1 V r.m.s.

Power for the current amplifier is provided by a heavy-duty toroidal transformer, Tr_2, with two secondary windings, each delivering 40 V, a bridge rectifier and four 10 000 µF electrolytic capacitors. The open-circuit supply voltage is around ±57 V. At full load, this drops to about ±51 V.

Power for the voltage amplifier is provided by a smaller transformer, Tr_1 which, after rectification and smoothing, supplies a voltage of ±70 V. This is brought down to ±60 V by discrete voltage regulators T_{12}–T_{15} and T_{16}–T_{19}. In each of the regulators, a differential stage compares the output voltage with a reference voltage; if necessary, the difference is nullified by a darlington series regulator in the supply line. The output voltages are set with P_2 and P_3.

The output relay is fitted on the current amplifier board and not on the protection circuit board so as to keep the loudspeaker leads as short as possible. A transistor, T_{27} and T_{30}, across one of the emitter resistors in each half of the amplifier measures the current through that resistor and actuates, if necessary, the protection circuit via T_{28}, T_{29}. The protection circuits come into action (output relay de-energized) when the output current becomes 10 A or higher.

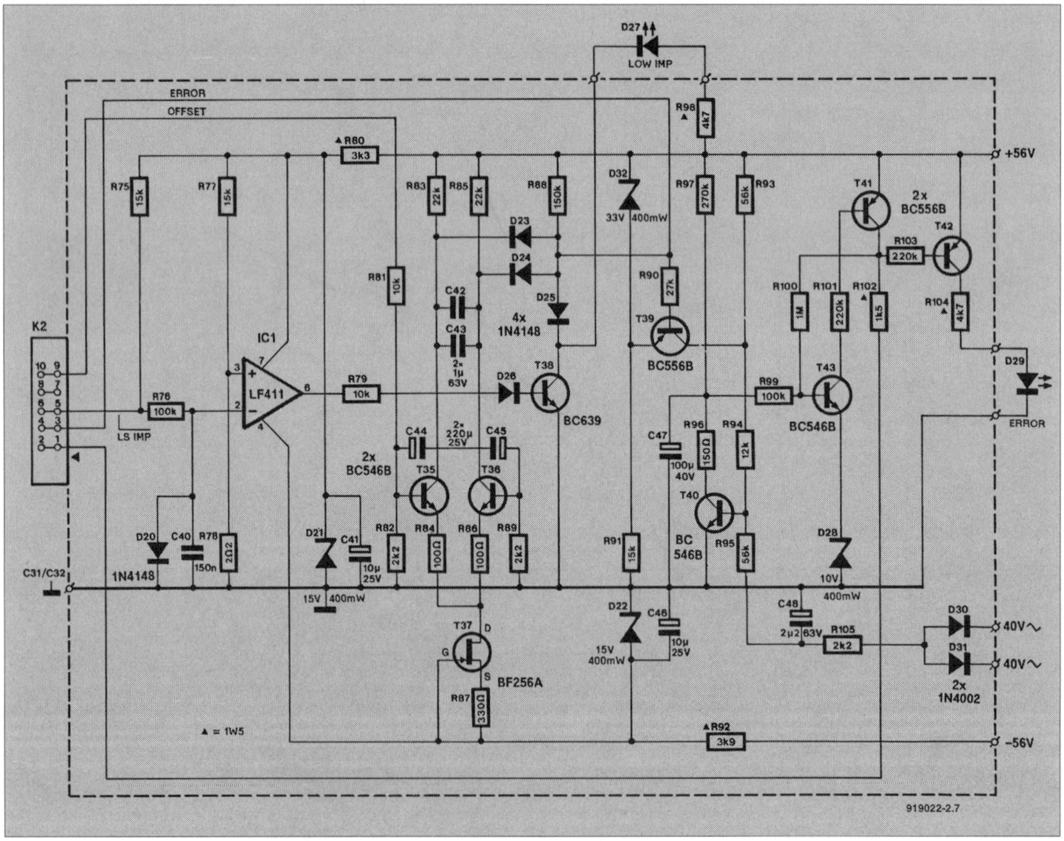

Fig. 7.4. Diagram of the protection circuits.

Protection

The protection circuits provide safeguards as follows.

- When the amplifier is switched on (power on), it takes a few seconds before the output relay is energized.
- On power-on, the resistance of the loudspeaker system is measured; if this is lower than 2.2 Ω, the output relay is not energized.
- If there is a direct voltage greater than 1 V at the output, the output relay is disabled.
- If the peak current through the output transistors exceeds 10 A, the output relay is disabled.
- If one or both of the secondary transformer voltages fail, the output relay is disabled. This protection also ensures that the loudspeakers are disconnected from the amplifier output the instant the amplifier is switched off.

Note that there is no short-circuit protection; this means that the connection between the amplifier outputs and the loudspeakers MUST be made before the amplifier is switched on. From then on, the protection circuits will ensure that nothing catastrophic can happen.

The diagram of the protection circuits is shown in Fig. 7.4.; this is not complete because the output relay and the peak current detection section are located on the current amplifier board.

The 24 V output relay is controlled by T_{43} and T_{41}. Since these transistors form a Schmitt trigger, the relay is energized when the potential across C_{47} is about 12 V and de-energized when that potential has dropped to around 6 V. The hysteresis is determined by R_{99} and R_{100}. Inverter T_{42} in the collector line of T_{41} causes an LED to light when the protection circuit is actuated. After the power supply is switched on, C_{47} is charged via R_{97}. When the potential across the capacitor has risen to about 12 V, T_{43} is switched on. whereupon T_{41} begins to condcut and the output relay becomes energized.

Capacitor C_{47} is shunted by a transistor via which the capacitor is discharged rapidly if a fault arises. The values of resistors R_{93}, R_{94} and R_{95} that form a voltage divider in the

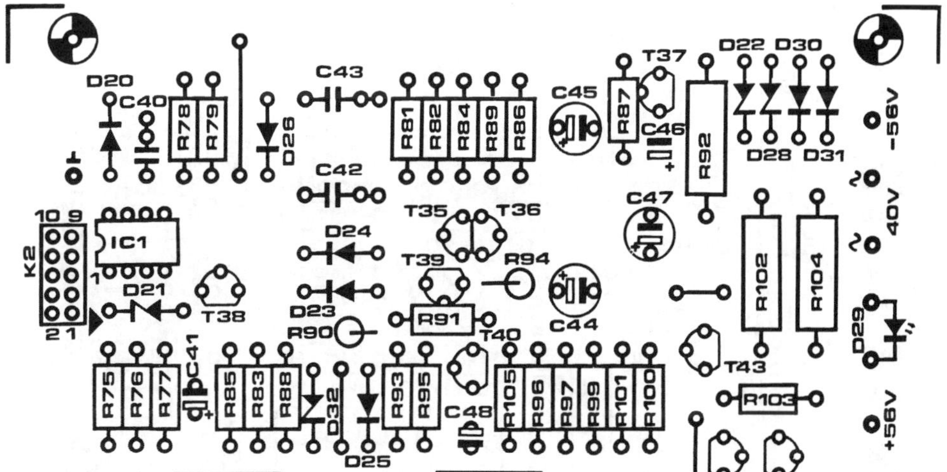

Fig. 7.5. Component layout of the PCB for the protection circuits.
The track layout is given on page 241.

base circuit of T_{40} are such that if one or both of the supply voltages fail, the output relay is disabled instantly.

Network T_{39}-D_{32}-R_{91} at the junction of R_{93}-R_{94} forms a kind of comparator. All protection networks are linked to the base of T_{39} via R_{90}. When the base potential of the transistor drops below about 23 V (56 V less the zener voltage of 33 V), T_{40} is switched on and the output relay is de-energized.

The (ohmic) loudspeaker resistance is measured by IC_1, whose inputs are linked to the junctions of a resistive bridge. One branch of this consists of R_{77}-R_{78} and the other of R_{75} and the loudspeaker coil (via pins 5, 6 of K_2). Measurements can be carried out only when the output relay is not energized. Network R_{76}-C_{40} prevents error indications if ambient noise should be picked up by the loudspeaker and converted into electrical signals (measurements are carried out with very small – millivolts – direct voltages). The potential across C_{40} is limited by D_{20}. Should the loudspeaker resistance be lower than the value of R_{78} (2.2 Ω), the output of IC_1 changes state, whereupon T_{38} is enabled. The base potential of T_{39} is then pulled down via D_{25} to nearly 0 V, so tht the output relay can not be energized; this state is indicated by the lighting of D_{27}. If the loudspeaker resistance is higher than 2.2 Ω, the output relay is energized a few seconds after the supply has been switched on. The loudspeaker coil is then no longer connected to pins 5, 6 of K_2, so that measurements can no longer take place. A new measurement can be effected only when the amplifier is switched on again or when the output relay is re-energized after an error.

The supply for IC_1 is obtained from the ±56 V line via zener diodes D_{21}, D_{22} and series resistors R_{80}, R_{90}.

Direct voltages at the amplifier output are measured by differential amplifier T_{35}-T_{36}. The direct-coupled output signal of the amplifier is applied to T_{35} via a potential divider

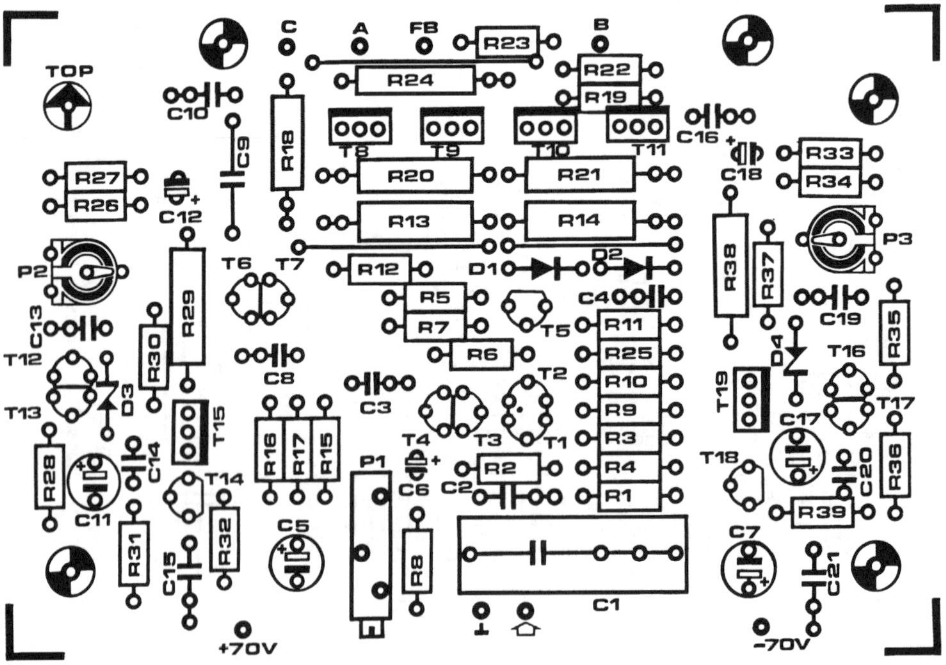

Fig. 7.6. Component layout of the PCB for the voltage amplifier.
The track layout is given on page 241.

and to T_{36} via bipolar electrolytic capacitor C_{44}-C_{45}. The amplified difference signal is applied to the collectors of the transistors, from where it is passed through low-pass filter R_{83}-R_{85}-C_{42}-C_{43}. If the (positive or negative) direct voltage exceeds 1 V, the collector voltage of T_{35} or T_{36} (depending on the polarity of the direct voltage) drops sufficiently for T_{39} to be enabled via D_{23} or D_{24}, whereupon the output relay is de-energized by T_{41}. The d.c. setting of the differential amplifier (about 2.5 mA) is provided by constant-current source T_{37}-R_{87}.

Transistors T_{27} and T_{30} measure the peak voltage across the emitter resistor of one of the output transistors in either half of the amplifier. The values of the resistors in the potential dividers in the base circuits of the transistors are such that either T_{27} switches on T_{28} or T_{30} enables T_{29}; T_{39} is then switched on via D_{11} or D_{12} respectively, whereupon the output relay is de-energized.

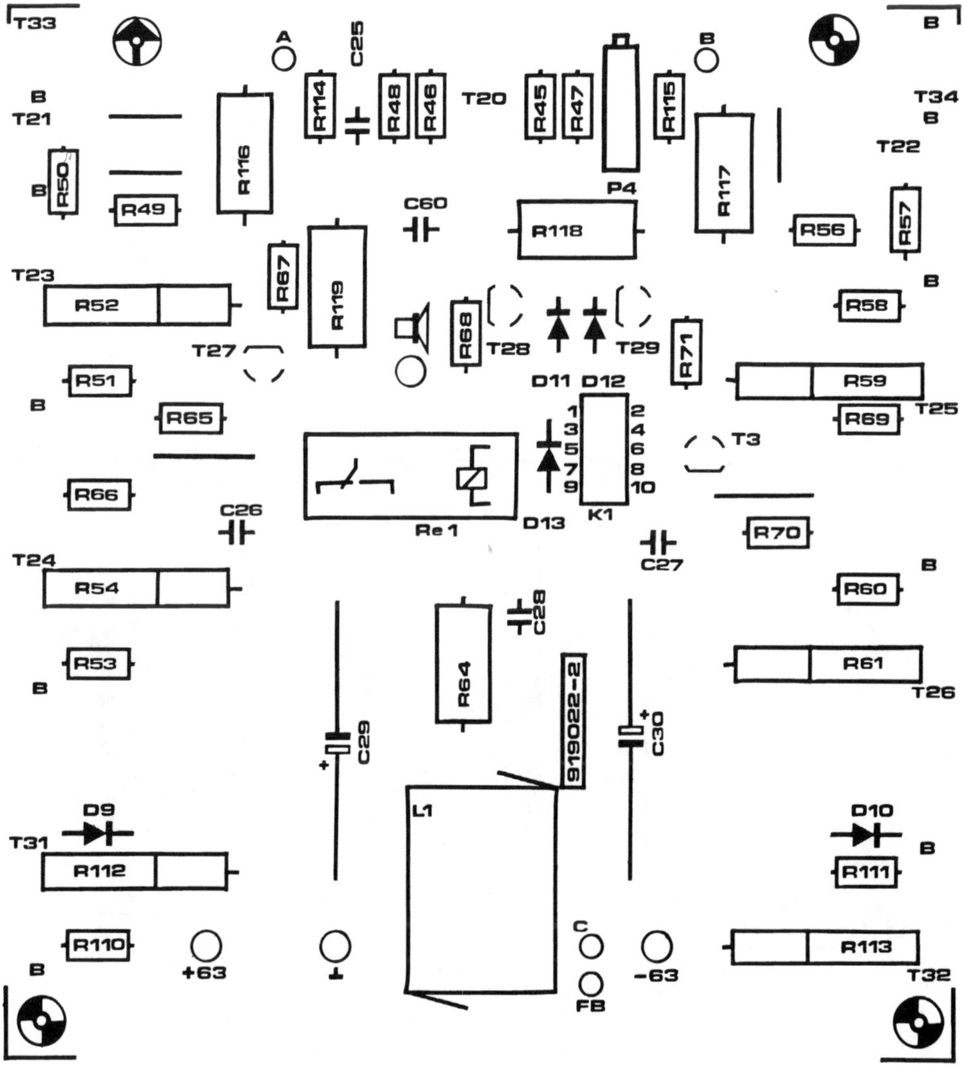

Fig. 7.7. Component layout of the PCB for the current amplifier.
The track layout is given on page 243.

Power for the protection circuits is taken from the ±56 V supply line.

Construction

In the following description it is assumed that the enclosure will contain both amplifiers of a stereo setup. The enclosure may be of a type whose side panels form heat sinks or it should be of a size that accommodates the required heat sinks. The heat sinks must be drilled with fixing holes (tapped afterwards) for the six output transistors, two drivers, two transistors of the feed forward stage, the 'zener' transistors, and for fixing the board. A drilling template is given in Fig. 7.12., while Fig. 7.13 shows an early stage of the construction on to the heat sink. Use a 2.5 mm drill and an M3 tap.

The two mains transformers, the large bridge rectifiers, and the four electrolytic capacitors are fitted in a lying position to the bottom of the enclosure. A mounting plate is fitted above these components on M4×80 spacers on to which all boards, except those for the current amplifier are mounted. The boards for the current amplifier are fitted at right angles to the heat sink—see Fig. 7.9.

Start the construction with populating the boards – first that for the current amplifier, then the voltage amplifier, followed by that for the protection circuits, and finally that for the power supply. Fit all components, except T_{20}, to the track side of the board a few millimetres above the surface.

Inductor L_1 consists of 20 turns 1.5 mm enalled copper wire with an internal diameer of about 20 mm. Fit the coil so that it floats a couple of millimetres above the copper tracks.

Four solder pins and a 10-way connector are used for linking the board to the remainder of the amplifier.

Because of the large currents, it is recommended to solder the connecting wires from

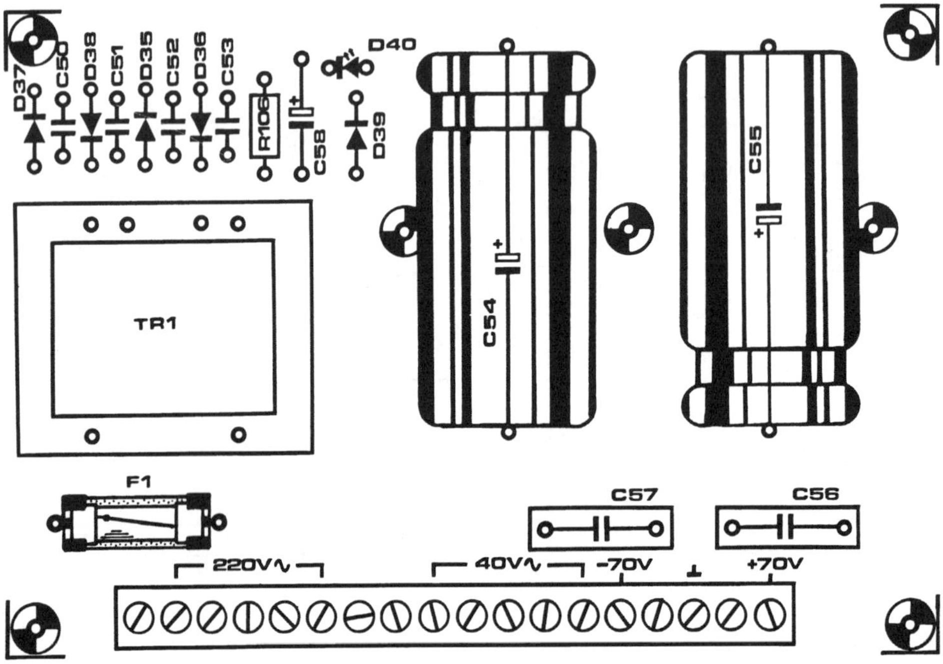

Fig. 7.8. Component layout of the PCB for the power supply.
The track layout is given on page 245.

Fig. 7.9. Completed current amplifier board shown in situ.

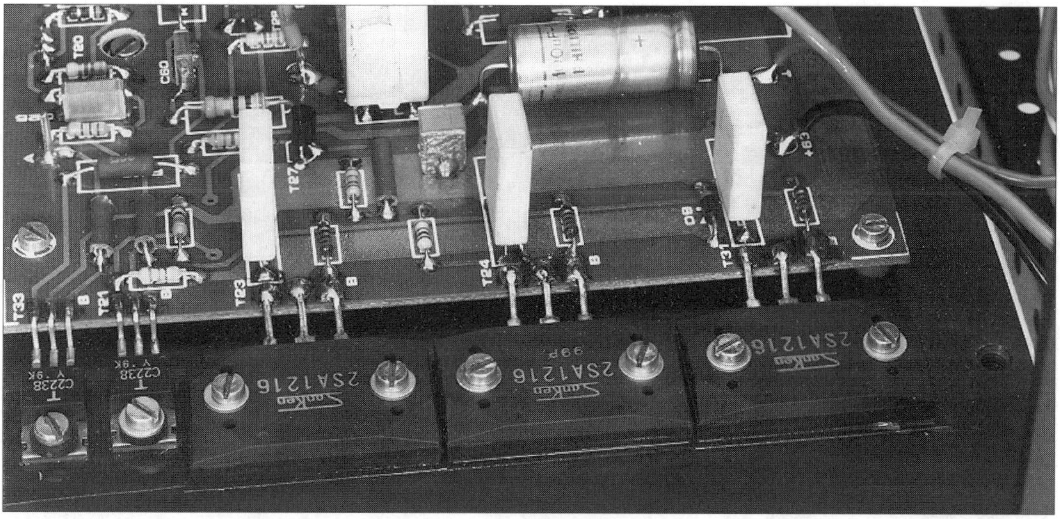

Fig. 7.10. Close-up of the current amplifier board, showing how the terminals of the output and driver transistors are bent and soldered in place.

the supply lines to earth, and to the loudspeaker output, directly on to the board.

Fit the output and driver transistors to the heat sink on insulating washers and with non-metallic screws; use plenty of heat-conducting paste. Make sure that the correct washers are used for the output transistors, since these have a rather unusual case.

Bend the terminals of T_{20} about 3 mm from the case at right angles and fix the transistor to the heat sink on insulating washers with non-metallic screws. Next, fit the current amplifier board to the heat sink on 5 mm spcacers in such a way that the terminals of T_{20} protrude through it. With small pliers, bend the terminals of the output and driver transistors as shown in Fig. 7.11. Finally, solder the terminals of all transistors to the board.

Several pairs of transistors on the voltage amplifier board need to be matched: T_3-T_4; T_6-T_7; T_8-T_9; and T_{10}-T_{11}. If matching is not possible, make sure that each pair is at least from the same production run and date (normally stated on the case).

Pairs T_3-T_4 and T_6-T_7 are fitted on to the board with the flat sides of the two transistors adjoining. Apply some heat conducting paste to these side and then fasten the transistors together with a nylon binder to ensure good thermal coupling.(the transistors must always be at the same temperature to ensure that their operating direct voltages remain identical).

Pairs T_8-T_9 and T_{10}-T_{11} must be fitted, using insulating washers, heat conducting paste and non-metallic fixing screws, on a small heat sink (see Fig. 7.14) which is subsequently fixed on the board on two small spacers.

All other components of the voltage amplifier are fitted on the component side of the board.

Before the boards can be interconnected, the power supply has to be built. The wiring is shown in the diagram in Fig. 7.15.

Note that there is only earthing point (at the electrolytic capacitors), which is connected

Fig. 7.11. Close-up of the final assembly of the boards and power supply in the enclosure.

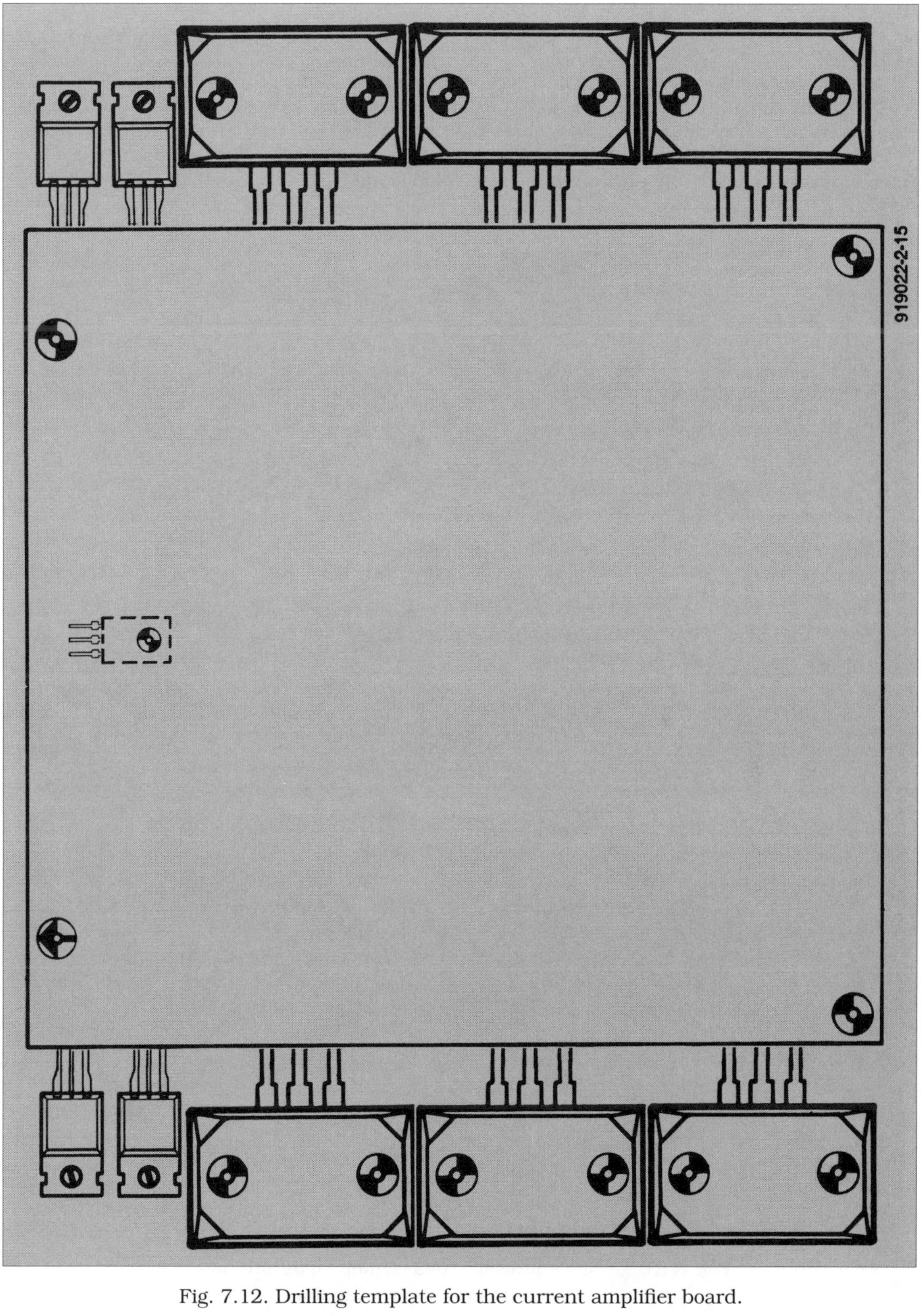

Fig. 7.12. Drilling template for the current amplifier board.

to the enclosure and mains earth.

Testing

Switch on the mains and check that there is a potential of about 70 V w.r.t. earth at the ±70 V terminals on the supply board. If a lower voltage is measured, say, ±45 V, switch off the mains and interchange the two mains connections to Tr_2. Switch on the mains again and recheck that there is 70 V at the appropriate terminals. Unplug the mains from the mains outlet and carefully discharge the electrolytic capacitors via a 470 Ω, 1 W resistor.

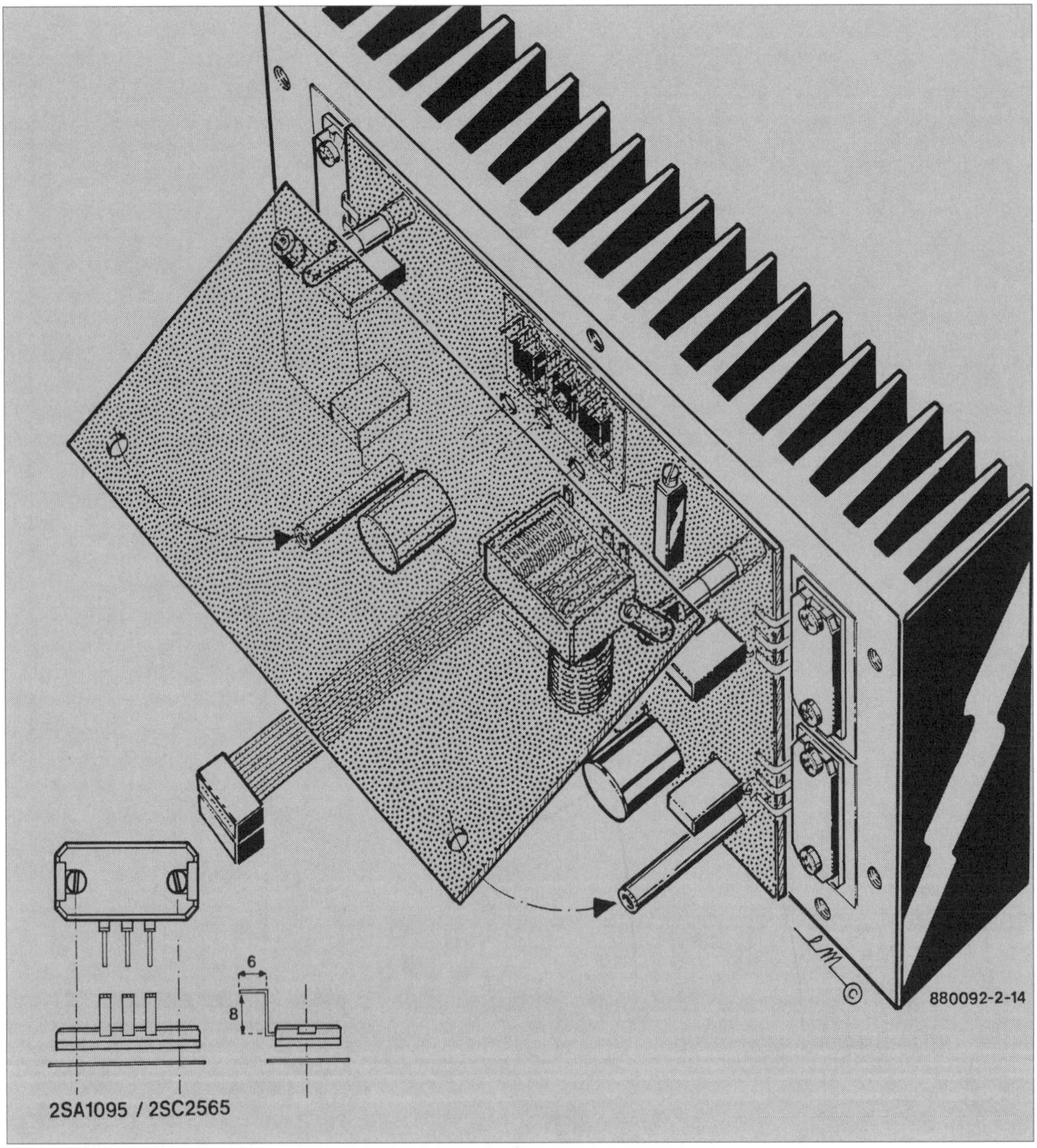

Fig. 7.13. Artist's impression of assembly of output and driver transistors on to the heat sink.

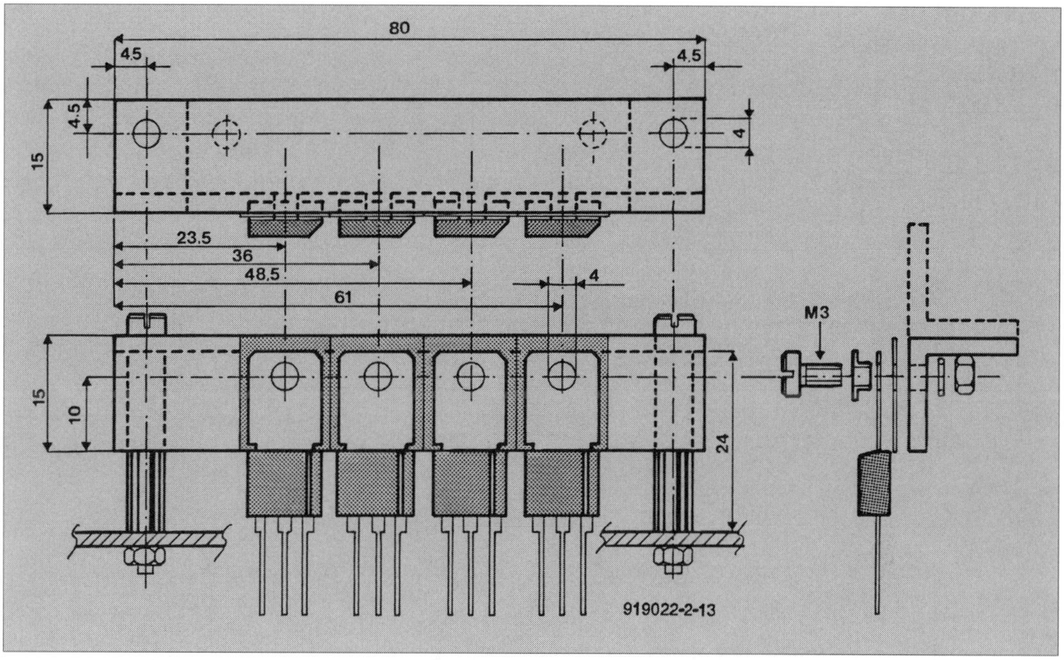

Fig. 7. 14. Heat sink for T_8–T_{11} in the voltage amplifier.

Link points A, B, C and FB on the current amplifier and voltage amplifier boards with short lengths of copper wire.

Connect the input socket with the input terminal of the voltage amplifier via a length of single screened cable.

Connect the supply input terminals on the current amplifier board to the terminals of the electrolytic capacitors with heavy-duty, insulated cable.

The links from the amplifier to the loudspeakers should also be heavy-duty cable.

Connect the supply terminals on the voltage amplifier board to the 70 V terminals on the supply board.

Connect the protection board to the current amplifier board via a length of 10-core flat-cable terminated at each end into a 10-way flatcable connector. Make sure that pin 1 on the protection board is linked to pin 1 on the current amplifier board.

Connect the supply voltage to the various boards as indicated in Fig. 7.15.

Wire up the three LEDs on the front panel as shown in Fig. 7.15.

Set P_1, P_2 and P_3 to the centre of their travel, and P_4 to maximum resistance. Reconnect the mains and switch on.

The supply voltages on the electrolytic capacitors should be around ±58 V w.r.t. to the central earthing point.

Measure the potential across R_{29} and R_{38} on the voltage amplifier board and adjust P_2 and P_3 for a reading of +60 V and –60 V respectively.

Connect a multimeter (direct voltage range) between junction L_1-R_{64} and earth and adjust P_1 until the meter reading is exactly 0 mV.

Connect a digital multimeter (direct voltage range) across R_{52} or R_{54} and slowly adjust P_4 until the meter reading is about 20 mV (which corresponds to a quiescent current of about 50 mA).

Parts list

VOLTAGE AMPLIFIER

Resistors:

R_1 = 100 kΩ*
R_2 = 1 kΩ*
R_3 = 33 kΩ*
R_4 = 560 Ω*

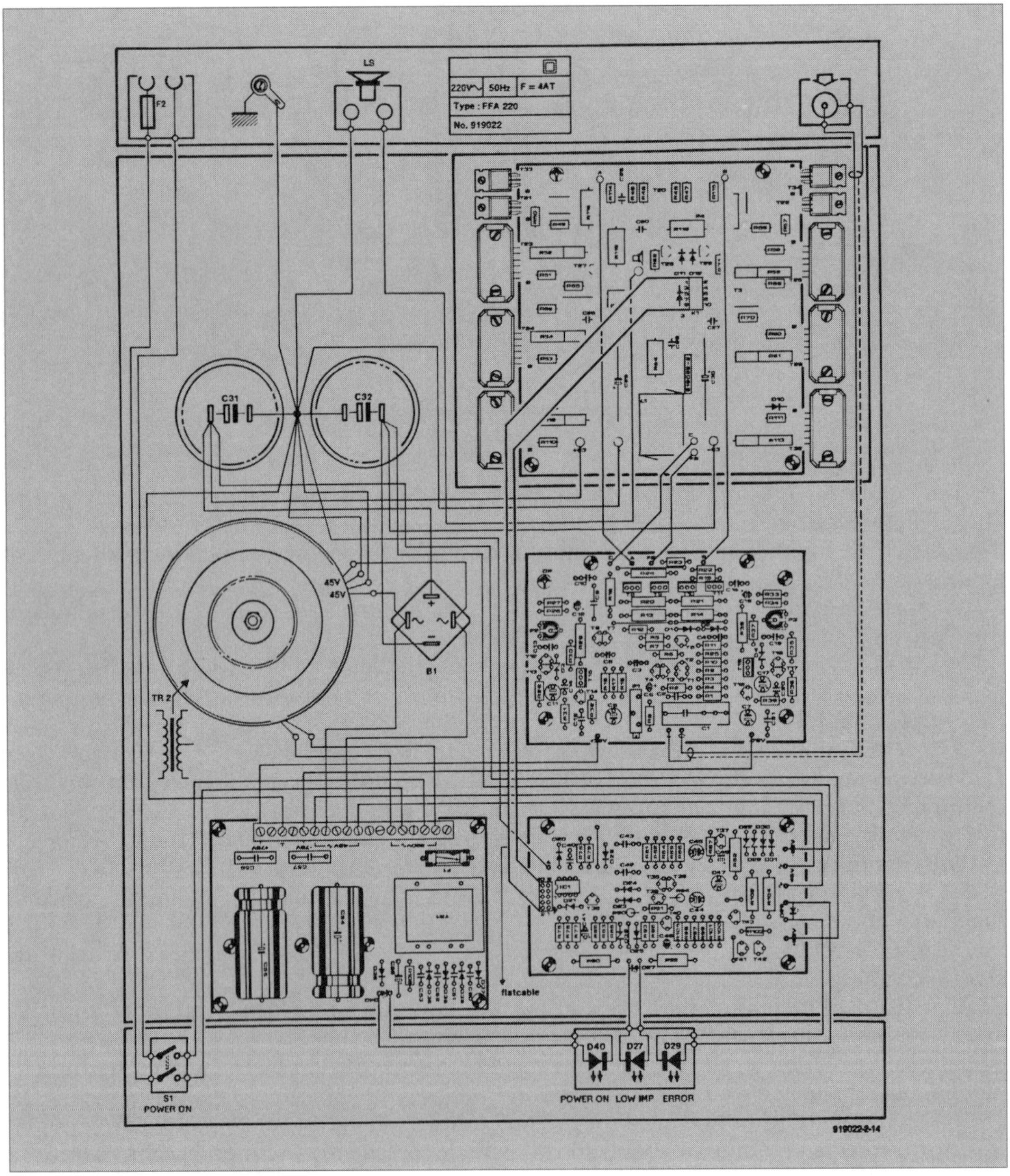

Fig. 7.15. Wiring diagram of (single-channel) 200 W power amplifier.

R_5 = 8.2 Ω
R_6 = 18 kΩ*
R_7 = 392 Ω
R_8 = 374 Ω*
R_9, R_{10} = 18.2 Ω*
R_{11} = 82 Ω
R_{12} = 1.8 kΩ
R_{13} = 15 kΩ, 1.5 W
R_{14} = 33 kΩ, 1.5 W
R_{15} = 47 Ω
R_{16}, R_{17} = 4.7 Ω
R_{18} = $2\times$2.2 kΩ, 1.5 W in series
R_{19}, R_{22} = 100 Ω*
R_{20}, R_{21} = 4.7 kΩ, 1.5 W
R_{23} = 2.2 kΩ
R_{24} = 10 Ω, 1.5 W
R_{25} = 3.32 kΩ*
R_{26}, R_{27}, R_{33}, R_{34} = 15 kΩ
R_{28}, R_{31}, R_{35}, R_{36} = 10 kΩ
R_{29}, R_{38} = 1 Ω, 1.5 W
R_{30}, R_{37} = 100 Ω
R_{32}, R_{39} = 5.6 kΩ
P_1 = multiturn potentiometer, 50 Ω
P_2, P_3 = 2.5 kΩ preset
* = metal film

Capacitors:
C_1 = 4.7 µF polypropylene
C_2 = 820 pF polyester
C_3 = 22 nF
C_4 = 100 nF
C_5, C_7 = 47 µF, 100 V
C_6, C_{11}, C_{17} = 10 µF, 40 V
C_8 = 10 nF
C_9 = 68 pF, 160 V, polyester
C_{10}, C_{16} = 680 nF
C_{12}, C_{18} = 1 µF, 100 V
C_{13}, C_{19} = 47 nF
C_{14}, C_{20} = 100 pF
C_{15}, C_{21} = 680 nF, 100 V

Semiconductors:
D_1, D_2 =1N4148
D_3, D_4 = zener diode, 33 V, 1.4 W
T_1, T_2 = 2SK146V
T_3, T_4 = BC550C
T_5 = BC639
T_6, T_7 = BC560C
T_8, T_9 = BF870
T_{10}, T_{11} = BF869
T_{12}, T_{13}, T_{14} = BC546B
T_{15} = BD139

T_{16}, T_{17}, T_{18} = BC556B
T_{19} = BD140

CURRENT AMPLIFIER
Resistors:
R_{45} = 39 Ω
R_{46} = 820 Ω*
R_{47} = 1 kΩ*
R_{48}, R_{49}, R_{56} = 47 Ω
R_{50}, R_{57} = 4.7 Ω, 1.5 W
R_{51}, R_{53}, R_{58}, R_{60}, R_{72}, R_{73} = 2.2 Ω
R_{52}, R_{54}, R_{59}, R_{61}, R_{74}, R_{75} = 0.22 Ω, 5 W, non-inductive
R_{55}, R_{62}, R_{63} = not used
R_{64}, R_{116}, R_{117} = 33 Ω, 1.5 W
R_{65} = 150 Ω
R_{66} = 100 Ω
R_{67} = 18 kΩ
R_{68}, R_{118} = 270 Ω
R_{69}, R_{70} = 120 Ω
R_{71} = 47 kΩ
R_{114}, R_{115}, R_{119} = 10 Ω
P_4 = multiturn potentiometer 1 kΩ
* metal film

Capacitors:
C_{25} = 100 nF
C_{26}, C_{27} = 680 nF
C_{28} = 27 nF, 250 V
C_{29}, C_{30} = 100 µF, 100 V
C_{60} = 12 nF, 250 V

Semiconductors:
D_9, D_{10} = 1N4002
D_{11}, D_{12}, D_{13} = 1N4148
T_{20} = BD139
T_{21} = 2SC3298
T_{22} = 2SA106
T_{23}, T_{24}, T_{31} = 2SA1216
T_{25}, T_{26}, T_{32} = 2SC2922
T_{27} = BC556B
T_{28}, T_{29}, T_{30} = BC546B

Miscellaneous:
Header, 10-way, male
Re_1 = relay, 24 V, 1 change-over contact
L_1 = 1 µH (see text)

PROTECTION CIRCUITS
Resistors:
R_{75}, R_{77}, R_{91} = 15 kΩ
R_{76}, R_{99} = 100 kΩ
R_{78} = 2.2 Ω

R_{79}, R_{81} = 10 kΩ
R_{80} = 3.3 kΩ, 1.5 W
R_{82}, R_{89}, R_{105} = 2.2 kΩ
R_{83}, R_{85} = 22 kΩ
R_{84}, R_{86} = 100 Ω
R_{87} = 330 Ω
R_{88} = 150 kΩ
R_{90}, R_{95} = 27 kΩ
R_{92} = 3.9 kΩ, 1.5 W
R_{93} = 56 kΩ
R_{94} = 12 kΩ
R_{96} = 150 Ω
R_{97} = 270 kΩ
R_{98}, R_{104} = 4.7 kΩ, 1.5 W
R_{100} = 1 MΩ
R_{101}, R_{103} = 220 kΩ
R_{102} = 1.5 kΩ, 1.5 W

Capacitors:
C_{40} = 150 nF
C_{41}, C_{46} = 10 μF, 25 V
C_{42}, C_{43} = 1 μF, 63 V
C_{44}, C_{45} = 220 μF, 25 V
C_{47} = 100 μF, 40 V
C_{48} = 2.2 μF, 63 V

Semiconductors:
D_{20}, D_{23}, D_{24}, D_{25}, D_{26} = 1N4148
D_{21}, D_{22} = zener diode, 15 V, 400 mW
D_{27} = LED, orange
D_{28} = zener diode, 10 V, 400 mW
D_{29} = LED, red
D_{30}, D_{31} = 1N4002
D_{32} = zener diode, 33 V, 400 mW
T_{35}, T_{36}, T_{40}, T_{43} = BC546B
T_{37} = BF256A
T_{38} = BC639
T_{39}, T_{41}, T_{42} = BC556B

Integrated circuits:
IC_1 = LF411

Miscellaneous:
Header, 10-way, male

POWER SUPPLIES
Resistors:
R_{106} = 1.2 kΩ

Capacitors:
C_{31}, C_{32} = 20 000 μF, 75 V (or each 2×10 000 μF)
C_{50}, C_{51}, C_{52}, C_{53} = 22 nF

C_{54}, C_{55} = 1000 μF, 100 V
C_{56}, C_{57} = 680 nF, 100 V
C_{58} = 22 μF, 25 V

Semiconductors:
B_1 = bridge rectifier BYW66
D_{35}, D_{36}, D_{37}, D_{38}, D_{39} = 1N4004
D_{40} = LED, green

Miscellaneous:
F_1 = fuse, 50 mA, slow with fuse holder for board monting
F_2 = fuse, 4 A, slow
Tr_1 = encapsulated PCB type mains transformer, secondary 2×15 V, 180 mA
Tr_2 = toroidal mains transformer, secondary 2×45 V, 5 A
S_1 = double-pole mains on/off switch
Mains entry with integral fuse holder
Heat sink, thermal resistance <0.55 K W^{-1}
3 off 6-way terminal blocks for board mounting

Part 3

Headphone amplifiers

Chapter 8

Headphone amplifier I

Headphone amplifiers normally consist of a simple dedicated output stage, but the one described in this chapter is rather more sophisticated. It is a good-quality miniature stereo amplifier whose volume control uses digital switches. Two push-button switches enable the loudness to be set in 16 discrete steps. It also has two switch-selectable inputs. The amplifier can be used with a variety of preamplifiers or integrated power amplifiers.

Circuit description

The circuit diagram in Fig. 8.1. shows both stereo channels. A logic level provided by electronic switches IC_{1b}–IC_{1d} enables switching between the tape and line inputs. The switching signal is provided by switches S_1 and S_2 via R-S bistable (flip-flop) IC_{2a}–IC_{2b}. This stage retains the last selected position and at the same time prevents chaos if S_1 and S_2 were accidentally pressed simultaneously. Network R_1-C_1 ensures that it is always set to position 'line' when the supply is switched on.

The input signal is also applied to a potential divider connected across the inputs of a 16-channel multiplexer. The divider for the left-hand channel consists of resistors R_{14}–R_{29} and that for the right-hand channel of R_{38}–R_{53}. The junctions of successive resistors are connected to the 16 inputs of IC_5 (IC_6). The common output of the multiplexer is con-

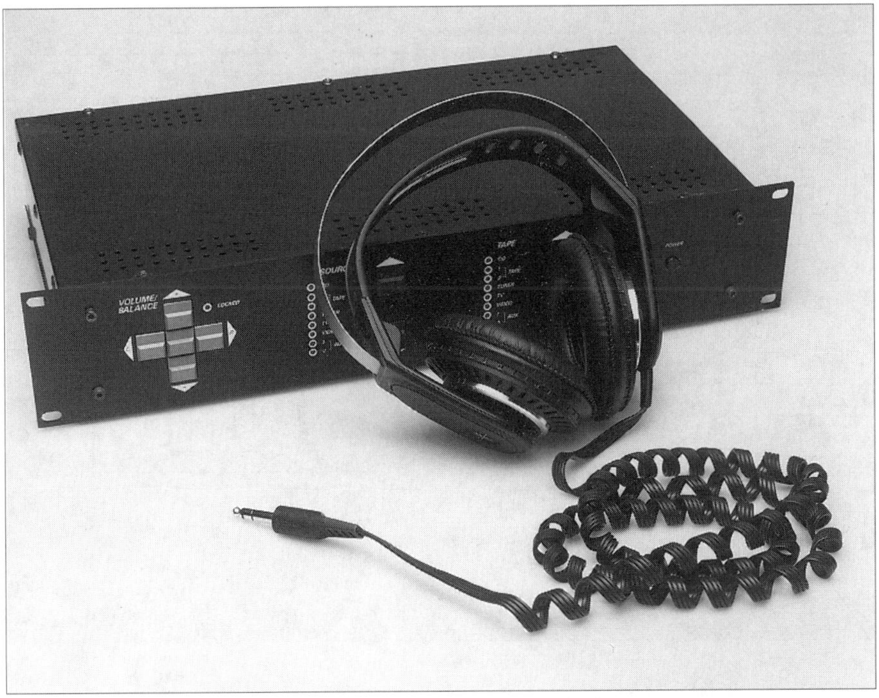

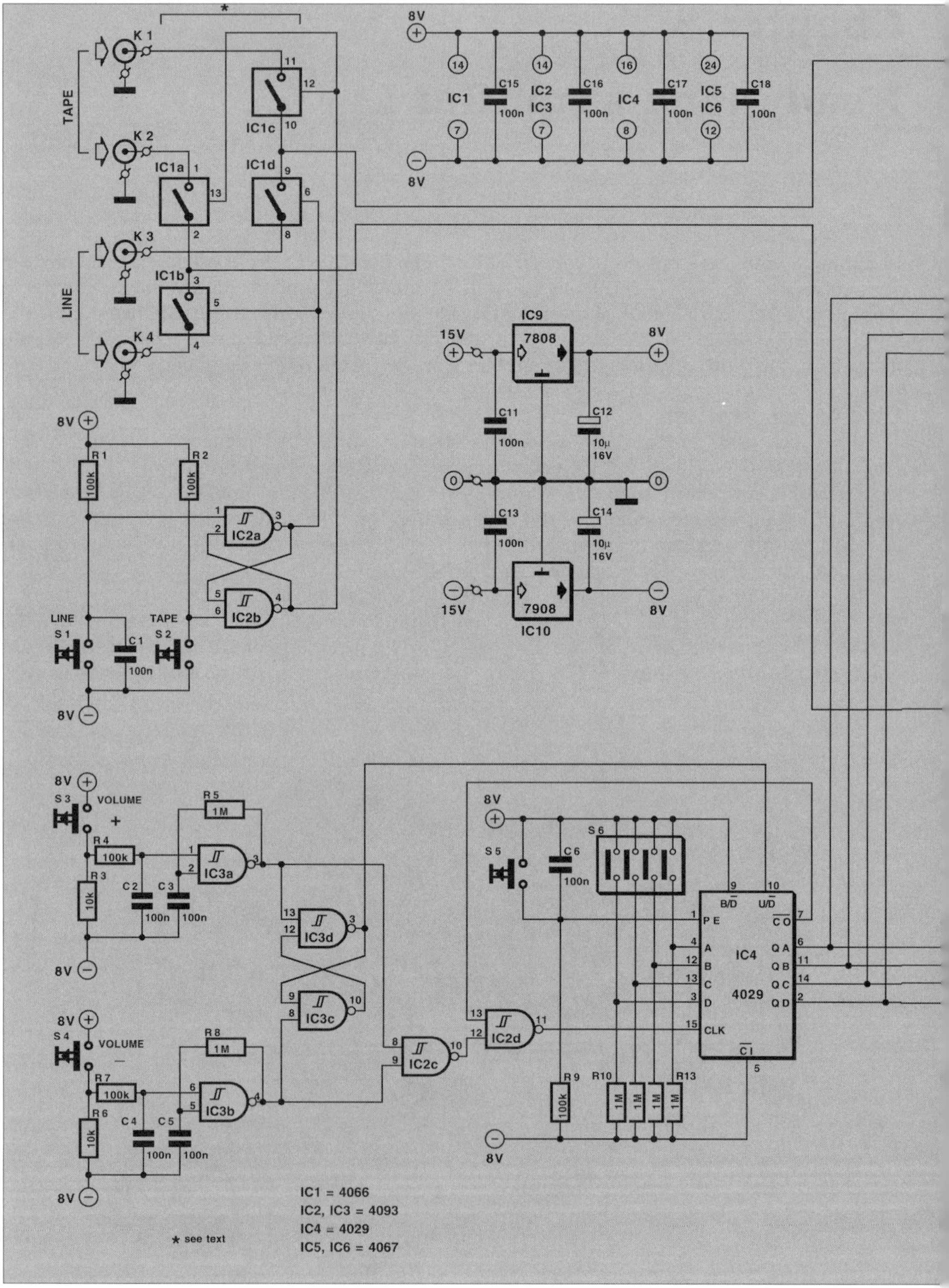

Fig. 8.1. Circuit diagram

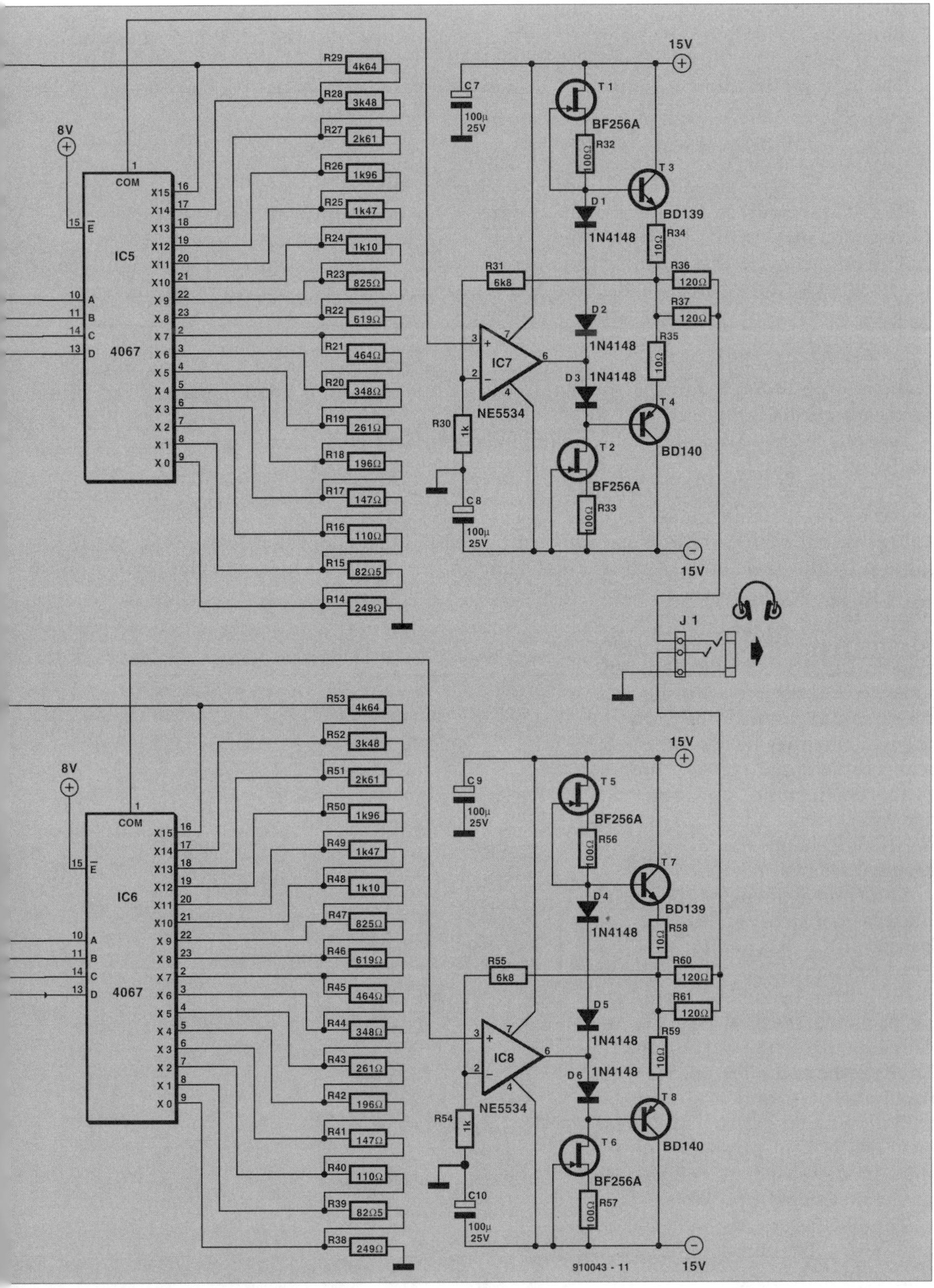

of headphone amplifier I.

nected to the non-inverting input of op amp IC_7 (IC_8).

Since the FET switches on board the 4067 are in series with the fairly high input resistance of the op amp, they introduce virtually no distortion.

The logic combination at inputs A–D of the multiplexer determines which of its inputs is connected to its output (pin 1).

The output stage consists of an op amp with two separate output transistors, T_3 and T_4 (left-hand channel). These transistors are driven by the output of IC_7 via quiescent-current diodes D_1–D_3. The quiescent-current circuit also contains current sources T_1-R_{32} and T_2-R_{33}. These sources provide sufficient current for the output transistors even in high-drive conditions, so that the distortion is kept low right up to the maximum drive level.

The output stage is a Class A type operating with a quiescent current of 50 mA. Negative feedback takes place via R_{30} and R_{31}. The stage uses no capacitors so that loss of quality caused by these components is avoided.

The output stage operates from a higher supply, ±15 V, to ensure a good dynamic range even if fairly insensitive headphones are used. This arrangement ensures that an output voltage of up to 8.5 V r.m.s. is available with a distortion not greater than 0.01%. The bandwidth is 400 kHz.

Any offset compensation for the op amp has been omitted on purpose. The offset voltage at the output varies by a couple of millivolts, which is caused by the relatively high bias current of the 5534 which flows through the feedback resistors. Therefore, compensation would be largely ineffectual.

The output of the amplifier is protected by two parallel-connected resistors of 120 Ω to give an output resistance of 60 Ω. This value was chosen to ensure that the power delivered to the headphones remains about the same, irrespective of whether a low- or high-impedance type is used. If it is felt that this arrangement does not provide optimum sound quality, the value of the resistors may be altered empirically to personal taste. It should be borne in mind, however, that the values depend to a large extent on the impedance characteristic of the headphones.

The control stage for the multiplexers consists of the circuit based on IC_3, IC_4 and (partly) IC_2. This section also contains two push-button switches, S_3 and S_4, to provide, respectively, an increase or reduction in the volume. Each of the switches operates an oscillator, IC_{3a} and IC_{3b} respectively. When either of these switches is kept pressed, the volume is increased or reduced in predefined steps—see Table 8-1.

The oscillators are followed by an R-S bistable (flip-flop), IC_{3c} and IC_{3d}, which arranges for counter IC_4 to count up or down depending

Table 8-1

	step size			
	2.5 dB	3.0 dB	4.0 dB	5.0 dB
Resistors	Value in Ω			
R_{29}, R_{53}	4640	3650	4220	11000
R_{28}, R_{52}	3480	2610	2670	6190
R_{27}, R_{51}	2610	1820	1690	3480
R_{26}, R_{50}	1960	1270	1050	1960
R_{25}, R_{49}	1470	909	681	1100
R_{24}, R_{48}	1100	649	422	619
R_{23}, R_{47}	825	464	267	348
R_{22}, R_{46}	619	332	169	196
R_{21}, R_{45}	464	232	105	110
R_{20}, R_{44}	348	162	68.1	61.9
R_{19}, R_{43}	261	115	42.2	34.8
R_{18}, R_{42}	196	82.5	26.7	19.6
R_{17}, R_{41}	147	56.2	16.9	11.0
R_{16}, R_{40}	110	40.2	10.5	6.19
R_{15}, R_{39}	82.5	28.7	6.81	3.48
R_{14}, R_{38}	249	69.8	11.5	4.42
R_{total} (abt)	18.5 kΩ	12.0 kΩ	11.5 kΩ	25.0 kΩ

on which of the switches is pressed.

The signal generated by IC_{3a} or IC_{3b} is also applied via gates IC_{2c} and IC_{2d} to the clock input, pin 5, of IC_4. Circuit IC_{2d} ensures that when the highest or lowest volume is reached the counter is disabled.

The counter may also be preset to a given position with the DIP switches contained in S_6. Whenever the amplifier is switched on afterwards, the counter automatically assumes the preset position. The position is read via R_9 and C_6. Push-button switch S_5 enables the preset position to be selected during operation.

The power supply has been kept simple, since normally the headphone amplifier will be integrated into an existing power amplifier, which almost certainly has a regulated ±15 V line available. All that is necessary to obtain a ±8 V supply for the present amplifier, with the exception of the output stages, is the addition of regulators IC_9 and IC_{10} as shown in Fig. 8.1.

Some practical points

No printed-circuit board has been designed for the amplifier, but its construction on a prototype board should not present any undue difficulties.

The electronic input switches cause very slight distortion. If the switching between line and tape is not required, it is therefore advisable to omit IC_1 and its associated control circuitry to reduce the distortion. This may also be achieved by the use of Texas Instruments Type TLC4066 LinCMOS input switches instead of the standard 4066 types. Note that the TLC4066 ICs require 6 V instead of 8 V regulators.

The values of the resistors in the voltage dividers as shown in the circuit diagram result in 2.5 dB steps in volume. This gives an overall volume range of 40 dB. If steps of different magnitude are required, the values of the resistors must be altered as shown in Table 8-1.

Although the BD139 and BD140 output transistors get pretty hot during normal operation, they do not need a heat sink, but there is, of course, no harm in using one.

Circuit IC_7 may be a Type NE5534 or a Type OP37. The OP37 is better, but also much dearer, than the NE5534.

Chapter 9

Headphone amplifier II

A first-class electroacoustic transducer, whether this is a headset or a loudspeaker, can only perform to its true specification if the amplifier driving it is also first class. It is, of course, not necessary to connect a headset to a power amplifier; after all, it needs only little energy. Yet, in practice, the headphone output is often taken from the power amplifier output via a potential divider. It is, however, far better to connect it to the output of

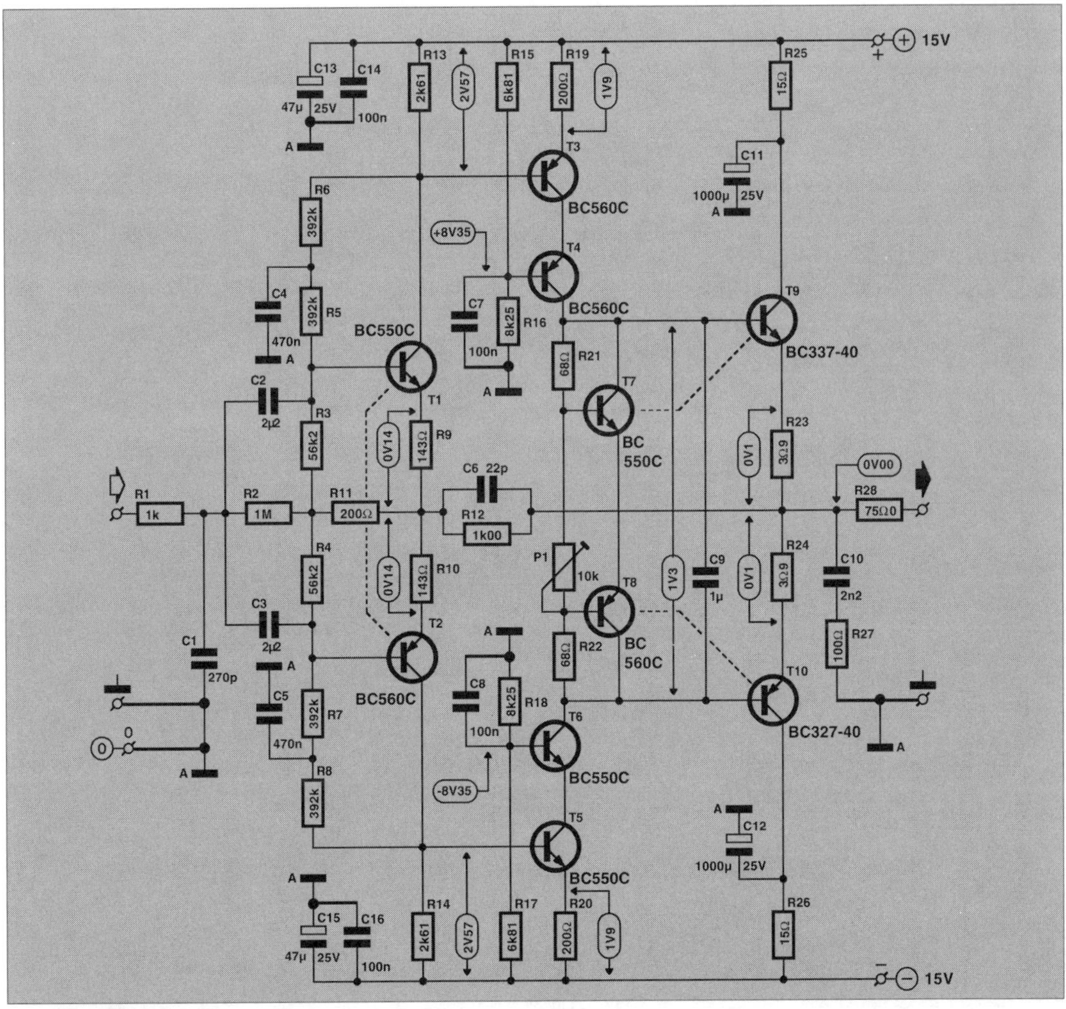

Fig. 9.1. Circuit diagram of the left-hand channel of headphone amplifier II.

the preamplifier via a dedicated headphone amplifier as described in this article.

A headphone amplifier is, strictly speaking, a sort of line amplifier with a power output. Its amplification need not be high, since the sensitivity of most headsets is usually fairly good. If we assume that the output of the preamplifier is 1 V r.m.s., an amplification of a few times is quite sufficient: some tens of milliwats is fine. Average good-quality headphones provide a sound pressure of 90–100 dB for an input of about 1 mW. It should be borne in mind that modern headphones have a fairly high impedance. Until not so long ago, this impedance was 8 Ω or so, but nowadays good-quality headphones have an impedance of hundreds of ohms (typically 600 Ω). The present amplifier can deliver up to 40 mW into 600 Ω. Never use this full power for listening long: it may permanently and irreversibly damage your hearing. Too many young people are going deaf prematurely because they listen to headphones at too high a volume!

Note that the amplifier is not suitable for use with electrostatic headphones. These need far more energy and are, therefore, normally driven straight from the power output amplifier.

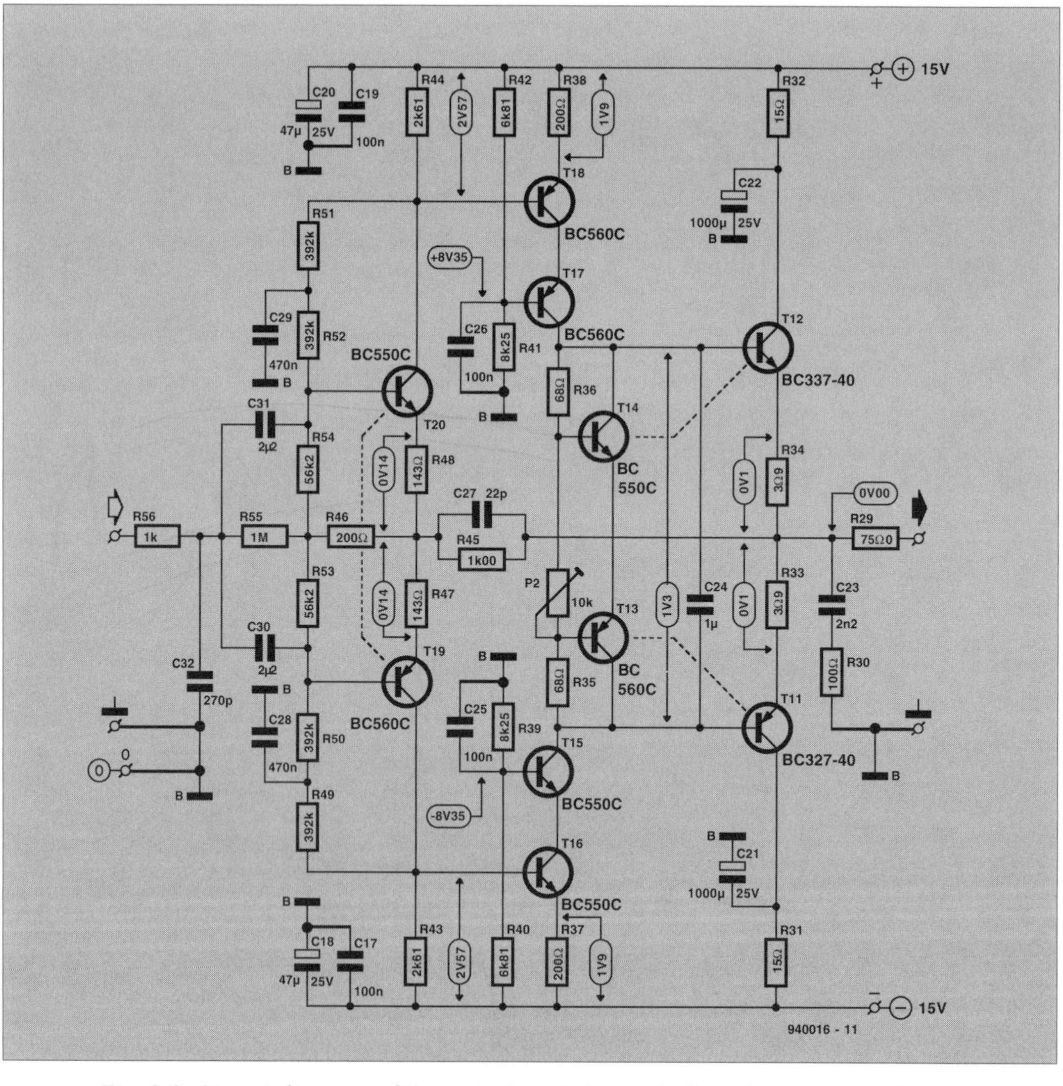

Fig. 9.2. Circuit diagram of the right-hand channel of headphone amplifier II.

Circuit description

The design—see Fig. 9.1 and 9.2— is a pure symmetric one which can be split into an amplifier that works from a positive supply and one that operates with a negative supply. This design, used in the past in preamplifiers for moving-coil pick-ups, is known for its excellent performance at low drive levels. However, it appears to do well also at higher signal levels. Its only drawback is the requirement for two input capacitors: one for the n-p-n stage and one for the p-n-p stage. But, since in a headphone amplifier the input impedance may be fairly high (here about 20 kΩ), these 2.2 μF capacitors can be kept fairly small so that good-quality ones (polythene or polypropylene) can be used.

The signal from the preamplifier is applied to R_1. This resistor and C_1 form a low-pass filter that limits the bandwidth of the incoming signal to about 400 kHz for a preamplifier output impedance of 600 Ω. The signal is then applied to amplifiers T_1 and T_2 via capacitors C_2 and C_3. The amplification of these stages depends on the values of R_9, R_{11}, R_{13} (in the case of T_1) and R_{10}, R_{11}, R_{14} (in the case of T_2). Resistor R_{11} is also part of the feedback loop of both tran-

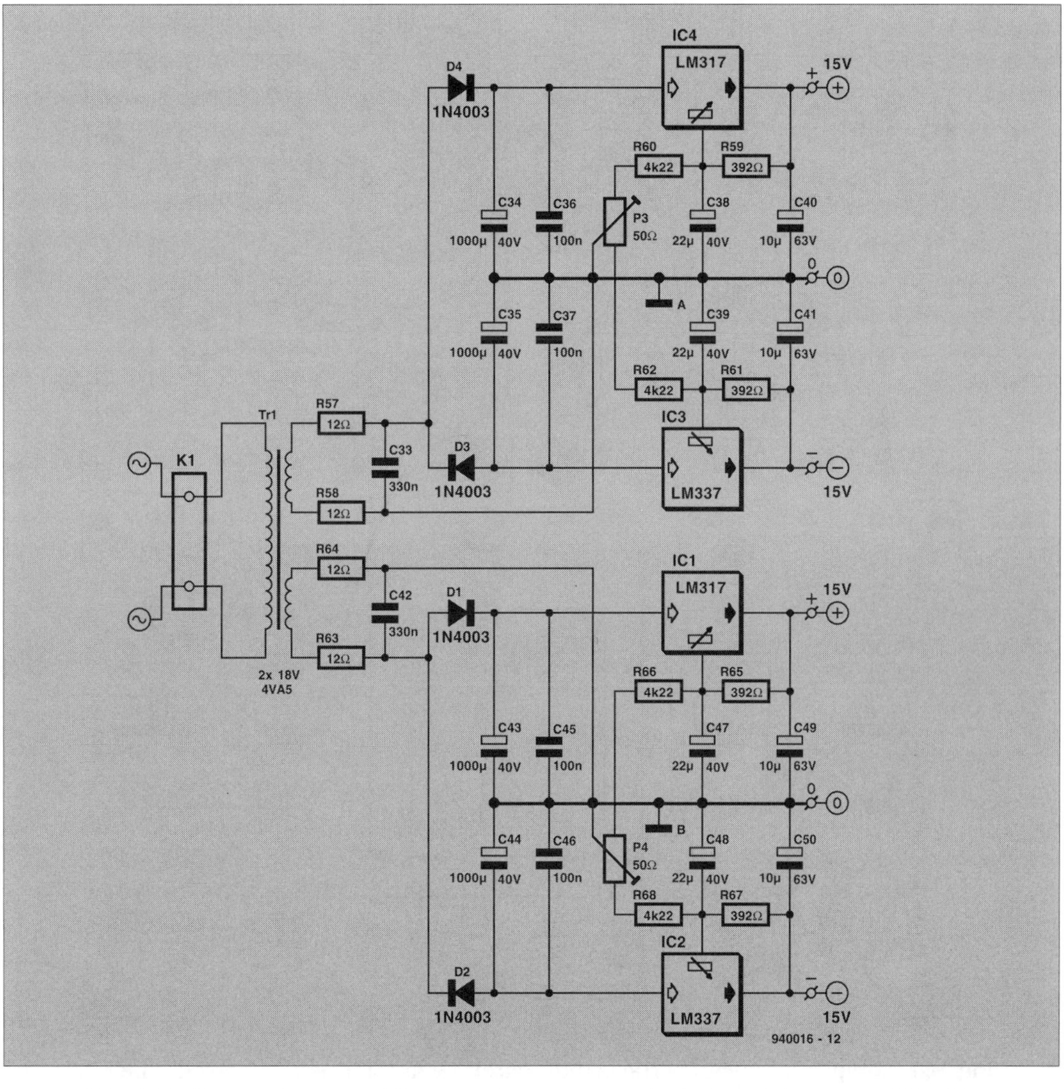

Fig. 9.3. Circuit diagram of the double power supply for headphone amplifier II.

sistors. The d.c. operating point is set by R_3-R_5-R_6 (T_1) and R_4-R_7-R_8 (T_2). These resistors also provide some local feedback.

The signals at the collector of T_1 and T_2 are fed to two cascode stages, T_3-T_4 and T_5-T_6 respectively. These stages provide wide-band amplification and impedance matching between the input transistors and the 'output' stage. A drawback of this arrangement is that output transistors can not be driven up to the supply voltage level, but that is not so important in the

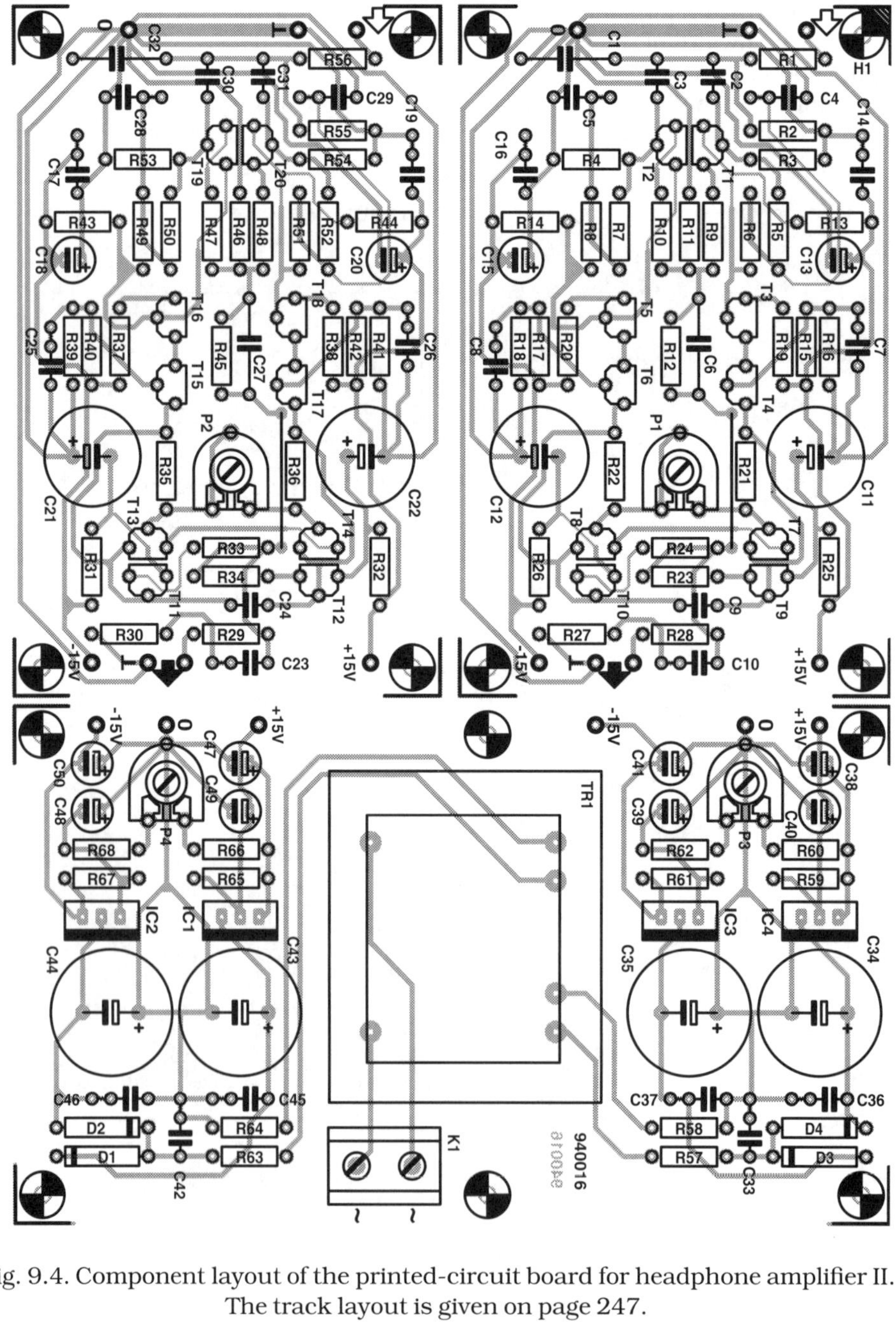

Fig. 9.4. Component layout of the printed-circuit board for headphone amplifier II. The track layout is given on page 247.

case of a headphone amplifier. The output transistors are driven by the collector signals of T_4 and T_6.

Transistors T_7 and T_8 form an adjustable 'zener' stage that sets the quiescent current. They are thermally coupled with T_9 and T_{10} to ensure that the quiescent current remains reasonably stable during temperature variations in the output transistors.

The current through T_9 and T_{10} is relatively high (25 mA), which is typical of Class A operation. The transistors are connected to the power supply via 15 Ω resistors, and buffered by electrolytic capacitors C_{11} and C_{12}.

Overall feedback is provided by R_{12}. Together with R_{11}, this resistor determines the amplification of the complete amplifier.

Network R_{27}-C_{10} ensures a constant load at high frequencies.

The output impedance is 75 Ω (R_{28}). Power amplifiers normally have a low output impedance, but headphone amplifiers need a higher one. On the one hand, a low impedance ensures that the varying impedance of the headphones does not influence the output characteristic, and on the other hand, a resistor is necessary to protect the output stage against short circuits that occur every time the jack plug of the headphones is inserted into the socket. The specified value of R_{28} was found to be a good compromise.

Since a stereo amplifier draws a fairly high current, the amplifier is given a dedicated power supply—see Fig. 9.3. To ensure good separation of the channels, the mains transformer uses two secondary windings, each of which provides a symmetrical voltage of ±15 V. In this way, only two windings are needed for two isolated symmetrical voltages. Deriving symmetrical voltages from a single winding is possible with half-wave rectification: one diode uses the positive half-periods to charge an electrolytic capacitor, and another diode rectifies the negative half-periods. The use of relatively large electrolytic smoothing capacitors ensures that the ripple is kept small in spite of the half-wave rectification. Resistors R_{57}, R_{58}, R_{63} and R_{64} limit peak currents.

Integrated circuits IC_1–IC_4 regulate the output voltages, which are held at ±15 V with the aid of resistance networks. Presets P_3 and P_4 enable setting exactly symmetrical voltages and setting the output of each output stage to exactly zero.

Construction

The complete amplifier is best built on the printed-circuit board in Fig. 9.4. This board consists of three parts: left-hand channel amplifier, right-hand channel amplifier and power supply. It is advisable to cut the board into three parts, so that the power supply can be fitted some distance from the amplifiers.

Population of the boards is straightforward, but make sure that the flat sides of transistor pairs T_1-T_2, T_7-T_9, T_8-T_{10}, T_{19}-T_{20}, T_{12}-T_{14} and T_{11}-T_{13} fit snugly together. Make two small rings of copper or aluminium and use these to clamp T_1 and T_2, and T_{19} and T_{29}, securely together for good thermal contact (for an example, see Fig. 4.4). Fit lobe-finned heatsinks as used for TO-39 transistors around the other four pairs. These heatsinks should be slightly flattened before being fitted—see Fig. 9.5.

Interwiring is minimal: six wires for the supply voltages.

It is advisable to build the amplifier in a dedicated case, since most preamplifier will not have the space to house it. Fit the mains entry at the back of the case and the audio input sockets and the (6.3 mm) jack socket at the front: its terminals are shown in Fig. 9.6.

When the construction is completed, set P_3 and P_4 to the centre of their travel. Connect a multimeter (200 mV direct voltage range) across R_{23}. Adjust P_1 until the voltage across R_{23} is 100 mV. Do the same in the other channel with P_2 and R_{34}.

Connect the multimeter to the output of each amplifier in turn and adjust P_3 and P_4 respectively for zero reading of the meter (50 mV range).

Fig. 9.5. This photograph shows how the pairs of transistors should be thermally coupled with the aid of a TO39 size slightly flattened lobe-finned heat sink.

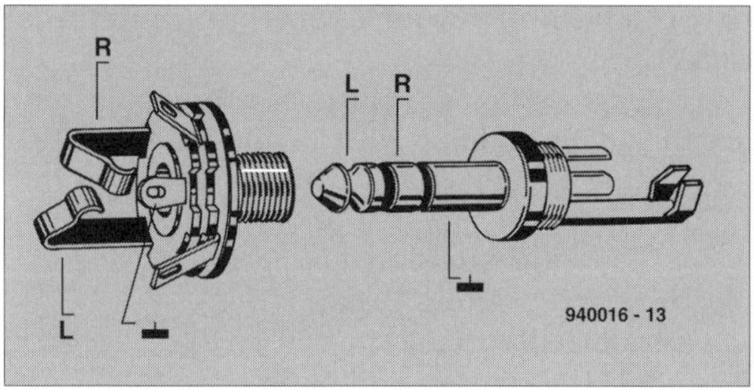

Fig. 9.6. Terminal layout of a 6.3 mm jack plug and socket.

Main parameters

Supply voltage	±15 V
Curren drain (each output stage)	about 30 mA
Quiescent current (each output stage)	25 mA
Output power	40 mW into 600 Ω
THD + noise (at 1 mW into 600 Ω)	<0.0015% (20 Hz–20 kHz)
THD at 1 kHz/1 mW	<0.0005%
Signal-to-noise ratio	>112 dB (A-weighted)
Input impedance	about 20 kΩ
Output impedance	75 Ω
Bandwidth	400 kHz
Slew rate (without R_1-C_1)	350 V μs^{-1}
Allowable loads	33–600 Ω

Parts list

Resistors:

R_1, R_{56} = 1 kΩ

R_2, R_{55} = 1 MΩ

R_3, R_4, R_{53}, R_{54} = 56.2 kΩ, 1%
R_5–R_8, R_{49}–R_{52} = 392 kΩ, 1%
R_9, R_{10}, R_{47}, R_{48} = 143 Ω, 1%
R_{11}, R_{19}, R_{20}, R_{37}, R_{38}, R_{46} = 200 Ω, 1%
R_{12}, R_{45} = 1.00 kΩ, 1%
R_{13}, R_{14}, R_{43}, R_{44} = 2.61 kΩ, 1%
R_{15}, R_{17}, R_{40}, R_{42} = 6.81 kΩ, 1%
R_{16}, R_{18}, R_{39}, R_{41} = 8.25 kΩ, 1%
R_{21}, R_{22}, R_{35}, R_{36} = 68 Ω
R_{23}, R_{24}, R_{33}, R_{34} = 3.9 Ω
R_{25}, R_{26}, R_{31}, R_{32} = 15 Ω
R_{27}, R_{30} = 100 Ω
R_{28}, R_{29} = 75.0 Ω, 1%
R_{57}, R_{58}, R_{63}, R_{64} = 12 Ω
R_{59}, R_{61}, R_{65}, R_{67} = 392 Ω, 1%
R_{60}, R_{62}, R_{66}, R_{68} = 4.22 kΩ, 1%
P_1, P_2 = 10 kΩ preset
P_3, P_4 = 50 Ω preset

Capacitors:
C_1, C_{32} = 270 pF, polystyrene
C_2, C_3, C_{30}, C_{31} = 2.2 µF, 50 V, polythene or polypropylene, pitch 5 mm
C_4, C_5, CC_{28}, C_{29} = 470 nF
C_6, C_{27} = 22 pF, polystyrene
C_7, C_8, C_{14}, C_{16}, C_{17}, C_{19}, C_{25}, C_{26}, C_{36}, C_{37}, C_{45}, C_{46} = 100 nF
C_9, C_{24} = 1 µF, pitch 5 mm
C_{10}, C_{23} = 2.2 nF
C_{11}, C_{12}, C_{21}, C_{22} = 1000 µV, 25 V, radial
C_{13}, C_{15}, C_{18}, C_{20} = 47 µF, 25 V, radial
C_{33}, C_{42} = 330 nF
C_{34}, C_{35}, C_{43}, C_{44} = 1000 µF, 40 V, radial
C_{38}, C_{39}, C_{47}, C_{48} = 22 µF, 40 V, radial
C_{40}, C_{41}, C_{49}, C_{50} = 10 µF, 63 V, radial

Semiconductors:
D_1–D_4 = 1N4003
T_1, T_5–T_7, T_{14}–T_{16}, T_{20} = BC550C
T_2–T_4, T_8, T_{13}, T_{17}–T_{19} = BC560C
T_9, T_{12} = BC337-40
T_{10}, T_{11}, = BC327-40

Integrated circuits:
IC_1, IC_4 = LM317
IC_2, IC_3 = LM337

Miscellaneous:
K_1 = 3-way terminal block, pitch 7.5 mm
Tr_1 = mains transformer, secondary 2×18 V, 4.5 VA

Chapter 10

Headphone amplifier III

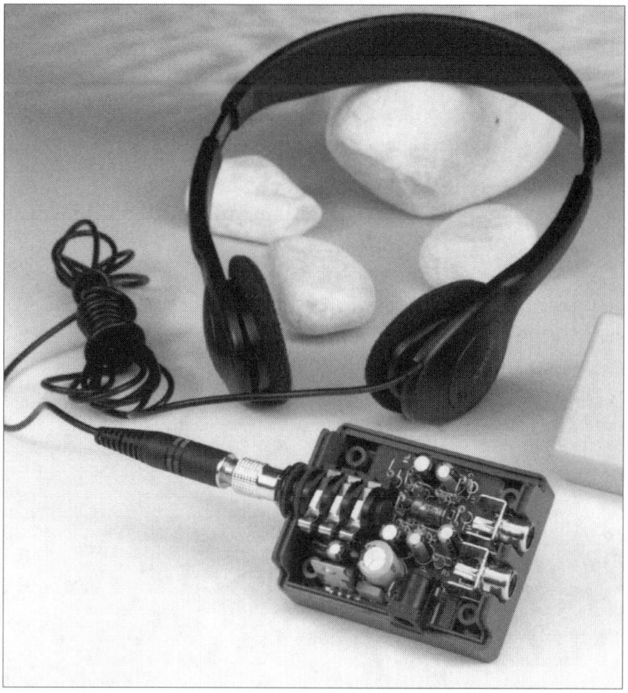

On much audio equipment, the headphone output is simply derived from the loudspeaker output via a series resistor: not exactly hi-tech! The present circuit describes a 'real' headphone amplifier that can be added to most equipment, but may also be used as a stand-alone unit.

Although many audiophiles still believe discrete components are best, the relentless progress of integrated circuits can not be stopped: these devices get better and better. Even top-quality commercial equipment is now loaded with ICs and no one can doubt their quality and reliability. In many modern CD players, preamplifiers and digital-to-analogue converters (DACs) there is hardly a transistor to be found. Only the design of power amplifiers often still relies on discrete devices. The present amplifier is based on an IC: a surface mount device (SMD) Type TDA1308T from Philips Components.

The IC was developed specifically for use as a headphone driver: the enthusiastic claims of the manufacturer as to its qualities appear to be rather less exaggerated than is often the case (see page 126: Performance). A signal-to-noise ratio of 110 dB and a distortion factor of <0.009% (with a 5 kΩ load) are undeniably good.

The IC can be used to good effect in CD players, DCC players, keyboards, laser disc systems, multimedia amplifiers, and more. It draws a quiescent current of only 3 mA and can work from supplies of 3–7 V. The latter makes it suitable for use in either battery-powered circuits or in standard mains-operated equipment. Its dynamic range is good, its bandwidth is 5.5 MHz and its slew rate is 5 V μs^{-1}.

The (simplified) internal design of the IC is shown in Fig. 10.1. The differential input stage uses MOSFETs, M_1, M_2, is provided with current mirrors and is powered by a current source, J_1. The input stage is followed by two operational amplifiers, A_1 and A_2, that drive output stages M_3 and M_6, which are also MOSFETs. The advantage of MOSFETs is that the necessary input bias current is very small: typically 10 pA; moreover, the swing of the amplifier with high impedance loads is nearly equal to the supply voltage.

The inverting and non-inverting inputs of the op amps have an excellent common mode

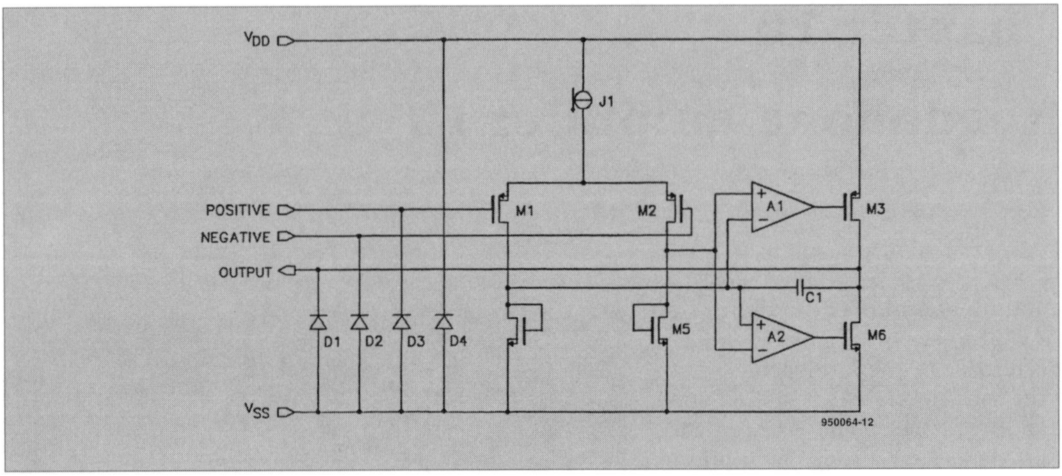

Fig. 10.1. Diagram of internal circuitry of the TDA1308T.

suppression that ranges from the negative supply voltage to 1.5 V under the positive supply voltage. The IC can be fed from single as well as bipolar supplies. The closed-loop gain can be set with two external resistors.

The outputs are short-circuit-proof and totally free of switching noise. The hum suppression is 90 dB.

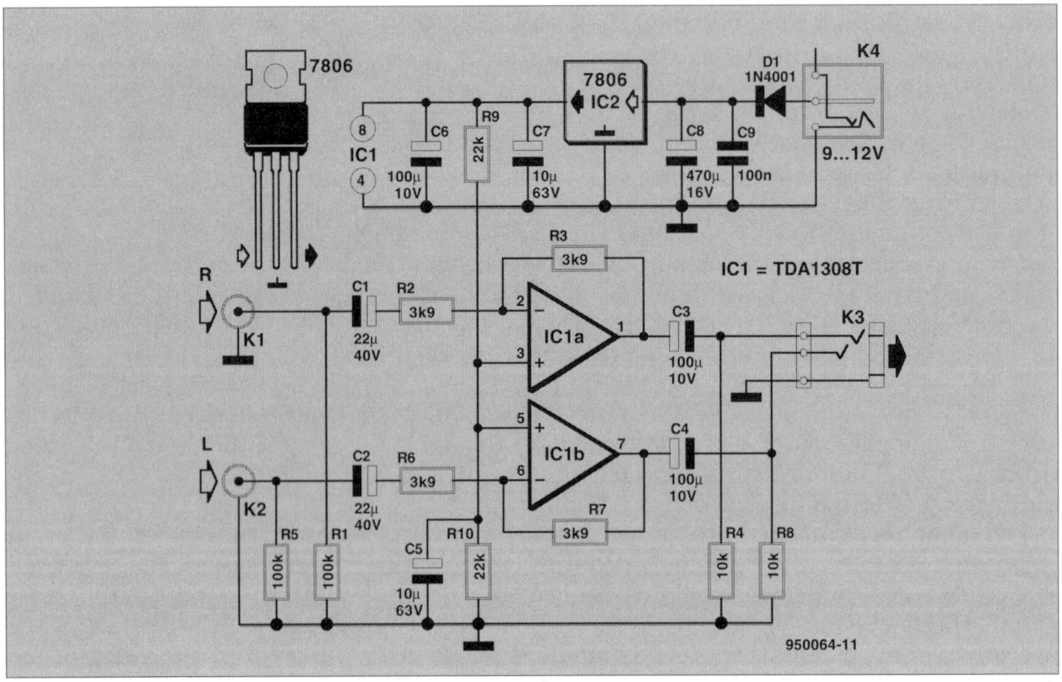

Fig. 10.2. Circuit diagram of headphone amplifier III.

Circuit description

In the circuit diagram (Fig. 10.2), values of components are generally those recommended by the manufacturer. Power is derived from a single, standard mains adaptor, which should output at least 9 V. The adaptor output is smoothed by C_8 and regulated by IC_2. Diode D_1 protects the circuit against wrong polarity.

The input impedance is determined largely by R_2 (R_6). The value of 3.9 kΩ presents no problems to any preamplifier. The amplification factor is set by the ratio R_2:R_3 (R_6:R_7). As is seen, the factor here is unity, so that the name 'amplifier' is, strictly, a misnomer; 'driver' would have been more appropriate. There is no need for amplification, because the usual line level of 1 V (nominal) is more than enough to drive any headphone. However, a standard line output can not provide sufficient current for driving low-impedance headphones. The present amplifier remedies this.

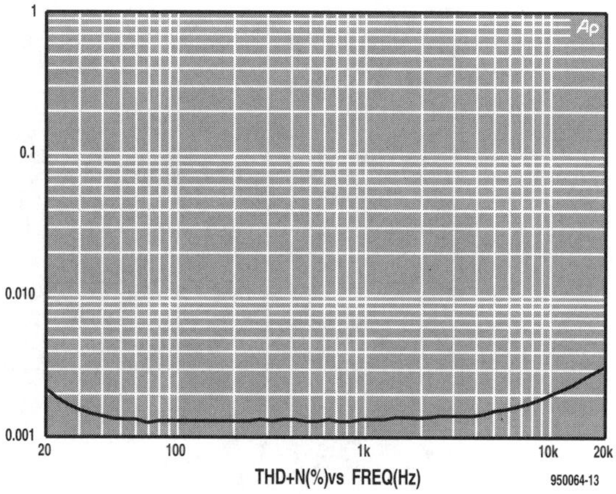

Fig. 10.3. The THD+N characteristic for 1 V input and a 600 Ω load.

Resistors R_9 and R_{10}, and capacitor C_5, set the IC for operation from half the supply voltage. Capacitor C_6 provides additional decoupling of the amplifier. Since the supply is asymmetrical, input capacitors C_1 and C_2 (C_3, C_4) are essential. Some audiophiles will raise their eyebrows at this, but in this application no adverse effects of these capacitors have been detected. Resistors R_1 and R_4 (R_5, R_8) make sure that these capacitors are charged even when the input and output are open.

Construction

The design has been kept as compact as feasible as can be seen from the drawings of the printed-circuit board in Fig. 10.4. In spite of the small dimensions, the construction is no more testing than most, at least not as far as the standard components are concerned. Soldering the surface-mount IC into place (not at the component side of the board, but at the track side as is usual with SMDs) is a tedious job. But, with patience and a small soldering iron with a fine tip, even relatively inexperienced constructors should be able to make a good job of it. Lightly tin the pads on the board and the pins of the IC and place the device in position. Take great care with the positioning of the IC: it is so small that mistakes are easily made. The side of the device where pins 1–4 are located is marked by a chamfered edge on the case. If you can not see this properly, use a magnifying glass. The chamfered edge should point in the

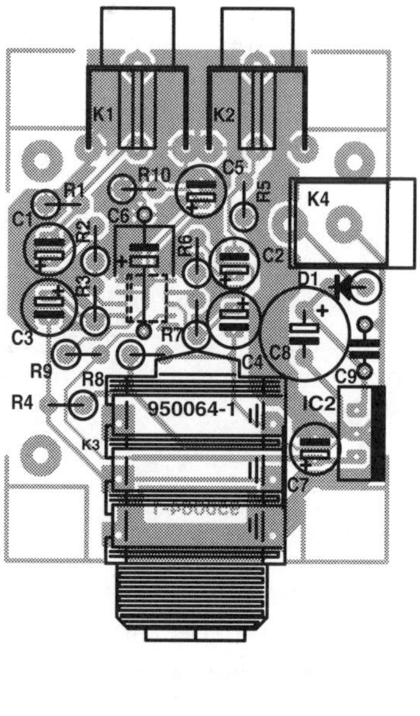

Fig. 10.4. Component layout of the PCB for headphone amplifier III. The track layout is given on page 249.

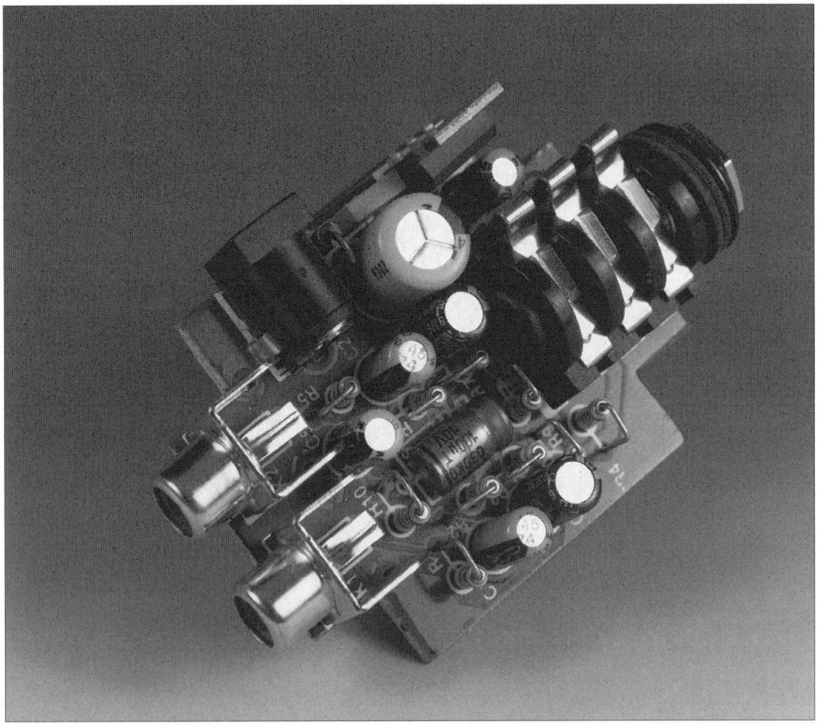

Fig. 10.5. Completed prototype of Headphone amplifier III.

direction of R_2 and R_3. Since it is very small, pressing with a fingernail will do. Gently solder one of the pins into place and make sure that everything is as it should be. If so, carefully solder the other pins on to the board.

The top of the finished board is shown in the photograph in Fig. 10.5. Note that the various connectors are soldered directly to the board: two phono plugs for inputs K_1 and K_2; an adaptor socket for K_4 and a 6.3 mm stereo jack for K_3. These connectors are, of course, required only if the amplifier is to be used as a stand-alone unit. If the amplifier is built into an equipment, the connectors can all be replaced by soldering pins from where the various connections are made. The signal lines should be in screened cable. Moreover, a mains adaptor will normally not be required, since power can invariably be derived from the main equipment: the amplifier draws only a very small current. If the voltage in he main equipment is too high, it can be dropped by a series resistor and zener diode (9 V or 12 V).

Performance

The performance of the Philips chip is typified by the distortion characteristic in Fig. 10.3. This shows that the THD+N is, as claimed, low: with a 1 kHz input signal at a level of 1 V and an output load of 600 Ω, the measured value was about 0.0015%. With a load of 32 Ω (Walkman-type headphones), it rose to 0.028%, which is still impressive for such a simple IC.

The channel separation measured at K_3 hovered around 90 dB with a 600 Ω load and 70 dB with a 32 Ω load (frequency range 20 Hz to 20 kHz). These values depend largely on the internal wiring of the headphones: a common earth wire leads to worse channel separation, but this can not really be attributed to the amplifier.

The maximum output voltage is 2 V r.m.s. across 560 Ω and 1.5 V r.m.s. across 32 Ω.

Parts list

Resistors:
R_1, R_5 = 100 kΩ
R_2, R_3, R_6, R_7 = 3.9 kΩ
R_4, R_8 = 10 kΩ
R_9, R_{10} = 22 kΩ

Capacitors:
C_1, C_2 = 22 µF, 40 V, radial
C_3, C_4 = 100 µF, 10 V, radial
C_5, C_7 = 10 µF, 63 V, radial
C_6 = 100 µF, 10 V
C_8 = 470 µF, 16 V, radial
C_9 = 100 nF, pitch 5 mm

Semiconductors:
D_1 = 1N4001

Integrated circuits:
IC_1 = TDA1308T (SMD)
IC_2 = 7806

Miscellaneous:
K_1, K_2 = audio socket for PCB mounting
K_3 = 6.3 mm stereo jack for PCB mounting
K_4 = Inlet for mains adaptor (for PCB mounting)
Enclosure (optional): 65×50×30 mm (e.g., Bopla E406 from Phoenix Mecano Ltd, 6–7 Faraday Road, Aylesbury HP19 3RY, Great Britain. Telephone +44 (0)1296 398855)

Part 4

Surround sound

Chapter 11

Dolby surround

A description of Dolby Surround must of necessity start with its background, which is in film making. The production of film sound is different from music-studio practice, if only because it evolved separately and, for the most part, earlier. The film industry developed facilities for recording, synchronization and complex mixes from several sound sources long before the first multi-track sound only reproduction equipment was developed. In fact, many of these techniques antedated the advent of magnetic recording. When filmmakers could not find equipment to implement those techniques, they invented and built it themselves.

One of the most complex systems in the film industry was Cinerama (an early form of widescreen cinematography shot with three adjacent cameras for presentation with three projectors on overlapping panels to form a continuous picture and using seven sound tracks). However, for financial and production reasons, film makers looked for alternative systems.

Involvement of the audience is what films endeavour to achieve and it does not matter whether the means to that end is dramatic or technical. Placing sound sources around the audience has been used universally for almost a decade. But the history of surround sound started with the 35 mm four-track format, which had, however, a narrow and therefore slightly noisy surround track. It was adequate for loud sounds, however, and when it was not in use, a pilot signal on the track shut it off. Films in 70 mm gave full-width quality on the surround track, making it possible to use subtler sounds and effects.

One of the systems to evolve in the mid 1970s was Dolby Stereo™, which was first used in

Ray Dolby was born in Portland, Oregon, in 1933, and received a B.S. degree in electrical engineering from Stanford University in 1957. From 1949 to 1952, he worked on various audio and instrumentation projects at Ampex Corporation, and from 1952 to 1957 he was mainly responsible for the development of the electronic aspects of the Ampex video tape recording system. After he was awarded a Marshall Scholarship, followed by a National Science Foundation graduate fellowship, he left Ampex in 1957 for further study at Cambridge University in England where he received a Ph.D. degree in physics in 1961, and was elected a fellow of Pembroke College. During his last year at Cambridge, he was also a consultant the the United Kingdom Atomic Energy Authority.

In 1963, he took up a two-year appointment as a United Nations adviser in India, and returned to England in 1965 to establish Dolby Laboratories in London. Since 1976 he has lived in San Francisco, where his company has established further offices and laboratories.

Dr Dolby holds a number of patents and has written papers on video tape recording, long wavelength X-ray analysis, and noise reduction. He is a fellow and past president of the AES (Audio Engineering Society), and a recipient of its Silver Medal Award. He is also a fellow of the British Kinematograph, Sound and Television Society, the SMPTE (Society of Motion Picture and Television Engineers), and a recipient of its Samuel L. Warner Memorial Award and Alexander M. Poniatoff Gold Medal. In 1979 he and his colleagues received the Scientific and Engineering Award of the Academy of Motion Picture Arts and Sciences.

the film *Star Wars*. In this system, the stereophonic reproduction is extended by a centre channel at the front and a surround channel at the rear of the audience. The centre channel serves to close the gap between the left-hand and right-hand channels; it makes dialogue more intelligible and more natural. The surround channel gives the sound a spatial effect. In cinemas the surround channel is usually reproduced by a number of small loudspeakers that are placed in the form of a U at the back and sides of the theatre. This system is widely used, because it places the information of the four channels on only two tracks of the film, which makes copying simple (which is, of course, important for the film industry). Moreover, it makes Dolby Stereo fully compatible with normal stereo, which means that films made in Dolby Stereo can be run by a standard stereo installation (special effects are, of course, not heard then).

Thousands upon thousands of feature films use Dolby Stereo. Since these films can be viewed at home, either on TV or by video cassette, the Dolby Laboratories introduced, in 1982, a system for consumer applications: Dolby Surround™. This early system was followed in 1987 by Dolby's Pro Logic decoder, which is similar to the professional Dolby Stereo system as far as characteristics and quality are concerned.

How does it work?

A simplified block schematic of a Dolby Stereo encoder, used in the film industry, is shown in Fig. 11.1. It shows how the four channels are combined into two film sound tracks. As far as signals in the centre channel are concerned, this is fairly simple: they are attenuated by 3 dB and then added, in phase, to the left-hand and right-hand channels. The signal in the surround channel is also attenuated by 3 dB and then passed through a bandpass filter, which limits

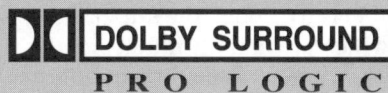

The Dolby Stereo trademark is used in prints and movie advertisements to denote a Dolby Stereo motion picture and is found only on pre-recorded VHS cassettes that employ B-type decoding on the two standard linear tracks. Furthermore, Dolby Stereo only appears on videocassettes or films that are licensed by Dolby Laboratories. Dolby System may be found on B-encoded cassettes of non-Dolby Stereo films that were released theatrically in four-track magnetic stereo or Academy mono. Therefore, no Dolby logo will be found on a video disc; although the LaserDisc is a high-quality medium, it does not employ Dolby noise reduction. Similarly, while the linear tracks on a VHS cassette might say 'Dolby Stereo', the 'VHS Hi-Fi' logo is not accompanied by any Dolby trademark.

Home surround decoders licensed by Dolby Laboratories have carried the 'Dolby MP Matrix', whose exact meaning was a mystery to most people. To remove this mystery, and to indicate the presence of Dolby Laboratories on the growing number of non-Dolby B home video formats, a new trademark, Dolby Surround, was introduced. The new logo appears on licensed decoders, replacing Dolby MP Matrix, as well as on all forms of video release, including the non-Dolby B stereo formats such as VHS and Beta Hi-Fi. Thus, Dolby Surround indicates the presence of stereo audio with surround information recorded and encoded with a Dolby DS-4.

Obtaining the Dolby Surround logo means that a manufacturer adheres to the basic outline provided by Dolby Laboratories. In general, licensed decoders have to contain:

- Metered input calibration to optimize levels for different VCR and videodisc output levels.
- A basic *L-R* surround matrix to extract the out-of phase surround information.
- A delay line, not only for time coherence of information that is both in the surround and front speakers, but also to reduce the perceptibility of unintentional surround leakage, especially the sibilant 'splatter' sometimes caused by azimuth errors. The recommended delay range for home decoders is 10–30 milliseconds (Dolby Cinema Processors are adjustable from 30 to 100 ms because of the larger front-to-back size of motion picture theatres).
- A 7 kHz low-pass filter. This is the high-frequency cut-off chosen by Dolby Laboratories to prevent bothersome hiss coming from surround speakers in theatres during quiet or inactive periods. The relatively steep HF roll-off also helps reduce delay-line noise and rear-channel sibilant splatter.
- A modified Dolby B-type decoder. *L-R* masters contain this modified B-type decoding on the surround-channel information to aid the low-pass filter in noise reduction and masking of sibilant breakthrough. It should be noted that the two tracks on a Dolby Stereo *LR* always employ A-type noise reduction that is decoded at some stage prior to video cassette duplication. The modified B-type decoding remains on the surround channel.
- An output stage with a ganged master level control.

Even though dialogue (speech or music) in home stereo playback seems natural coming from the phantom centre, many believe that the addition of a centre speaker helps to lock the dialogue to the image on the television receiver, with benefits observable even on 19-inch (48 cm) sets. In large living rooms, especially, a centre channel can help stabilize the dialogue (which is almost always in the centre) for those seated next to a front (left-hand or right-hand) speaker.

the frequency band to 100 Hz to 7 kHz. It is subsequently applied to a Dolby B compressor and then, phase shifted by –90°, to the right-hand channel, and by +90° to the left-hand channel. The chopping of frequencies below 100 Hz in the surround channel is to protect the surround speakers, which are usually much smaller than the front speakers and thus unable to handle these low tones. The limiting to 7 kHz and the compressing provide the required noise reduction. Both measures ensure that any sibilant splatter caused by phase and amplitude errors in the centre channel are reproduced via the surround channel (which would be both unnatural and annoying).

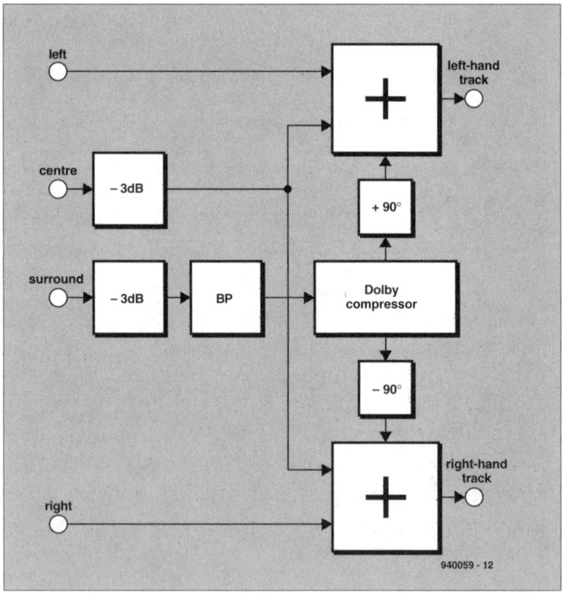

Fig. 11.1. Simplified block diagram of a Dolby Stereo™ encoder. This is used only in the film industry.

Matrix decoder

A decoder is required to derive the original four channels from the two film sound tracks. This may be an active device (to which will be reverted) or a passive, so-called matrix, decoder. In the latter, the two missing channels are regained mainly by application of sum and difference processes. The basic design of such a decoder is shown in Fig. 11.2. It is seen that the left-hand and right-hand channels are derived directly from the two sound tracks, while the centre channel is formed by adding the two sound tracks together.

The process of obtaining the surround channel is rather more complex. The difference of the two sound tracks is delayed by 20–60 ms, so as to make it impossible for the listener to determine the exact (sound) location of surround speakers close to him. This is essential, because the first wave front must come from ahead, since that must remain the direction of orientation.

The signal is then passed through a bandpass filter, after which a Dolby expander restores the original dynamics.

The main advantage of a passive decoder is the simplicity of its design. A drawback is, however, that the design does not provide good channel separation. The maximum attainable separation between the four channels is shown in Fig. 11.4.

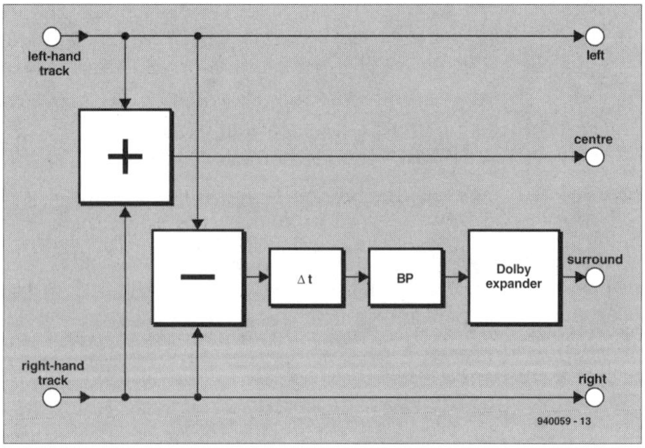

Fig. 11.2. Block diagram of a Dolby Matrix decoder. This is also for use with film sound tracks only.

Pro Logic (active) decoder

Analysis of the output signals of a matrix decoder shows various weaknesses: the two main channels contain information not only from left and right tracks, but also constituents from the centre and sur-

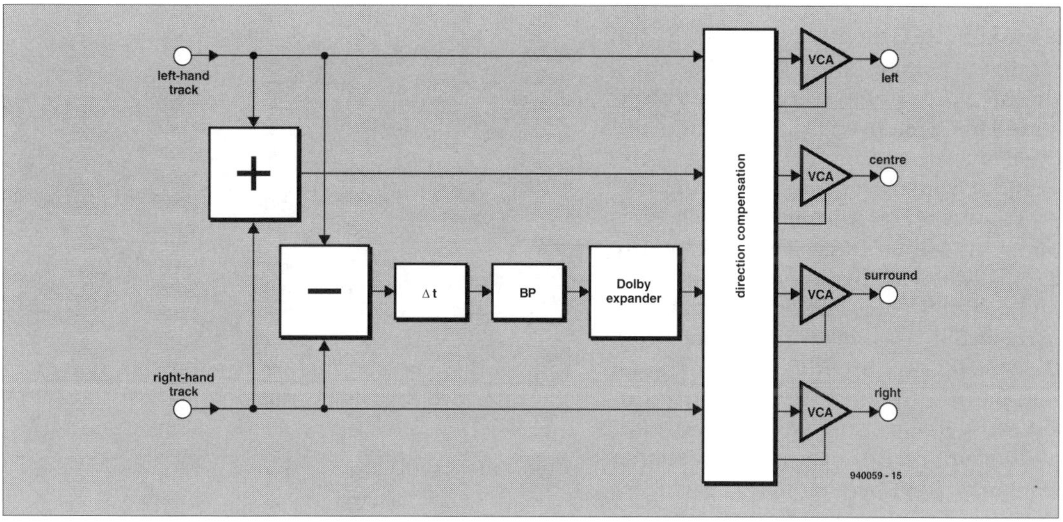

Fig. 11.3. A Pro Logic decoder provides dynamic direction compensation, which analyses the signal and continuously adjusts the output levels of the four channels via voltage-controlled amplifiers (VCAs).

round channels, since these have not been filtered out. Moreover, the surround channel contains constituents from the difference signal, while the centre channel contains parts of both the left and right track.

It is clear that the main task of an active decoder is to improve appreciably the channel separation. This is why it is provided with dynamic direction compensation. The Pro Logic decoder analyses the composition of the signal as far as phase, amplitude and frequency are concerned and generates the necessary correction signals. If, for instance, at a certain instant the left-hand and right-hand signals are identical in amplitude and phase, the decoder rightly acts as if this mono signal belongs in the centre channel. It then increases the gain of the centre channel and lowers that of the left-hand and right-hand channel to ensure that the total sound volume remains the same.

Active decoders usually have four voltage controlled amplifiers, whose gain is adjusted by the programme material. This selective amplification increases the channel separation between the main and auxiliary channels from 3 dB to 35 dB. Although some perfectionists feel that this is still a low figure, practice has shown that greater channel separation is not necessary, since the four channels form a unified sound source.

The block diagram of the Pro Logic decoder is given in Fig. 11.3. It shows that the basic matrix and preliminary processing of the surround channel signals are identical to those of Fig. 11.2. An addition is the dynamic direction compensation which controls the four voltage-controlled amplifiers (VCAs). In the very latest types of decoder, the entire signal processing and compensation are effected by digital levels. The audio signals are converted into digital levels by an ADC (analogue to digital converter) and processed by a signal processor and suitable algorithms. Filtering and gain control are also by digital levels. The signal delay of the surround channel is effected by DRAMs (dynamic random access memories).

The channel separations obtained with a Pro Logic decoder are shown in Fig. 11.5.

Domestic surround

As mentioned before, since Dolby Stereo makes use of only two sound tracks, the sound may be reproduced by any standard stereo installation to which a decoder and an additional pair of loudspeakers have been added. This compatibility with stereo is perhaps the strongest point of the system: it enables the pleasure of listening to surround sound to be removed from the cinema to the living room.

Apart from the fact already mentioned that thousands of feature films have been made in Dolby Stereo, there are many CDs available with original Dolby Stereo film sound tracks. And, of course, many of these film are shown on television. As the TV stations broadcast the original film sound, this can be reproduced in the living room. Consequently, there is a plethora of suitable decoders available from all self-respecting manufacturers: Denon, JVC, Pioneer, Sony, Toshiba, Yamaha, and many others. Prices have been coming down, too.

Most of these decoders have a number of additional features. Often there is an integral noise generator which is useful for adjusting the levels of the four channels. An auto balance compensates for differences in output level of the TV. And, of course, there is the inevitable remote control. Some decoders also have provision for connecting a subwoofer: the necessary cross-over filter is already fitted in the decoder. Whether a subwoofer is required depends, of course, on what speakers are already available with the existing audio system.

When buying a decoder or decoder/amplifier for Dolby Surround, make sure that the unit carries the original Dolby logo preceded by the two mirrored Ds and Pro Logic underneath it as shown on page 133. Whatever the salesman tells you, equipment that is not so fitted or which carries an incomplete logo is suspect and should not be bought (at least not for the present purpose).

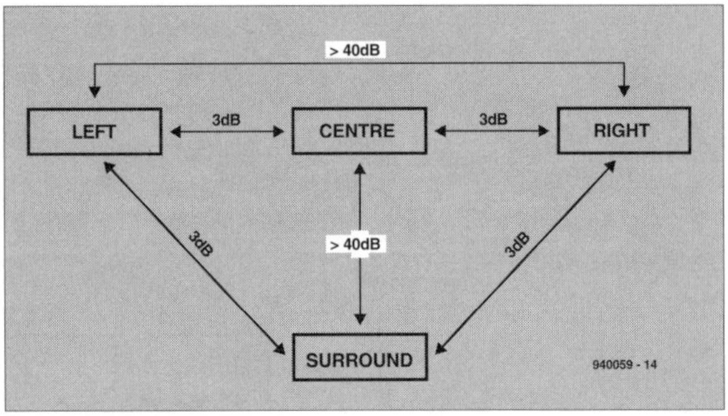

Fig. 11.4. The channel separation provided by a passive decoder is not particularly good.

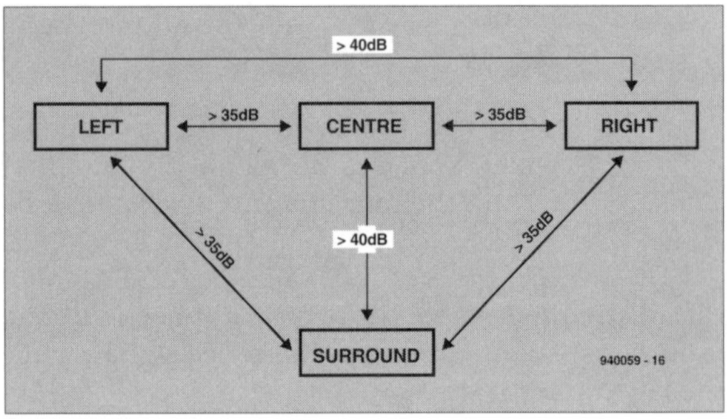

Fig. 11.5. The control system of an active decoder substantially improves the channel separation.

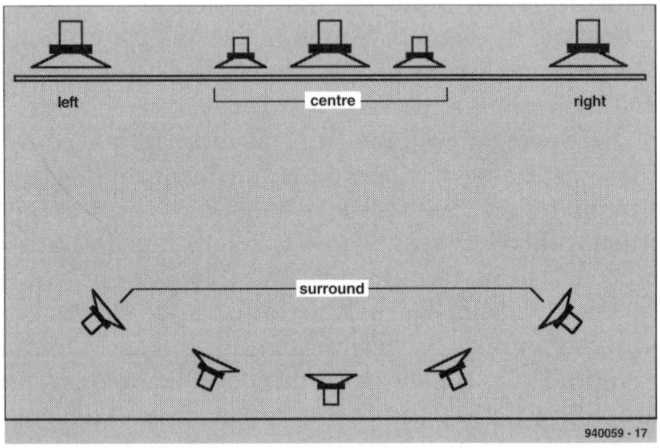

Fig. 11.6. In a cinema the additional channels are usually reproduced by a number of loudspeakers.

Loudspeakers

A typical loudspeaker array as used in cinemas is shown in Fig. 11.6. To the main speakers for the left-hand and right-hand channel have been added one or more speakers for the centre channel and a series of smaller speakers for the surround channel at the rear and sides of the theatre.

The setup in a domestic living room is rather simpler as shown in Fig. 11.7. Two small speak-

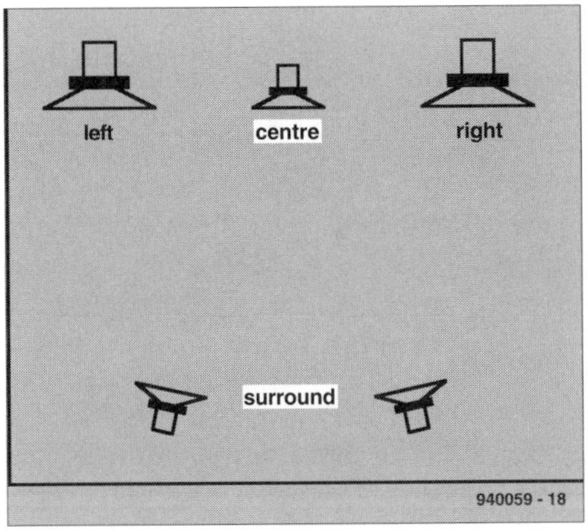

Fig. 11.7. In a domestic living room, an additional speaker for the centre channel and two small speakers for the surround channel can be added to the existing loudspeakers.

ers for the surround channel and one for the centre channel in addition to the existing stereo speakers are normally sufficient.

There may be some confusion as to what type of loudspeaker should be used for the surround systems. A number of manufacturers offer special sets of speakers and many people will wonder whether there is a need for these. The answer is yes and no. No, emphatically no, as regards the main speakers which should be standard hi-fi units: normally these will be the existing ones. As far as the surround channel is concerned, the usual hi-fi speaker is really too good for this, since the frequency range is limited to 100 Hz to 7 kHz (why use an expensive speaker that goes down to 40 Hz or even 20 Hz?)

The importance of the centre channel must not be underestimated, because this is a real pillar of the Dolby system. Film makers correctly maintain that the centre of the sound of a feature film contains much information, since virtually all dialogue between actors and actresses, as well as much other action, takes place there. It is at this centre that a standard stereo installation provides no reproduction. Add to this that in most cinemas the left-hand and right-hand speakers are (of necessity) far apart, and it is clear that many viewers find correct localization of the sound impossible.

All this is, of course, not terribly important in the living room, but it illustrates the point that the speaker for the centre channel must be of reasonable quality, although it need not be as large as the main speakers, because it does not handle so much bass power (–3 dB with respect ton the main speakers—see Fig. 11.4). The best place for the centre channel speaker is near, but not too close to, the TV set. If too close, the magnet of its drive unit may adversely affect the TV picture. Note, however, that there are speakers available with magnetic screening.

Chapter 12

Surround sound processor

Creating surround sound in a domestic room can be approached in two ways: by a processor that generates the four signals required: left-hand, right-hand, centre and surround, or by one that adds the two missing channels, that is, centre and surround, to the existing stereo sound. The first is the most elegant, but also the most complicated and most expensive. Moreover, it requires an additional line to return the left-hand and right-hand signals to the input of the TV/audio equipment. The second way is much more straightforward and has proved in practice to give an excellent spatial effect. Moreover, it can be accomplished in a compact and fairly inexpensive unit. The design in this chapter is of the second kind.

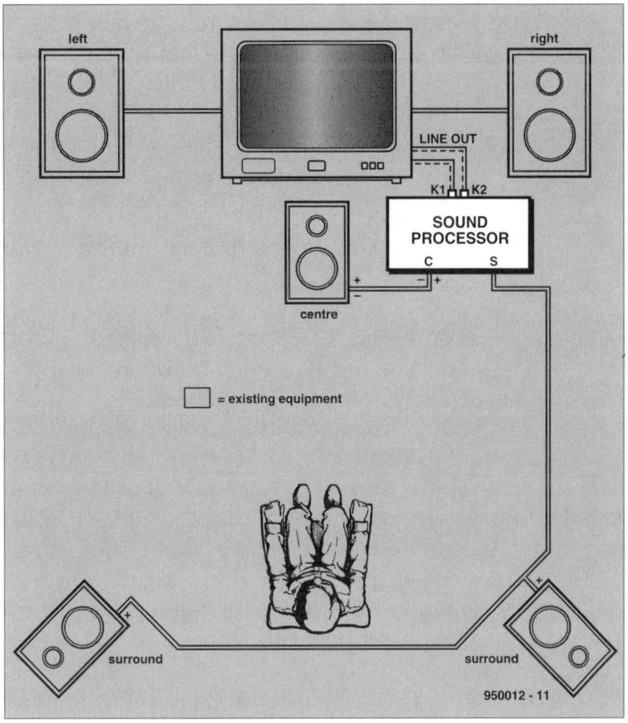

Fig. 12.1. Basic setup of a surround sound system in a living room.

The basic setup in a domestic room is shown in Fig. 12.1. The left-hand and right-hand channels are reproduced as before, that is, via the loudspeakers in the TV receiver, as shown, or by those of the audio installation to which the TV receiver is connected. The extra items are the processor and three loudspeakers. The inputs of the processor are linked, possibly via the SCART connector, with the line out terminal of the TV receiver, or audio amplifier, while the extra loudspeakers are connected to the outputs of the processor. The processor contains two integral amplifiers each of which provides 20 W output into 4 Ω: quite sufficient for the centre and surround loudspeakers.

As briefly discussed in Chapter 11, the additional loudspeakers need be no more than compact (bookcase type) hi-fi types that are not too expensive. If possible, however, choose types whose efficiency is about the same as that of the main loudspeakers: this affords rather more freedom when the system is set up as a whole.

Basic design

A block diagram of the basic design is given in Fig. 12.2. The design of the processor is

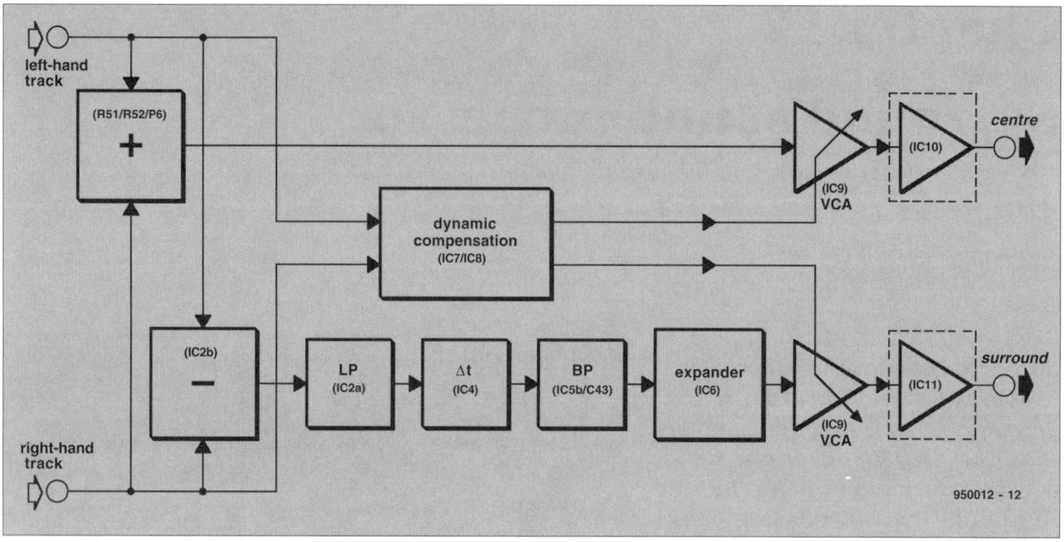

Fig. 12.2. Block diagram of the surround sound processor.

similar to that of the active decoder discussed in the (previous) chapter on Dolby surround sound. There are some differences between the two, since the present processor not only functions as a decoder, but also provides the signals for the centre and surround channels and contains two output amplifiers.

As explained in Chapter 11, the two additional channels are processed (whence the name of the unit) from the sum and difference of the two stereo channels. The centre channel is the simpler to produce, because it suffices to add the left-hand and right-hand channels together and apply the sum to a suitable output amplifier via a voltage-controlled amplifier (VCA).

To produce the surround signal, the right-hand signal is deducted from the left-hand signal (or vice versa) and the resulting signal is applied to a delay network via a low-pass filter. The delay can be preset between 10 ms and 30 ms. The signal is then applied to an expander via a band-pass filter. The expander is essential because the surround signal is compressed during recording. The output of the expander is applied to a second output amplifier via a VCA.

The dynamic compensating network, in conjunction with the VCAs, reflects the difference between an active and a passive (matrix) design. In this network, the correlation between the two stereo channels is analysed continuously. The results of the analysis are converted into control signals for the VCAs which constantly adjust the levels of the centre and surround signals. This arrangement ensures a much larger channel separation than possible with a passive design.

Centre channel

From the inputs of the left-hand and right-hand channels, K_1 and K_2 in Fig. 12.4, the signals are taken via buffer amplifiers IC_{1a} and IC_{1b} to R_{51} and R_{52}, and then summed in preset P_6. From there, the signal is applied to pin 5 of IC_9. This IC contains four electronic potentiometer circuits (of which only two are used) which function as VCAs; their amplification is governed by a control voltage at pins 9 and 10. The output of one of the circuits is available at pin 7, from where it is applied to output amplifier IC_{10}. This circuit provides an output of up to 20 W into 4 Ω.

The output of IC_{10} is applied to the centre channel loudspeaker via relay contact Re_{1b}. The relay is controlled by a simple delay circuit, T_1, and obviates any clicks and plops in the speaker caused by the switching on and off of the processor.

Surround channel

The signals at the outputs of IC_{1a} and IC_{1b} are also applied to the inverting and non-inverting inputs of IC_{2b} respectively, so that the output of the op amp is the difference of the two stereo signals (L–R). The difference signal is applied to a 4th-order low-pass filter based on IC_{2a}, which limits its upper bandwidth to 7 kHz. This anti-aliasing filtering serves to obviate the formation of spurious mixing products of the signal and the clock of the following delay line, which is based on IC_4. This IC is a 2048-stage bucket brigade device. The rate at which the internal electronic switches are operated is determined by IC_3. This CMOS-IC is designed especially to generate a low-impedance, double-phase clock. The specified values of its frequency-determining components, R_{15}, R_{16}, C_9 and P_1 allow a delay between 10 ms and 30 ms to be set with P_1.

The outputs of IC_4 are applied via buffer IC_{5a} to a low-pass filter based on IC_{5b} (identical to that based on IC_{2a}) which filters out any residue of the clock signal. The cut-off frequency is 7 kHz. The signal is subsequently fed to compander IC_6, whose input network contains a high-pass filter, R_{55}-C_{43}, the specified values of which give a lower cut-off frequency of about 50 Hz. The overall effect of the low-pass and high-pass filters is, of course, that of a band-pass filter as shown in Fig. 12.2.

The compander IC contains two circuits each consisting of a rectifier, a variable gain cell and an op amp. In the present processor only one of these circuits is used and that as an expander. The values of external components R_{27}–R_{32} and C_{23}–C_{29} allow for an expansion factor of 1:1.3.

The surround signal is then applied to the second electronic potentiometer circuit in IC_9, whose output is available at pin 17. From there, the signal is fed to output amplifier

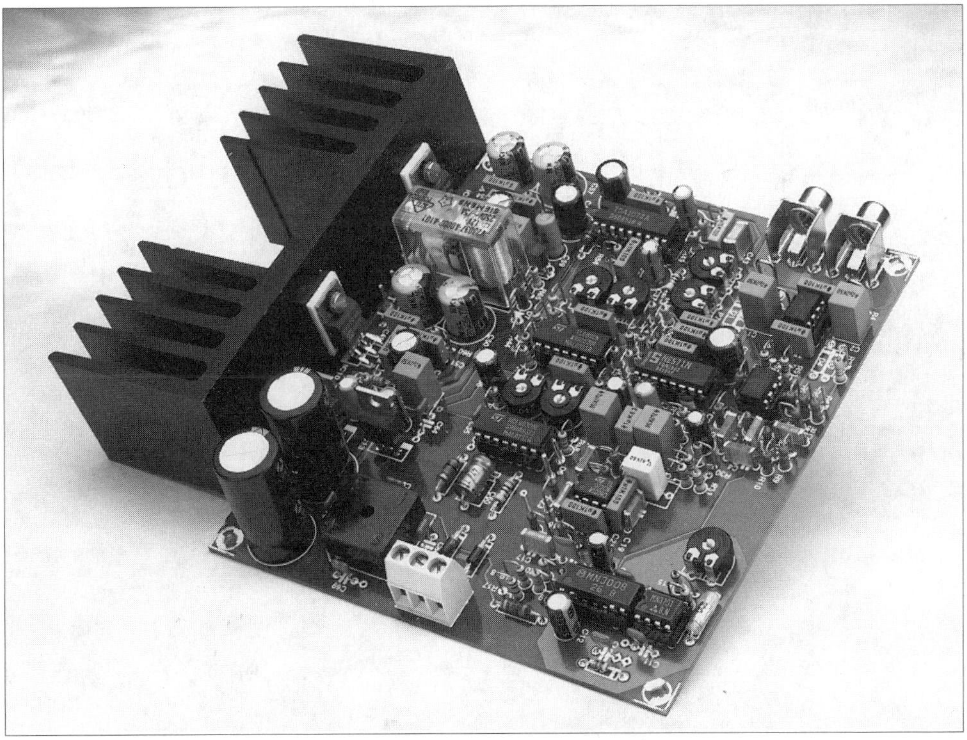

Fig. 12.3. Completed prototype board.

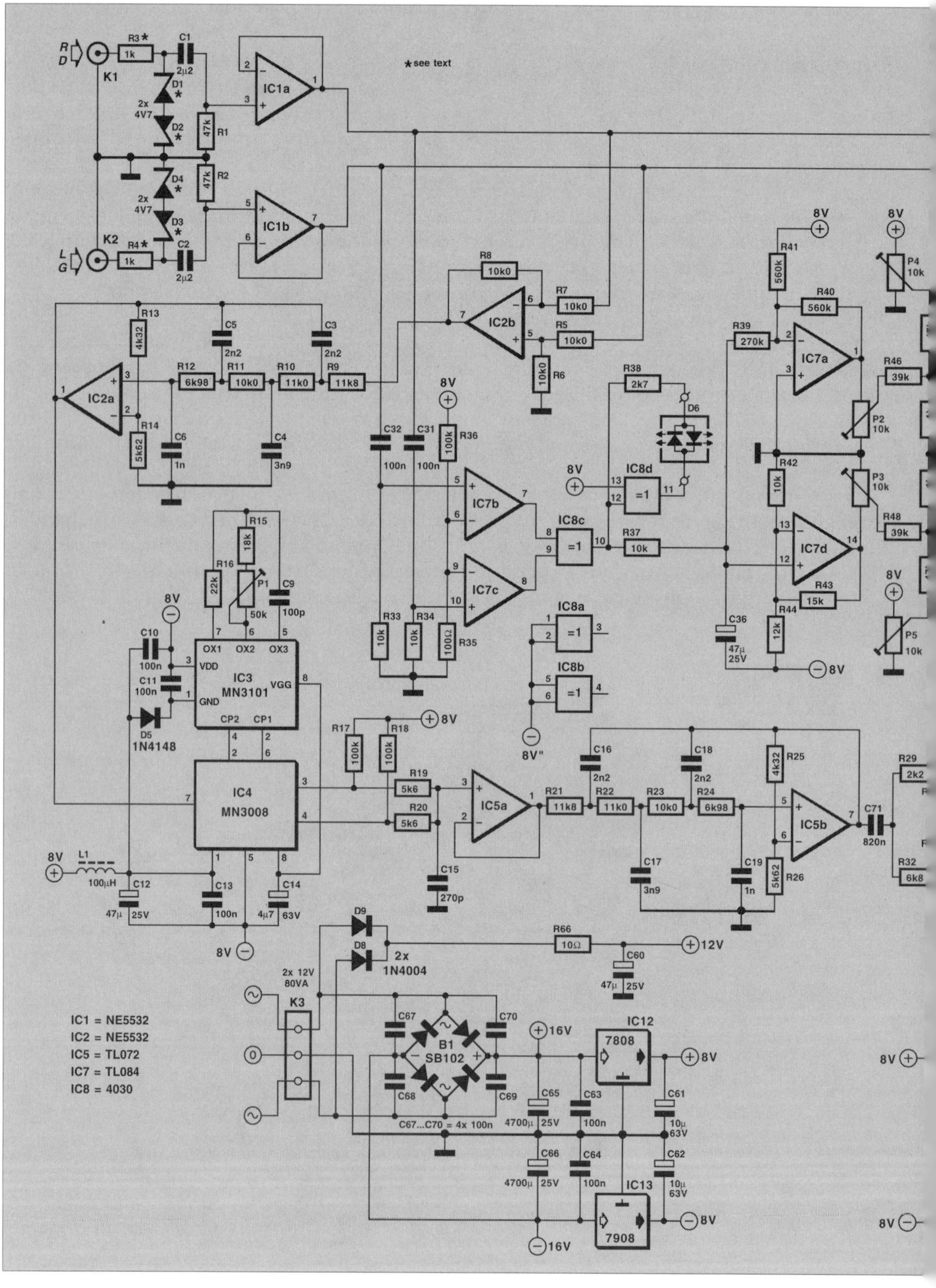

Fig. 12.4. Circuit diagram

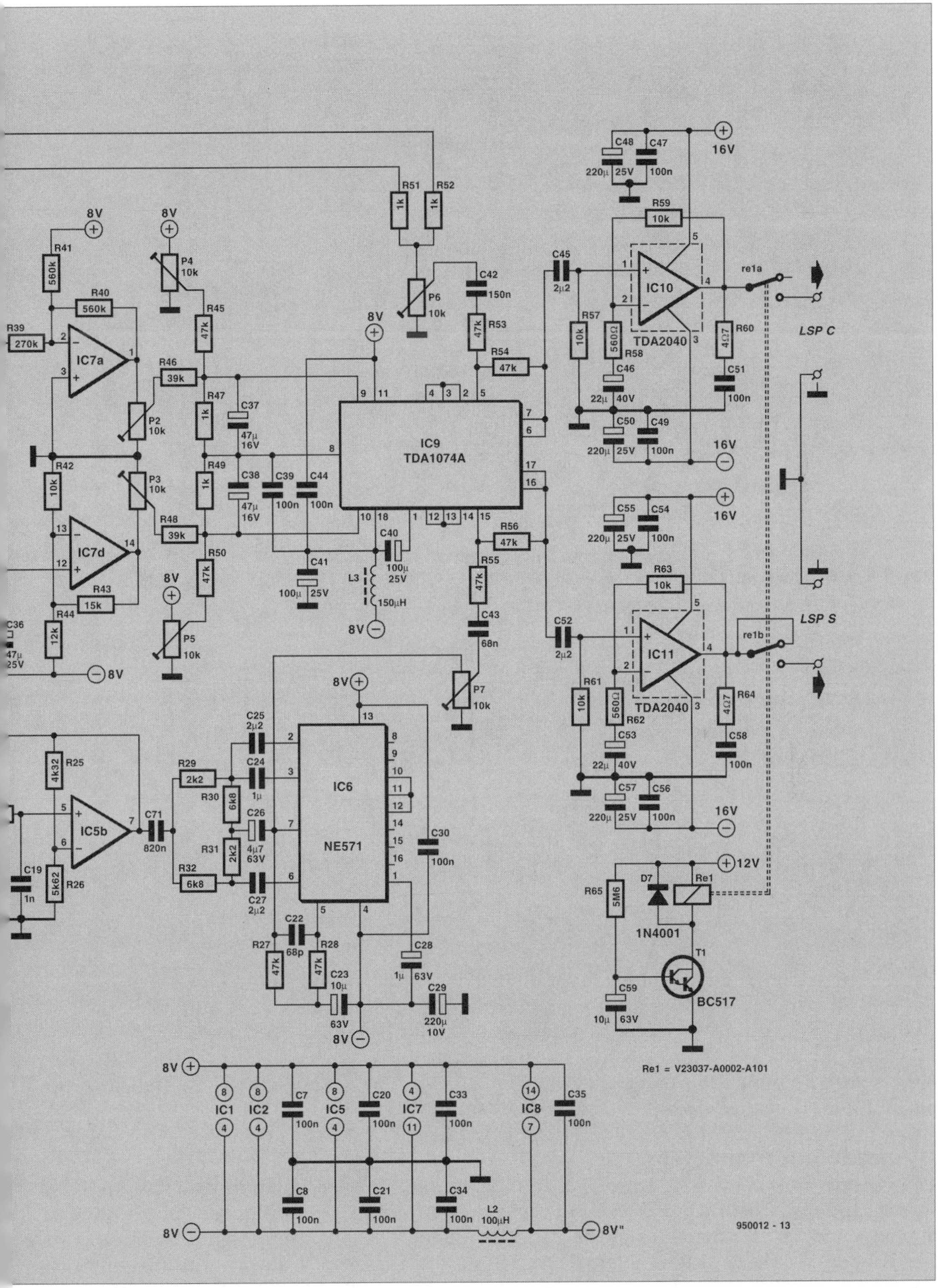

of the surround sound processor.

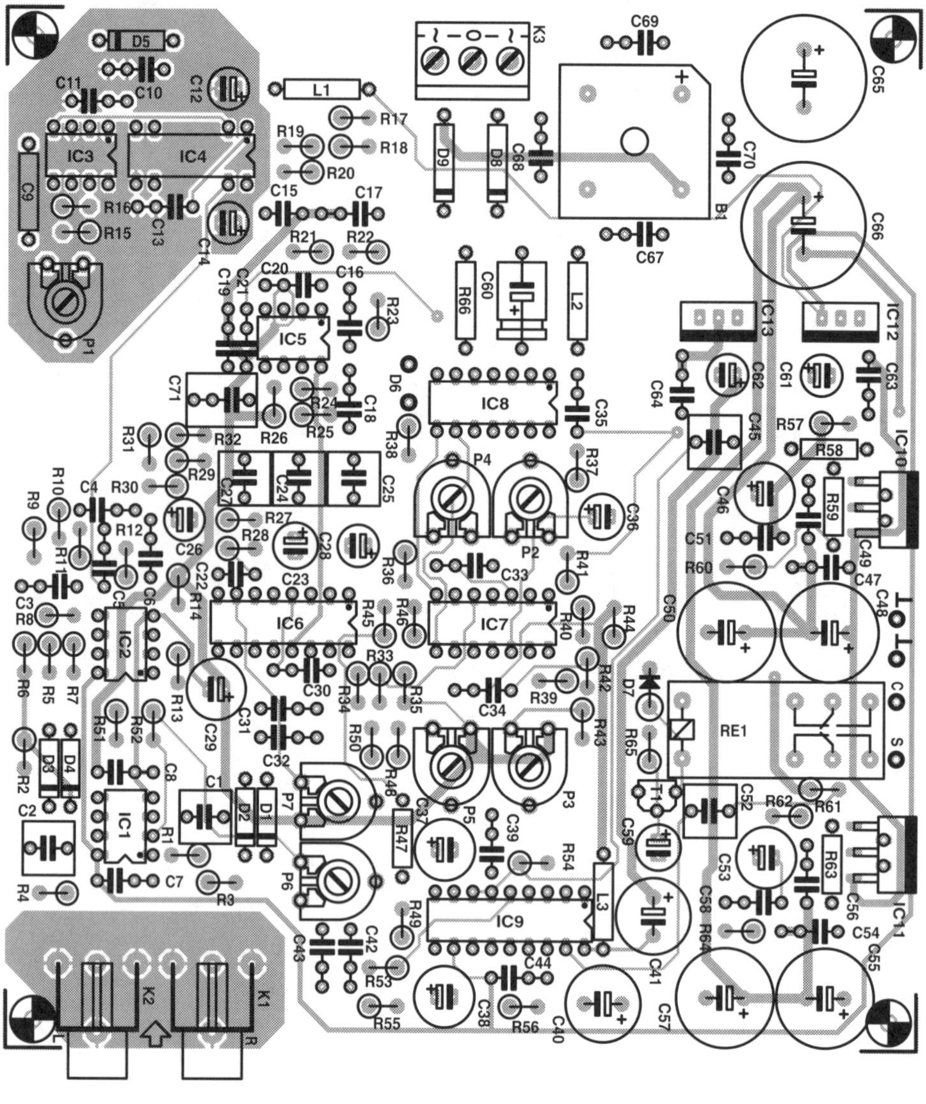

Fig. 12.5. Component layout of the PCB for the surround sound processor. The track layout and component overlay are given on pages 251 and 253.

IC_{11}, whose amplification is identical to that of IC_{10}. The output of IC_{11} is applied to the surround loudspeaker(s) via a second contact on Re_1.

Dynamic compensation

The outputs of buffers IC_{1a} and IC_{1b} are also applied to twin comparators IC_{7b} and IC_{7c} via C_{31} and C_{32}. The output of each of these comparators is a rectangular voltage the frequency of which is a measure of the variation in the relevant input signal. Both outputs are applied to XOR gate IC_{8c}. (Remember that an XOR gate has an output only when its inputs are dissimilar). Integration of the output pulses of the gate by R_{37}-C_{36} results in a direct voltage whose amplitude is a measure of the phase difference between the two stereo signals.

This direct voltage is applied via IC_{7a} (inverted) and IC_{7d} (non-inverted) to the control inputs (pins 9 and 10) of IC_9. This arrangement ensures that when a mono signal is present at the inputs (no or hardly any phase difference), the amplification of the VCA controlling the centre channel is raised. Conversely, when a surround signal is present (large phase difference), the amplification of the VCA controlling the surround channel is increased.

The degree to which the amplification of the VCAs is influenced by the control signals is preset by P_2 and P_3. When the wipers of these controls are at earth potential, the amplification is fixed; when they are at the opposite end of their travel, control is maximum.

Presets P_4 and P_5 serve to shift the operating point of the VCAs to some extent. They thus make the preset range wider and, in fact, support the operation of P_6 and P_7. If, for instance, P_7 has already set the surround level to maximum, P_5 enables this to be increased slightly. The same applies to P_4 insofar as the level of the centre channel preset with P_6 is concerned.

The currents through R_{45} and R_{46} and those through R_{48} and R_{50} are simply added together: there is, therefore, no interaction between P_2 and P_4 nor between P_3 and P_5.

Further circuit details

Resistors R_3 and R_4 and diodes D_1–D_4 limit the level of the input signal to a safe value and are imperative if the stereo signals are taken from the loudspeaker outputs of the TV receiver. Note that even line out terminals sometimes provide a signal at a level well above 1 V. If it is absolutely certain that the line output level is 1 V, and this is the only input, the resistors can be replaced by a wire bridge and the diodes may be omitted.

Bi-colour LED D_6 functions as a kind of signal monitor that shows the change from surround channel to centre channel and vice versa. In the case of a surround signal, the output of IC_{8c} is high. Since one output of IC_{8d} is at +8 V, both inputs of this XOR are then high, so that its

Fig. 12.6. Completed prototype processor with cover removed.

output is low. This results in the red segment of D_6 lighting. In the case of a centre signal, pin 12 of IC_{8d} is low, so that its output is high, resulting in the green segment of D_6 lighting. In practice, the changes between the signal are so rapid that the LED shows a fluent transition from red to green and back to red again.

The power supply provides three different voltages. The secondary of the mains transformer is connected to K_3. The 12 V input, after rectification and smoothing, results in a symmetrical supply of ±16 V, which is used to power output amplifiers IC_{10} and IC_{11}.

From the ±16 V lines, voltage regulators IC_{12} and IC_{13} derive a supply of ±8 V, which is used to power the remainder of the circuit.

The 12 V line for the relay is taken directly from K_3 and rectified by D_8-D_9.

Construction

The processor is best constructed on the PCB illustrated in Fig. 12.5. Since this board is double sided and through-plated, it is not possible to make it without special tools and equipment.

Populating the board is straightforward and should not present undue difficulties. Note that when the specified enclosure is used, capacitors C_{65} and C_{66} must be not higher than 38–40 mm.

Use gold-plated connectors for K_1 and K_2 to ensure good, lasting connections. The positions for these connectors are at the edge of the board so that all that is necessary when fitting the board into the enclosure is drilling a few holes in the back panel through which these sockets can protrude. Note that they must not touch the enclosure.

Output amplifiers IC_{10} and IC_{11} are located at the edge of the board so that they can be fitted readily to a heat sink. The ICs must be electrically isolated from the heat sink by ceramic washers and heat conducting paste.

The photograph in Fig. 12.3 shows the completed prototype board.

The main requirements of the enclosure are that it is made of metal and that it provides adequate space for the finished board.

Apart from K_1 and K_2, fit suitable sockets or spring-loaded terminals at the back of the enclosure for connecting the centre and surround loudspeakers. Link these terminals with heavy-duty insulated wire to points 'C' and 'S' and the adjacent earthing points on the board. The specified mains entry with integral fuse holder should also be fitted at the back of the enclosure.

The mains on/off switch and D_6 should be fitted at the front panel (for which a ready-made foil is not available). The diode should be connected to the relevant points on the board via lengths of flexible stranded wire.

Finally, link the centre pin of K_3 to the mass of the enclosure with the aid of a solder tag.

The completed prototype is shown in the photograph of Fig. 12.6.

The output amplifiers are suitable for operation with load impedances ≥4 Ω, but not lower ones. Thus, for the surround channel, two 8 Ω loudspeakers may be connected in parallel only if it is absolutely certain that the impedance is 8 Ω. If it is not, connect the speakers in series: this is safer. It is essential that the two speakers are in phase: the +terminals must go to the same terminal on the board: whether this is earth or 'S' does not matter.

The centre loudspeaker must be in phase with the main speakers. Since the relevant VCA functions as an inverter, the +terminal of this speaker must be connected to the earth point on the board; the –terminal to point 'C'.

Calibration

Start by setting presets P_2–P_7 to the centre of their travel, and P_1 to maximum (fully clockwise).

Inject a (mono) speech signal and adjust P_6 until the sound appears to come from the centre speaker. Wait for a surround signal (indicated by the red segment of D_6 lighting) and turn P_7 till sound emanates from the surround speakers. Do not set the level too high, because this leads quickly to an exaggerated effect. If, however, it is felt that the desired

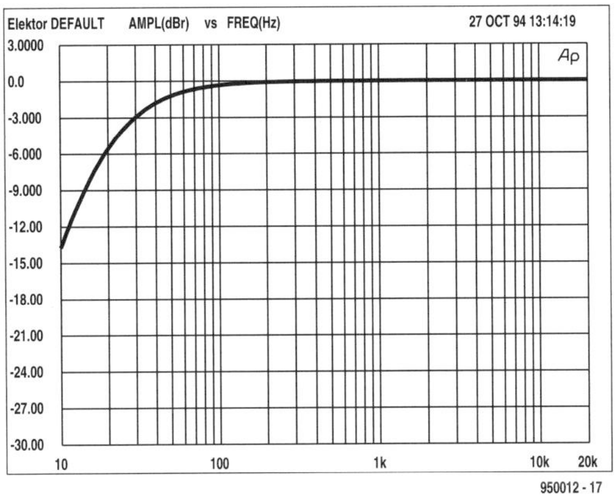

Fig. 12.7. Frequency response of the centre channel.

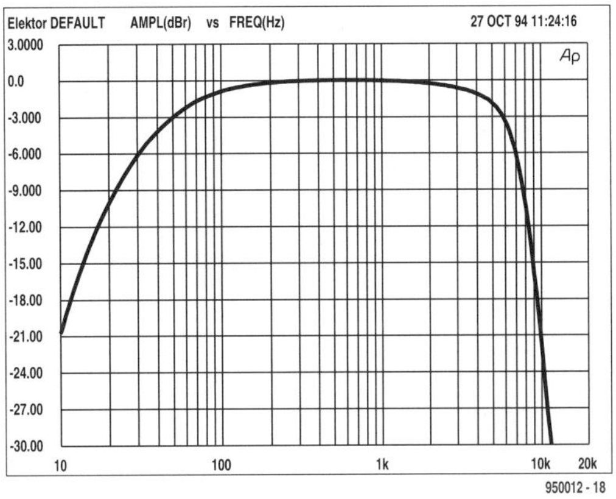

Fig. 12.8. Frequency response of the surround channel.

level can not be obtained with P_6 or P_7, as the case may be, adjust P_4 or P_5, or both, as required.

Next, create a spatial effect by slowly turning P_2 and P_3 until the centre channel and the surround channel seem well 'separated'. It is more than likely that P_6 and/or P_7 must then be readjusted. Note that these controls give an instinctive 'wrong' feel: turning them clockwise *reduces* the level.

If the surround speakers give exaggerated reverberation, reduce the delay with P_1. In the average living room, a delay of 25 ms appears correct: this corresponds to P_1 being almost at its maximum setting.

Do not be surprised if after watching and listening to a number of films, some readjustment of the controls is found desirable.

Characteristics

The curves in Fig. 12.7 and 12.8 show the amplitude vs frequency characteristics of the centre channel and surround channel respectively. The curves were obtained with an audio analyser. It is evident that they correspond closely with the descriptions. The –3 dB point of the centre channel is at 30 Hz. The –3 dB points of the surround channel are at 50 Hz and 6 kHz; the –6 dB bandwidth is roughly 30 Hz to 7 kHz.

Figure 12.8 shows that it does not make sense to use tweeters with a linear characteristic up to 20 kHz for the surround channels.

Both curves make it clear that a subwoofer must be connected to the main channels and not to the centre channel or surround channel.

Parts list

Resistors:

R_1, R_2, R_{27}, R_{28}, R_{45}, R_{50}, R_{53}–R_{56} = 47 kΩ
R_3, R_4 = 1 kΩ (see text)
R_5–R_8, R_{11}, R_{23} = 10.0 kΩ, 1%
R_9, R_{21} = 11.8 kΩ, 1%
R_{10}, R_{22} = 11.0 kΩ, 1%
R_{12}, R_{24} = 6.98 kΩ, 1%
R_{13}, R_{25} = 4.32 kΩ, 1%
R_{14}, R_{26} = 5.62 kΩ, 1%

R_{15} = 18 kΩ
R_{16} = 22 kΩ
R_{17}, R_{18}, R_{36} = 100 kΩ
R_{19}, R_{20} = 5.6 kΩ
R_{29}, R_{31} = 2.2 kΩ
R_{30}, R_{32} = 6.8 kΩ
R_{33}, R_{34}, R_{37}, R_{42}, R_{57}, R_{59}, R_{61}, R_{63} = 10 kΩ
R_{35} = 100 Ω
R_{38} = 2.7 kΩ
R_{39} = 270 kΩ
R_{40}, R_{41} = 560 kΩ
R_{43} = 15 kΩ
R_{44} = 12 kΩ
R_{46}, R_{48} = 39 kΩ
R_{47}, R_{49}, R_{51}, R_{52} = 1 kΩ
R_{58}, R_{62} = 560 Ω
R_{60}, R_{64} = 4.7 Ω
R_{65} = 5.6 MΩ
R_{66} = 10 Ω
P_1 = 50 kΩ preset
P_2–P_7 = 10 kΩ preset

Capacitors:
C_1, C_2, C_{25}, C_{27}, C_{45}, C_{52} = 2.2 µF,
polypropylene, pitch 5 mm
C_3, C_5,C_{16}, C_{18} = 2.2 nF
C_4, C_{17} = 3.9 nF
C_6, C_{19} = 1 nF
C_7, C_8, C_{20}, C_{21}, C_{30}–C_{34}, C_{39}, C_{44}, C_{47}, C_{49}, C_{51}, C_{54}, C_{56}, C_{58} = 100 nF
C_9 = 100 pF polystyrene, axial
C_{10}, C_{11}, C_{13}, C_{35}, C_{63}, C_{64}, C_{67}–C_{70} = 100 nF, ceramic
C_{12}, C_{36}, C_{60} = 47 µF, 25 V, radial
C_{14}, C_{26} = 4.7 µF, 63 V, radial
C_{15} = 270 pF
C_{22} = 68 pF
C_{23}, C_{59}, C_{61}, C_{62} = 10 µF, 63 V, radial
C_{24} = 1 µF, polypropylene, pitch 5 mm
C_{28} = 1 µF, 63 V, radial
C_{29} = 220 µF, 10 V, radial
C_{37}, C_{38} = 47 µF, 16 V, radial
C_{40}, C_{41} = 100 µF, 25 V, radial
C_{42} = 150 nF
C_{43} = 68 nF
C_{46}, C_{53} = 22 µF, 40 V, radial
C_{48}, C_{50}, C_{55}, C_{57} = 220 µF, 25 V, radial
C_{65}, C_{66} = 4700 µF, 25 V, radial
C_{71} = 820 nF

Inductors:
L_1, L_2 = 100 µH
L_3 = 150 µH

Semiconductors:
D_1–D_4 = zener diode, 4.7 V
D_5 = 1N4148
D_6 = bi-colour LED (green/red)
D_8, D_9 = 1N4004
B_1 = SB102, 10 A, 100 V, for PCB mounting
T_1 = BC517

Integrated circuits:
IC_1, IC_2 = NE5532
IC_3 = MN3101
IC_4 = MN3008
IC_5 = TL072
IC_6 = NE571
IC_7 = TL084
IC_8 = 4030
IC_9 = TDA1074A
IC_{10}, IC_{11} = TDA2040
IC_{12} = 7808
IC_{13} = 7908

Miscellaneous:
K_1, K_2 = audio socket for PCB mounting
K_3 = 3-way terminal block, pitch 5 mm
Re_1 = relay 12 V, 5 A, 270 Ω
Heat sink SK57, 37.5 mm high*
Ceramic washers Type AOS220*
Enclosure 300×45×210 mm (W×H×D)
($11^7/_8 \times 1^3/_4 \times 8^1/_4$ in)
Mains transformer with 2×12 V, 80 VA secondary
Mains entry with integral fuse holder and 500 mA slow fuse
Loudspeaker terminals (spring loaded) or suitable audio sockets
Mains on/off switch

* Available from Dau (UK) Ltd, 70–75 Barnham Road, Barnham PO22 0ES, telephone (01243) 553 031

Part 5

Loudspeakers

Chapter 13

Active 3-way loudspeaker

Designing an active loudspeaker system is a complex (but fascinating) matter because each drive unit needs its own output amplifier and cross-over filter. But, of course, an active system has several advantages over a passive one. The present circuit, comprising three filter sections, active bass corrective network, and three output stages, fits on just one printed-circuit board.

Two of the advantages of an active loudspeaker system over a passive one are that neither loudspeaker cables nor a passive cross-over filter are required. The lack of loudspeaker cables saves money,too, since good quality ones are expensive. A passive filter dissipates energy in the inductors and capacitors, and this causes some deterioration in the quality of sound reproduction. An active filter uses no inductors and the capacitors have a much lower value, so that they can be of better quality. Moreover, in direct coupling there are no capacitors and inductors with loss resistances, and this means that the loudspeakers are under more direct control of the output stages.

Two of the drawbacks of an active system are its cost and higher complexity. These two go together, because it is the complexity of the electronic circuits (each loudspeaker needs it own output stage, for instance) that costs the money.

In the present design, the output stages have been kept small through the use of modules for the medium and high frequency sections.

Parameters

Three filters and three output stages on one PCB

Filters may be 1st order, 2nd order, or 3rd order as required

Optional facility for Linkwitz correction network

Power rating	-low	70 W into 4 Ω (U_b=±25 V)
(output stages)		70 W into 8 Ω (U_b=±35 V)
	- middle	30 W into 8 Ω
	- high	30 W into 8 Ω
Nominal sensitivity	1.1 V r.m.s.	
Input impedance	47 kΩ	

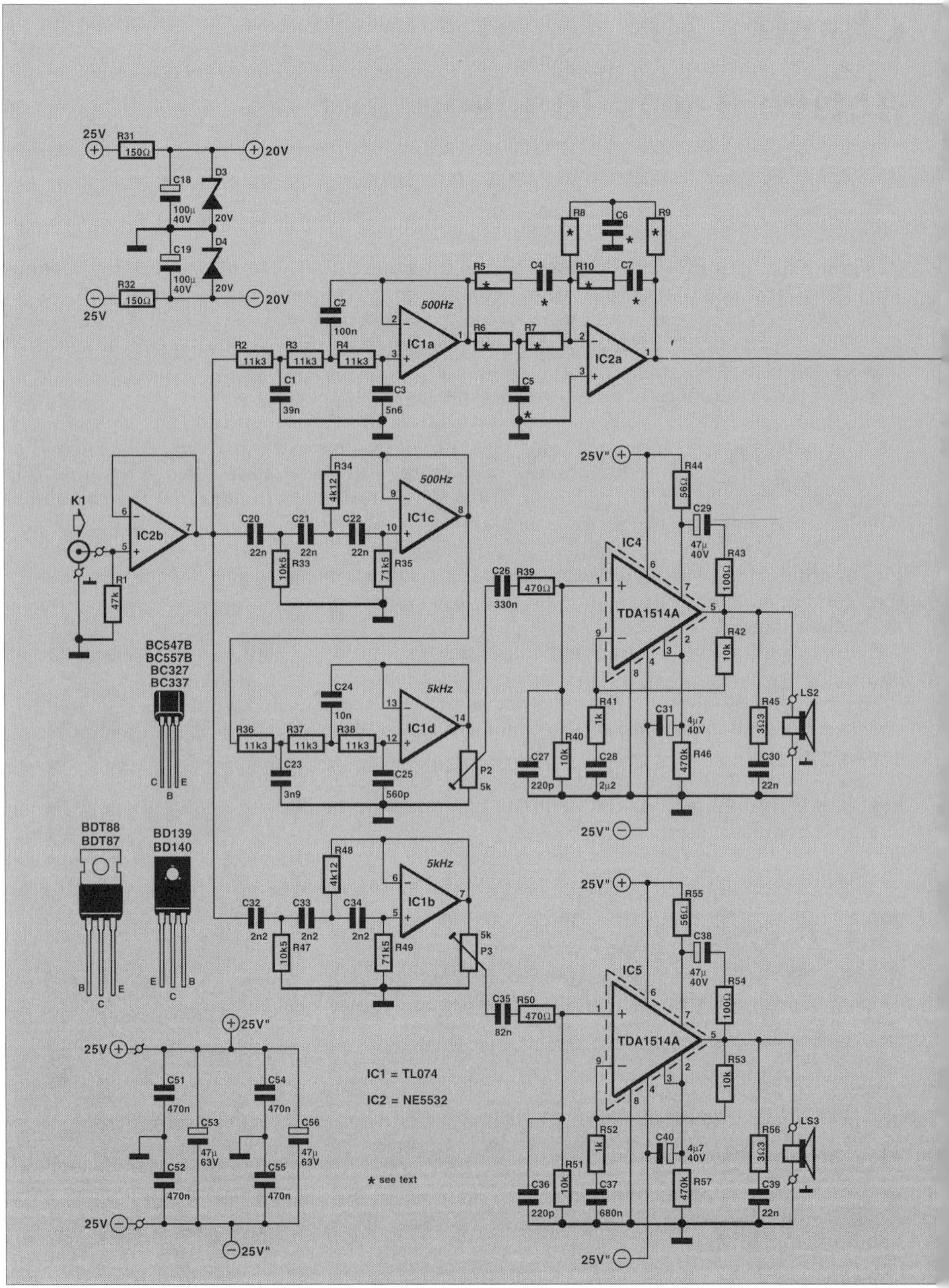

Fig. 13.1. Circuit diagram

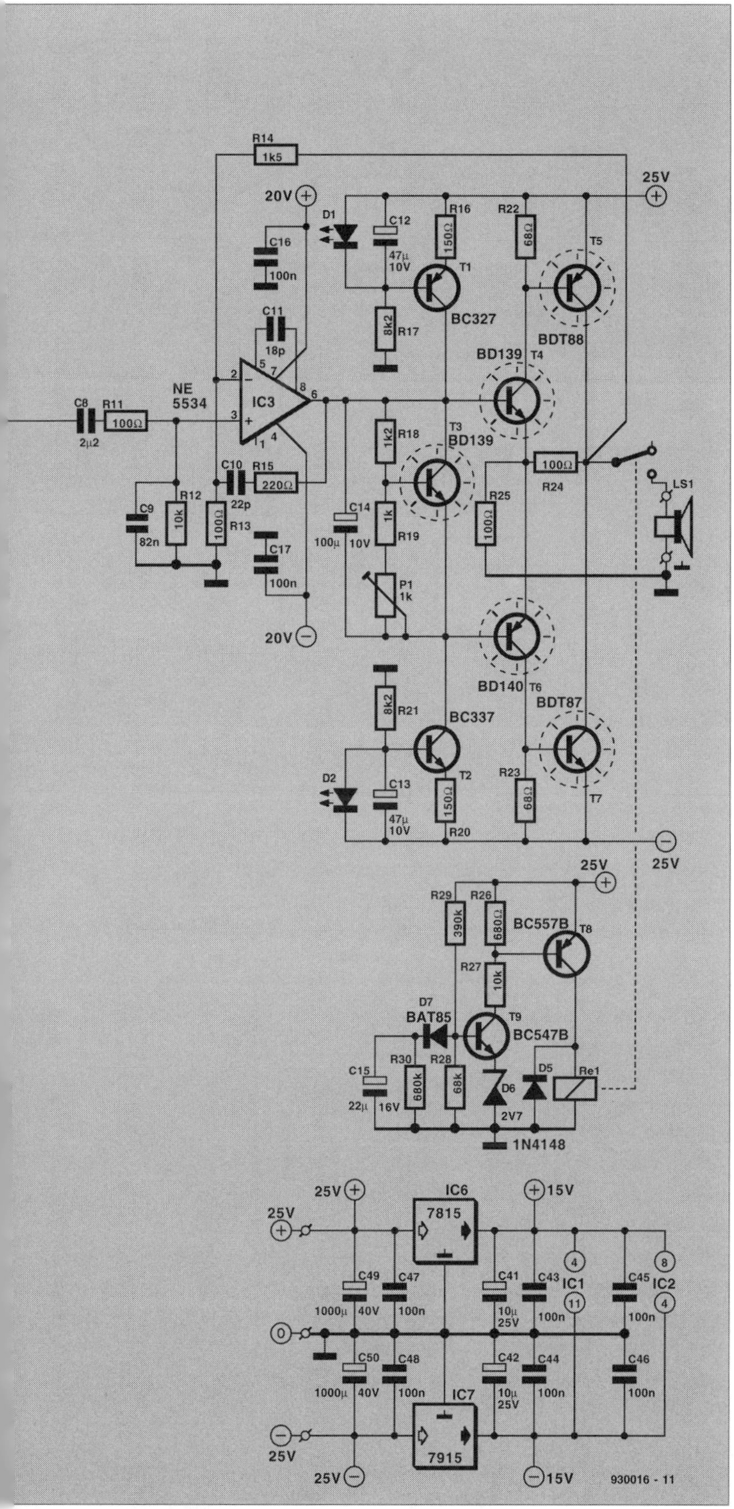

of the active 3-way loudspeaker system.

Some design considerations

The design (see Fig. 13.2) allows for various configurations of the cross-over networks. The audio signal from the preamplifier is applied to the three filters (bass, medium and treble) via a buffer stage. Each of the filters may be given a rolloff of 6 dB, 12 dB or 18 dB per octave depending on the value of certain components.

The low-pass filter is followed by a bass correction network. The original design of this network is due to Linkwitz. It is particularly useful to lower the response of the woofer in a closed box.

The formulas for the calculations of the correct component values and the rolloffs will be given later.

The outputs of the filters are applied to amplifiers. Note that the LF stage has more than twice the power output of the middle and high frequency one. Since the LF stage has been designed with discrete components, its output is applied to the drive unit via a power-on delay. This delay is an integral part of the medium and high frequency modules.

Practical design

Opamp IC_{2b} (see Fig. 13.1) buffers the applied audio signal to prevent this being loaded unnecessarily. The input impedance is determined by R_1. The output of IC_{2b} is split threeway to IC_{1a}, IC_{1c}, and IC_{1b}. Each of these circuits forms a third order filter, which may be converted into a first or second order type by omitting certain components. The middle-frequency section has two filters, IC_{1c} and IC_{1d}, because it needs a rolloff at the low frequency end and one at the high frequency end. With values as shown, the cut-off frequencies are at 500 Hz and 500 Hz respectively. The response is a Butterworth type.

It is possible to convert the present system into an active two-way one by omitting the entire middle-frequency section and the associated output stage (IC_4).

The low-pass filter is followed by the Linkwitz correcting network that matches the frequency response to the low cut-off point of the box. In this way, an octave is added to the lower portion of the response of the enclosure. It is, however, only possible to use the arrangement with boxes whose Q_{tc} and f_c are known. The calculations of the values of the network components are given later.

The supply voltages for the opamps are stabilized by IC_6 and IC_7. These voltages are derived from the ±25 V supply for the LF output amplifier (and for the other output stages if this is desired—more about this later).

Output modules IC_4 and IC_5 need only a few external passive components and a feedback loop. If the supply to them is ±25 V, they deliver an output of up to 30 W into 8 Ω. Noteworthy in the diagram of their internal circuitry, shown in Fig. 13.3, are the many protection cir-

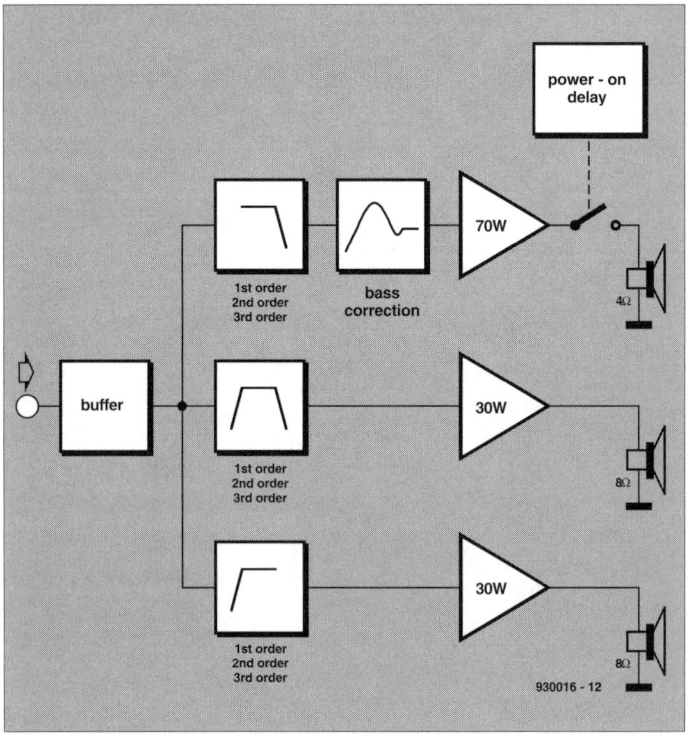

Fig. 13.2. Block diagram of the active 3-way loudspeaker system.

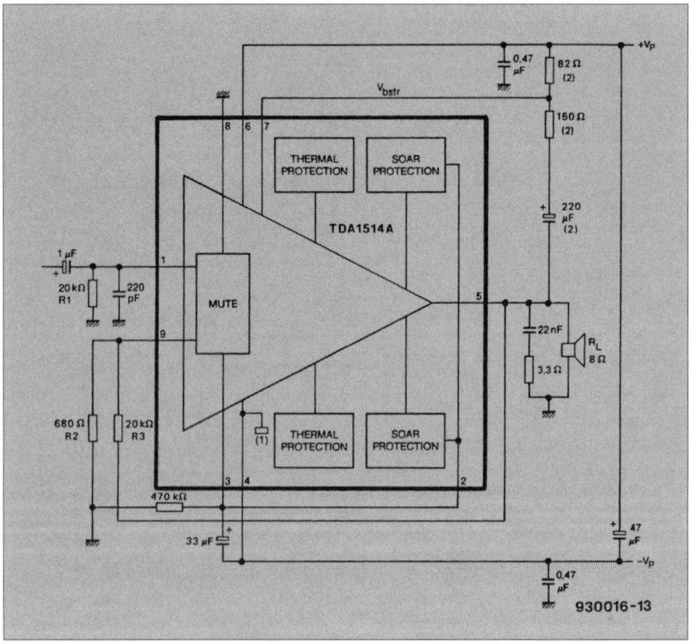

Fig. 13.3. Circuit diagram of the TDA1514A.

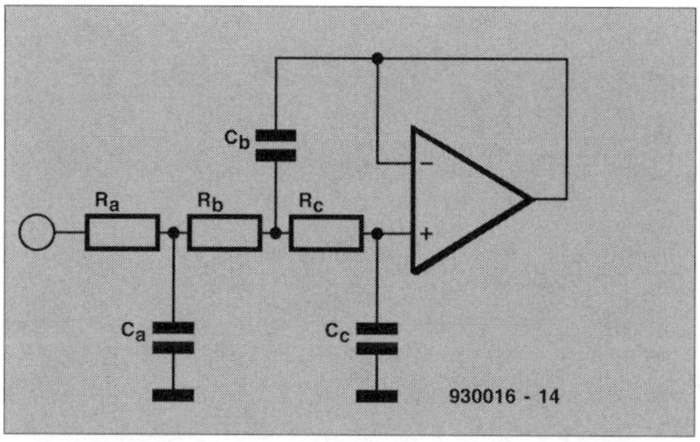

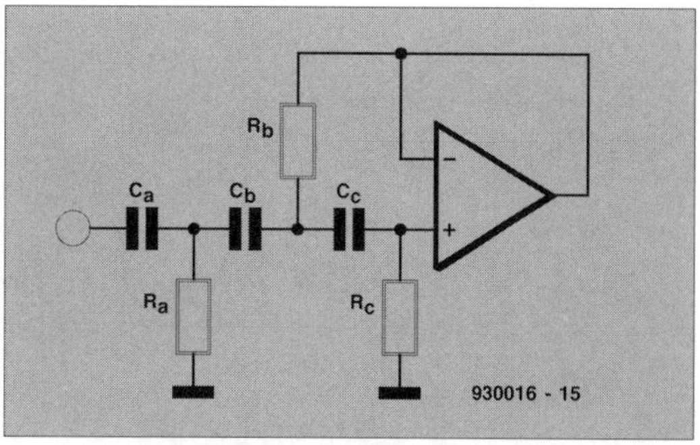

	Low-pass		**High-pass**	
	Butterworth	Bessel	Butterworth	Bessel
1st order	R_a=10 kΩ R_b=R_c=wire bridge C_b=C_c=not used C_a=0.1592/fR		C_a=4.7 nF C_b=C_c=wire bridge R_b=R_c=not used R_a=0.1592/fR	
2nd order	R_a = wire bridge C_a = not used R_b = R_c = 10 kΩ		C_a = wire bridge R_a = not used C_b = C_c = 4.7 nF	
	C_b=0.2251/fR_b C_c=0.1125/fR_b	C_b=0.1443/fR_b C_c=0.1082/fR_b	R_b=0.1125/fC_b R_c=0.2251/fC_b	R_b=0.1755/fC_b R_c=0.234/fC_b
3rd order	R_a = R_b = R_c = 10 kΩ		C_a = C_b = C_c = 4.7 nF	
	C_a=0.2215/fR_a C_b=0.5644/fR_a C_c=0.03221/fR_a	C_a=0.1572/fR_a C_b=0.2265/fR_a C_c=0.04039/fR_a	R_a=0.1125/fC_a R_b=0.04488/fC_a R_c=0.7864/fC_a	R_a=0.1611/fC_a R_b=0.1116/fC_a R_c=0.6272/fC_a

Table 13-1. Computations for low-pass and high-pass filters.

cuits. Their input has a mute stage to prevent on and off switching becoming audible.

Concentrating on IC_4, capacitor C_{26} prevents any d.c. components reaching the opamp. Low-pass filter R_{39}-C_{27} limits the bandwidth of the input signal to a degree that is suitable for the output amplifier. The input impedance of that amplifier is determined by R_{40}. The a.c. amplification of the module is set by feedback network R_{42}-R_{41}. The d.c. amplification is limited to unity by C_{28}. Network R_{43}-R_{44}-C_{29} forms a bootstrap that increases the full power available from the output stage slightly. Boucherot network R_{45}-C_{30} in parallel with the loudspeaker serves as load at high frequencies when the loudspeaker becomes inductive. Power-on delay is provided by R_{46}-C_{31}.

The presets between the filters and the loudspeakers serve to match the speakers as far as efficiency is concerned.

The LF output stage is a semi-discrete design to obtain a higher power since human hearing is less sensitive to low frequencies. The board has been designed to enable this

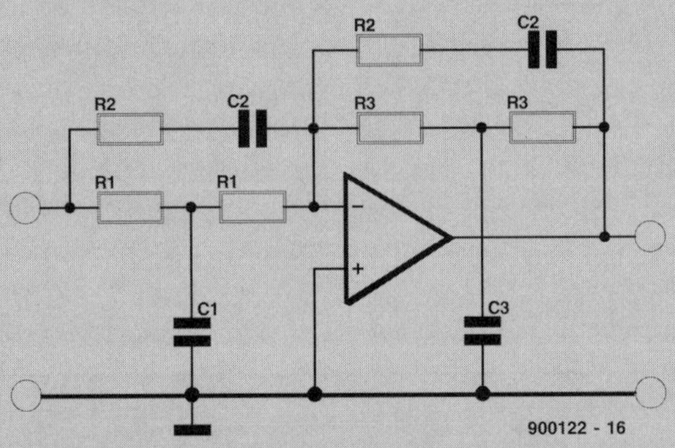

Parameters to be known

Q_{tc} = Q factor of loudspeaker in closed box

f_c = resonance frequency of loudspeaker in closed box

Required new parameters

Q_{tc}' = new Q_{tc} with correction network

f_c' = new f_c with correction network

The new parameters should be chosen such that the necessary correction is not too large to prevent the loudspeaker operating outside its linear performance.

Condition for chosen Q_{tc}' and f_c':

$$k = \frac{\dfrac{f_c}{f_c'} - \dfrac{Q_{tc}}{Q_{tc}'}}{\dfrac{Q_{tc}}{Q_{tc}'} - \dfrac{f_c'}{f_c}}$$

where k is the pole shifting factor.

Computation

Choose a value for R_1 and compute the remaining components as follows

$$R_2 = 2kR_1$$

$$R_3 = \left(\frac{f_c}{f_c'}\right)^2 \times R_1$$

$$C_1 = \frac{2Q_{tc}(1+k)}{2\pi f_c R_1}$$

$$C_2 = \frac{1}{4\pi f_c Q_{tc}(1+k)R_1}$$

$$C_3 = \left(\frac{f_c'}{f_c}\right)^2 \times C_1$$

Table 13-2. Computations for Linkwitz correction network.

amplifier operating from a power supply different from that for the middle and high frequency stages.

The supply for the differential input opamp, IC_3, is derived from the mains supply via networks R_{31}-D_3 and R_{32}-D_4. The input impedance of the opamp is determined mainly by R_{12}. The bandwidth of the input to the opamp is limited by R_{11}-C_9. Resistors R_{15} and capacitor C_{10} form a compensating network.

The output of IC_3 is applied to a compound output stage, which provides not only current amplification, but also voltage amplification. This allows the opamp, although its output is limited by the ±20 V supply, to fully drive the power stage, whose supply is ±25 V.

The output stage consists of T_4-T_5 for the positive half of the signal and T_6-T_7 for the negative half. The power stage is driven by IC_3 via two current sources, T_1-D_1 and T_2-D_2. The quiescent current through the output stage is determined by T_3, which functions as a

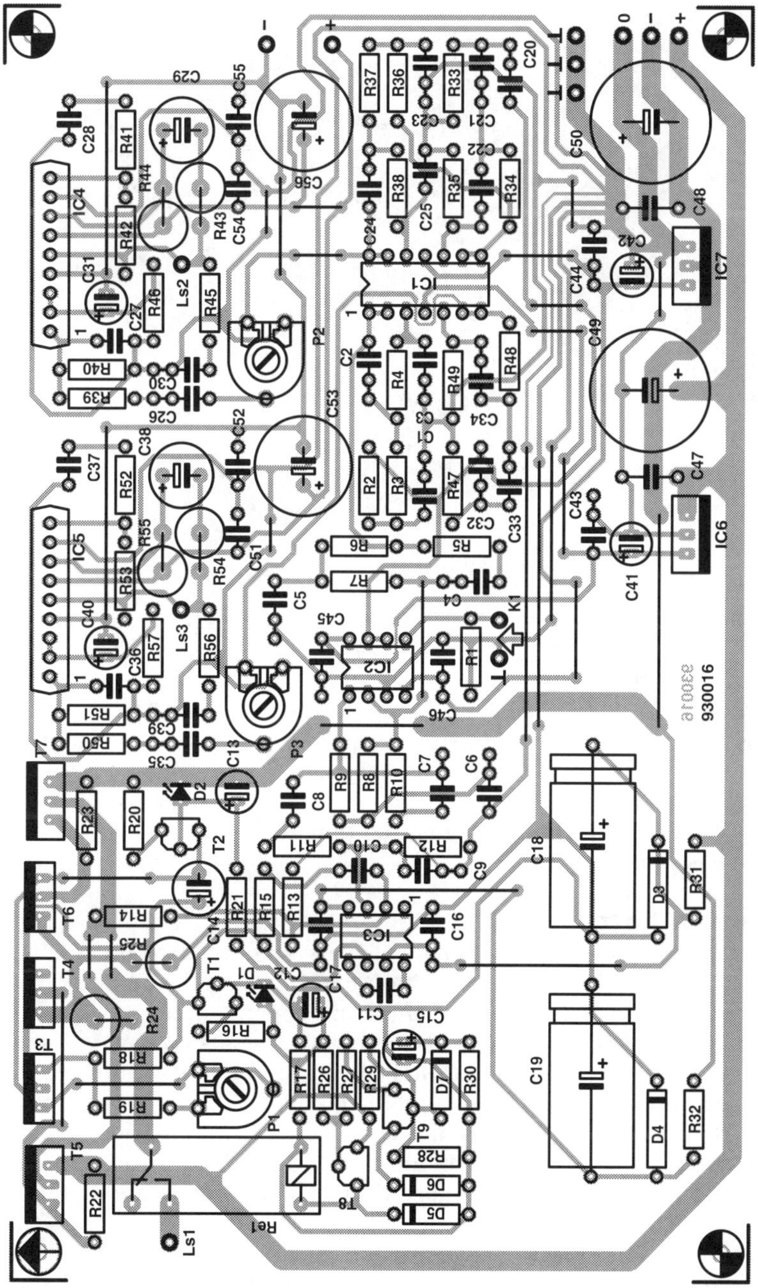

Fig. 13.4.Component layout of the PCB for the active 3-way loudspeaker system. The track layout is givenon page 255.

Fig. 13.5. The completed board assembled on to the heat sink.

variable zener diode. The overall feedback is provided by R_{13} and R_{14}.

The relay contact at the output of the low-frequency power stage prevents on/off clicks becoming audible. At power on, the relay is energized after a delay by T_8 and T_9. Before T_9 can conduct, C_{15} needs to be charged via R_{29} and D_7 to a voltage of 2.7 V (D_6) plus 0.6 V (base-emitter junction of T_9). Diode D_7 ensures that the relay is deenergized immediately the supply voltage is switched off.

Pre-construction notes

As already stated, the entire electronics part of the system is fitted on a single printed-circuit board—see Fig. 13.4. Before construction can be started, the gradients and cut-off points of the filters need to be known, and these are dependent on the loudspeaker enclosure. Component values of the bass correction network are also dependent on the bass enclosure used. Formulas for the computation of the component values for the filters and the correction network are given in Tables 13-1 and 13-2. The design of the circuit in Fig. 13.1 is based on third-order filters and cut-off frequencies of 500 Hz and 5 kHz. Different values may be needed, or it may be desired to imitate the performance of a passive filter (in an existing enclosure) in active form. All this is possible and will be discussed in more detail later.

The construction of the output stages depends to some extent on the loudspeakers used. The power distribution over the audio spectrum shows that in a multi-way system the woofer needs at least twice as much power as the other units combined. This requirement may be met by using a 4 Ω drive unit for the woofers and 8 Ω speakers for the other units. If that is not possible, for instance, because the low-frequency drive unit is not available in a 4 Ω version, the low-frequency output stage may be powered by a higher supply voltage. In the present design, this is achieved by connecting a separate 25 V supply to terminals + and – adjacent to capacitor C_{56}. The supply must not exceed ±35 V since regulators IC_6 and IC_7 cannot handle higher input voltages. A drawback of this arrangement is the higher price since two separate power supplies are needed.

Another aspect that needs bearing in mind is the polarity of the drive units. Owing to the bass correction network, the bass unit is out of phase with the other two units and its polarity must therefore be reversed. If the correction network is not used, all components around IC_{2a} with an asterisk must be omitted, except R_7 and R_{10}, whose value must be altered to 10 kΩ. Resistor R_6 and capacitors C_7 must be replaced by a wire bridge.

The prototype loudspeaker system is housed in a 90 cm (36 in) high enclosure, details of which are shown in Fig. 13.8. The drive units are a 200 mm (8 in) 4 Ω woofer, a 50 mm (2 in) 8 Ω mid-range unit, and a 25 mm (1 in) 8 Ω tweeter.

The frequency characteristics of the three cross-over networks are shown in Fig. 13.6. It may be noted that the characteristics do not conform to a pure Bessel, Butterworth, or similar, design. This is because in practice cross-over filters must have a slightly non-standard characteristic, since the frequency response of the drive units and the summing behaviour around the cut-off frequencies must be taken into account. Simulation programs such as PSPICE or MicroCap are of great help in the design.

As already stated earlier, if a 4 Ω woofer is not available, an 8 Ω type may be used, but the supply voltage to the bass frequency output stage should then be increased to ±35 V. The output of this stage is then 70 W, but that of the other two output stages remains 30 W. This is a costly solution, because two separate power supplies are required. Note that ±35 V is the absolute maximum if the specified voltage regulators are used. It is possible to connect a resistor in series with IC_6 and IC_7 to lower the voltage slightly, but it is better to replace the regulators by 20 V types: 7820 and 7920 respectively, which can handle input voltages up to 40 V. In that case, C_{49} and C_{50} must be 50 V types and IC_1 must be replaced by a type that can handle higher voltages than the TL074: for example, TL34074(A); TLE2144; LF147; LF444A (not LF444); MC34074; MC34084; OP11 (not OP11GR).

Construction

The PCB is populated in the usual manner: first the wire bridges, then the passive components, next the semiconductors (but see below for T_3–T_6 and IC_4 and IC_5), and finally the relay and electrolytic capacitors C_{49} and C_{50} (which must be mounted upright).

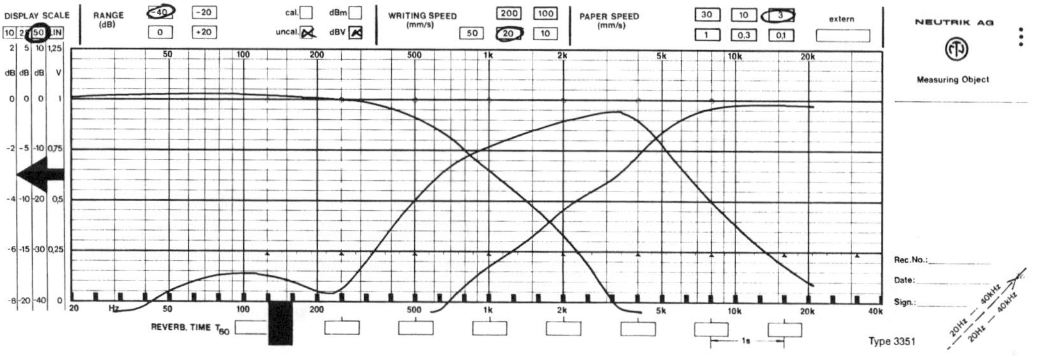

Fig. 13.6. Frequency response curves of the active filters.

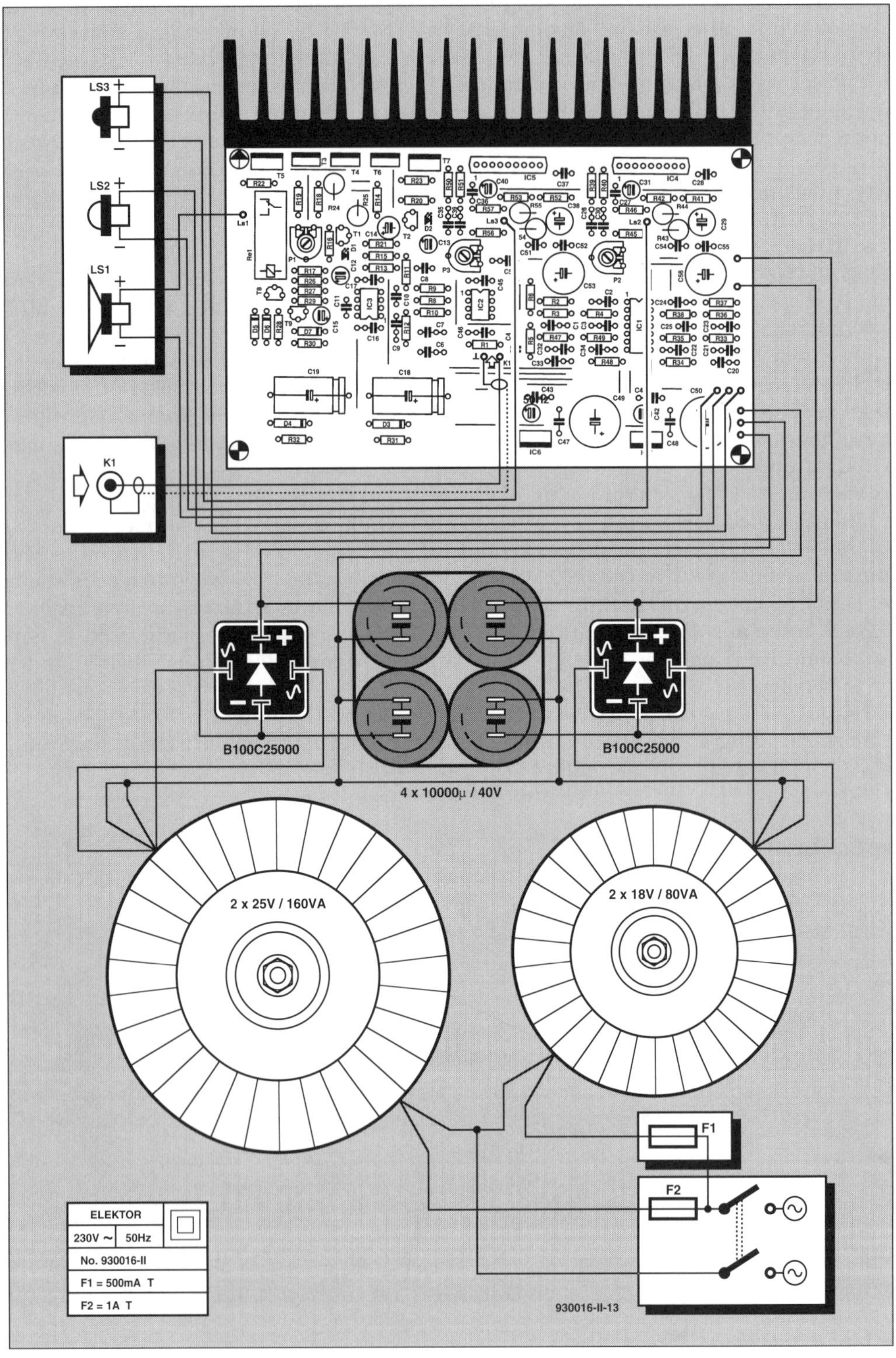

Fig. 13.7. Wiring diagram for one unit using two separate power supplies.

Drill and tap suitable screwholes in the heat sink for the power transistors and the modules. Fix the heat sink to a sheet of aluminium of about 230×260 mm ($9^1/_4 \times 10^1/_2$ in) with the aid of right-angle brackets.

Next, fit the power transistors and modules to the heat sink with the aid of heat conducting paste and insulating washers (these should be ceramic for the transistors). Bend the terminals of these devices (and give them a kink as well to allow for tension during temperature changes) so that they fit properly into the relevant holes on the PCB and solder them in place.

Mount the mains switch, fuse holder(s), transformers, electrolytic capacitors, and the audio input socket on the aluminium sheet (see Fig. 13.5). Fit the bridge rectifiers on the free area of the heat sink. The audio input socket should be an insulated type, which enables the best earthing point to be established: the aluminium sheets is strapped either to the socket earth or to the central earth of the electrolytic capacitors: use the position that gives the least hum.

After all parts have been fitted securely to the aluminium sheet, wire up the assembly as shown in Fig. 13.7. This diagram shows the value of the four buffer capacitors for the bass amplifier as 10 000 μF; this is a minimum value: use 20 000 μF if possible. Fit the mains cable with a strain relief sleeve.

Next, the quiescent current for the bass amplifier must be set. Start by turning P_1 fully anticlockwise. Remove the –35 V connection from the board and insert a milliammeter in series with this supply line. Switch on the mains and turn P_1 till the meter reads 50 mA. Switch off the mains, remove the meter and reconnect the –35 V line to the board.

A few seconds after the mains has been switched on again, the relay should change over. If it does not, it is probable that T_8 does not provide enough current for the relay (remember, a current source is used here). In that case, the value of R_{26} must be increased to the next or second higher E12 value. The value is correct when, a few seconds after the mains has been switched on, the potential across Re_1 is 24 V.

For safety's sake, measure all relevant supply voltages on the board and check that the direct voltage levels at the amplifier outputs are zero or very nearly so.

Finally, set P_2 and P_3 so that the resistance between wiper and ground is 55% of the total in the case of P_2 and 70% in the case of P_3.

The enclosure

The box is made from 18 mm thick medium density fibre board (MDF) or high-density chip board. The construction details are shown in Fig. 13.8. Basically, it is a rectangular box provided with two reinforcing cross members that prevent panel vibrations.

The one hole not shown in the drawing is that for the electronics assembly. It is best to fit this at the bottom of the enclosure, so that the board rests on spacers on the bottom panel. Saw a rectangular hole in the rear panel (see Fig. 13.5) whose width and height are 20 mm smaller than the aluminium sheet. On completion, the sheet is fixed across the hole with 10 or 12 suitable wood screws and sealed with draught-proofing tape.

Before that, however, fill the enclosure above the lower cross member evenly with the polyster wadding. The bottom third of the box, where the bass reflex duct and electronics assembly will be located, remains free of wadding.

Next, fit the drive units, provide them with cables and connect these to the board. Make sure that the polarity is as shown in Fig. 13.7.

Then, screw the electronics assembly across the rectangular hole at the rear as already discussed.

Next, push the bass reflex duct into the port on the front panel.

Finally, where deemed desirable, it is, of course, possible to use normal potentiome-

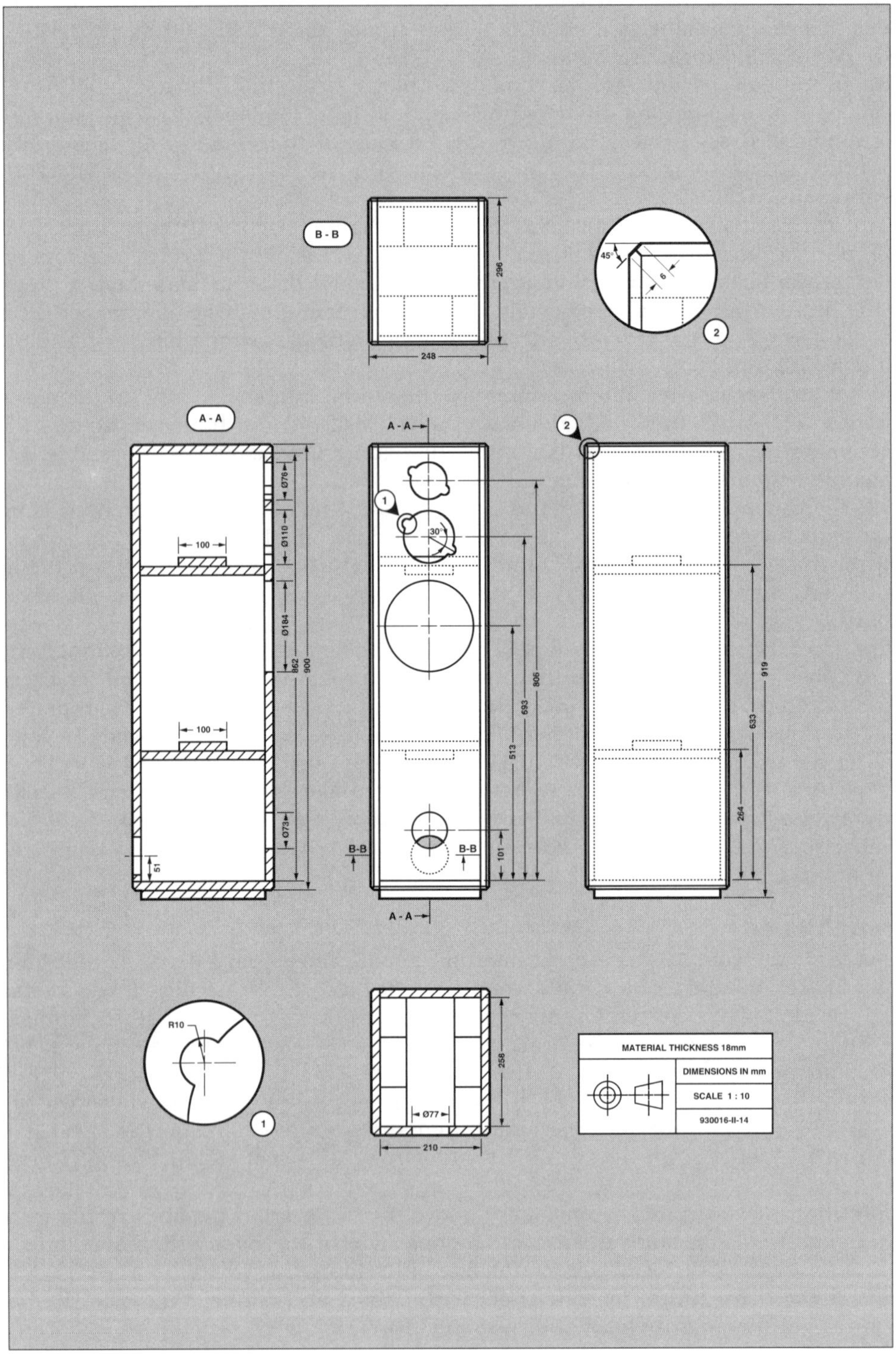

Fig. 13.8. Construction diagram of the enclosure. The aperture for moumting the electronics assembly is not shown – see Fig. 13.9.

Fig. 13.9. Detail of the prototype.

ters for P_2 and P_3 and fit these to the aluminium sheet, so that they can be readjusted, when required, without the need of removing the entire assembly from the box. In that case, the connections between these controls and the board should be by screened audio cable.

Parts list

Resistors:

R_1 = 47 kΩ
R_2–R_4, R_{36}–R_{38} = 11.3 kΩ, 1%
R_5–R_{10} = see text
R_{11}, R_{13} = 100 Ω
R_{12}, R_{40}, R_{42}, R_{51}, R_{53} = 10 kΩ
R_{14} = 1.5 kΩ
R_{15} = 200 Ω
R_{16}, R_{20}, R_{31}, R_{32} = 150 Ω
R_{17}, R_{21} = 8.2 kΩ
R_{18} = 1.2 kΩ
R_{19}, R_{41}, R_{52} = 1 kΩ
R_{22}, R_{23} = 68 Ω
R_{24}, R_{25} = 100 Ω, 2.5 W
R_{26} = 680 Ω
R_{27} = 10 kΩ
R_{28} = 68 kΩ
R_{29} = 390 Ω
R_{30} = 680 kΩ

R_{33}, R_{47} = 10.5 kΩ, 1%
R_{34}, R_{48} = 4.12 kΩ, 1%
R_{35}, R_{49} = 71.5 kΩ, 1%
R_{39}, R_{50} = 470 Ω
R_{43}, R_{54} = 100 Ω, 1.5 W
R_{44}, R_{55} = 56 Ω, 1.5 W
R_{45}, R_{56} = 3.3 Ω
R_{46}, R_{57} = 470 kΩ
P_1 = 1 kΩ preset
P_2, P_3 = 4.7 kΩ (5 kΩ) preset

Capacitors:
C_1 = 39 nF
C_2, C_{16}, C_{17}, C_{43}–C_{48} = 100 nF
C_3 = 5.6 nF
C_4–C_7 = see text
C_8, C_{28} = 2.2 µF, 50 V, polypropylene
C_9, C_{35} = 82 nF
C_{10} = 22 pF
C_{11} = 18 pF
C_{12}, C_{13} = 47 µF, 10 V, radial
C_{14} = 100 µF, 10 V, radial
C_{15} = 22 µF, 16 V, radial
C_{18}, C_{19} = 100 µF, 40 V
C_{20}–C_{22}, C_{30}, C_{39} = 22 nF
C_{23} = 3.9 nF
C_{24} = 10 nF
C_{25} = 560 pF, polystyrene
C_{26} = 330 nF
C_{27}, C_{36} = 220 pF
C_{29}, C_{38} = 47 µF, 40 V, radial
C_{31}, C_{40} = 4.7 µF, 40 V, radial
C_{32}–C_{34} = 2.2 nF
C_{37} = 680 nF
C_{41}, C_{42} = 10 µF, 25 V, radial
C_{49}, C_{50} = 1000 µF, 40 V, radial
C_{51}, C_{52}, C_{54}, C_{55} = 470 nF
C_{53}, C_{56} = 47 µF, 63 V, radial

Semiconductors:
D_1, D_2 = LED, 3 mm, red
D_3, D_4 = zener diode, 20 V, 1.5 mW
D_5 = 1N4148
D_6 = zener, 2.7 V, 400 mW
D_7 = BAT85
T_1 = BC327
T_2 = BC337
T_3, T_4 = BD139
T_5 = BDT88
T_6 = BD140
T_7 = BDT87
T_8 = BC557B

T_9 = BC547B

Integrated circuits:
IC_1 = TL074
IC_2 = NE5532
IC_3 = NE5534
IC_4, IC_5 = TDA1514A
IC_6 = 7815
IC_7 = 7915

Miscellaneous:
K_1 = audio plug for PCB mounting
Re_1 = relay with make contact, 24 V coil
Heat sink, 0.5 K W^{-1} (SK47)

POWER SECTION

Mains transformer, 80 VA, secondary 2×18 V
Mains transformer, 160 VA, secondary 2×25 V
2 off bridge rectifier B100C25000
4 off capacitor ≥10,000 μF, 40 V
Mains entry with integral fuse holder and 1 A slow fuse, and on/off switch
Fuse holder with 500 mA slow fuse

ENCLOSURE

Woofer, 100 W, 200 mm (8 in), 4 Ω
Mid-range drive unit, 50 W, 50 mm (2 in), 8 Ω
Tweeter, 30 W, 25 mm (1 in), 8 Ω
18 mm thick medium density fibre board or high-density chip board:
  2 off 864×210 mm
  2 off 900×260 mm
  2 off 210×296 mm
  2 off 210×100 mm
  2 off 260×100 mm
  1 off 210×260 mm
Bass reflex duct, 145 mm (5 11/16 in) long
3 metres mains cable
2 bags polyester wadding

Chapter 14

Active phase-linear cross-over network

The most serious drawback of ordinary cross-over filters is best illustrated with reference to a two-way system, which consists of a low-pass and a high-pass filter. One of the properties of a low-pass section is that it causes a time delay of the signal. A high-pass filter on the other hand causes an acceleration of the signal. These actions result in several complications at the cross-over point:

- the signals from the two sections partially cancel one another;
- the strongly varying phase shift between the two signals adversely affects the radiation efficiency of the overall system;
- the radiation pattern becomes frequency-dependent.

In the early 1980s, a series of papers[1, 2, 3] laid the foundations of the phase-linear cross-over network. This network uses a low-pass section that, in conjunction with a time-delay and subtraction circuit, also provides a high-pass characteristic. Although the time-delay is

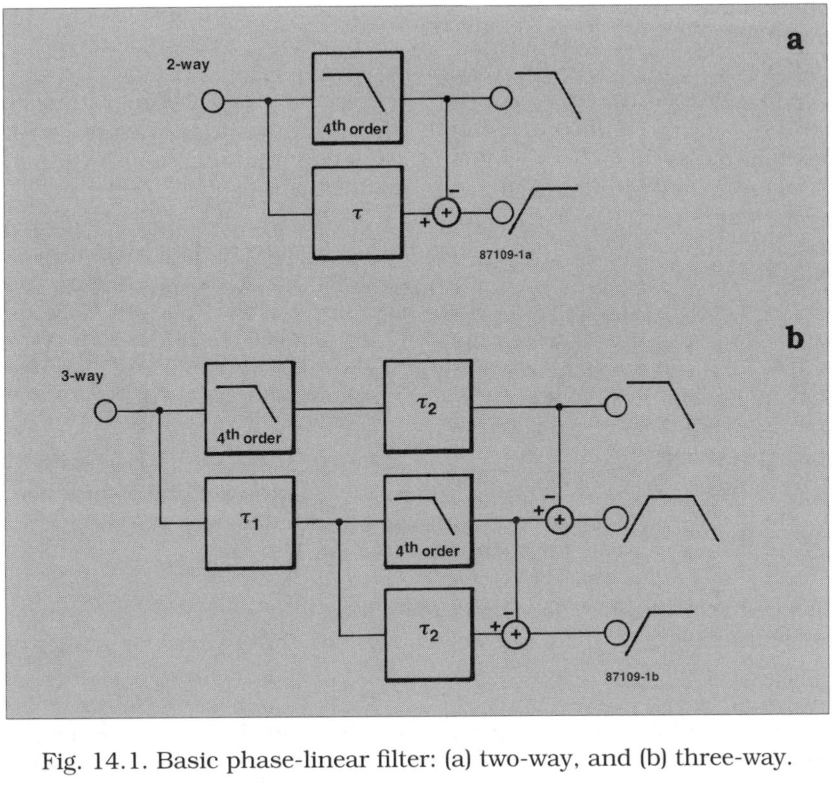

Fig. 14.1. Basic phase-linear filter: (a) two-way, and (b) three-way.

not constant over the entire frequency range, it varies only slowly. Moreover, there are no phase differences between the two output signals, even near the cross-over frequency.

A block diagram of a two-way and a three-way system based on these papers is shown in Fig. 14.1. It should be emphasized that the time-delay is an essential aspect of the design. There are filters that make use of the subtraction method only, but these do not exhibit phase linearity.

A standard fourth-order low-pass filter in the upper branch provides the usual low-pass performance. The delay, τ_1, is arranged such that it has exactly the same phase behaviour as the low-pass section and functions as an all-pass section. When the output signal of the low-pass section is subtracted from the delayed signal, the result is a high-pass characteristic that has the same phase behaviour as that of the low-pass filter. Adding the two signals together results in a perfectly straight line.

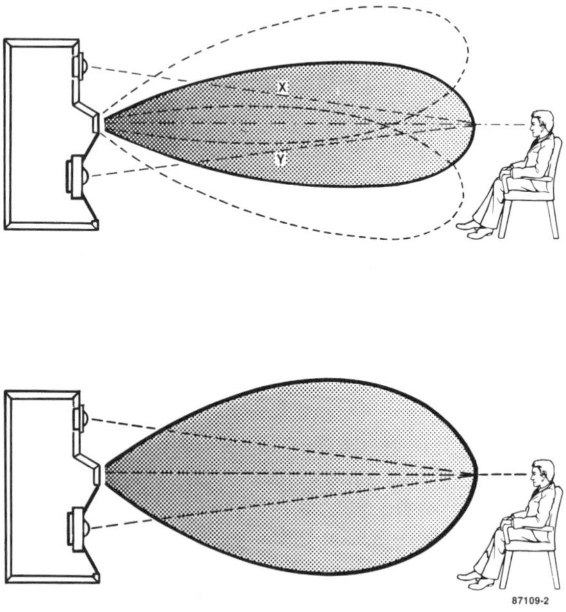

Fig. 14.2. Vertical radiation pattern:
(a) of system with conventional network, and
(b) of system with phase-linear network.

The set-up of a three-way system is rather more complicated, because an additional low-pass section is provided in the centre limb to obtain a bandpass characteristic for the middle-frequency loudspeaker. This additional section is compensated by a second delay, τ_2. Thus, in a three-way system, the τ_1 circuit simulates the time delay of the usual bass filter, while the τ_2 delay simulates the delay of the low-pass filter in the middle-frequency section.

The vertical radiation pattern (polar response) of a conventional loudspeaker system is shown in Fig. 14.2a. The dispersion is fairly small in the region where both loudspeakers provide a signal. The spread also varies with frequency, which causes the lobe to tilt or sag. The pattern of the phase-linear system in Fig. 14.2b shows that the lobe is much broader and points forward at all frequencies. In all this it is assumed that the acoustic centres of the loudspeakers lie on a vertical line, otherwise the pattern deteriorates.

A practical network

In a practical network, it is not possible (at least with an acceptable number of components) to simulate any phase behaviour with a delay circuit.

All-pass networks have some interesting properties:

- they cause a phase shift, but no signal attenuation over a given frequency range;
- the phase shift caused by them is twice as large as that caused by a filter of the same order.

From this, it is evident that the low-pass section should be an even-order type, that is, second-, fourth- or sixth-order. In the present network, fourth-order sections are used,

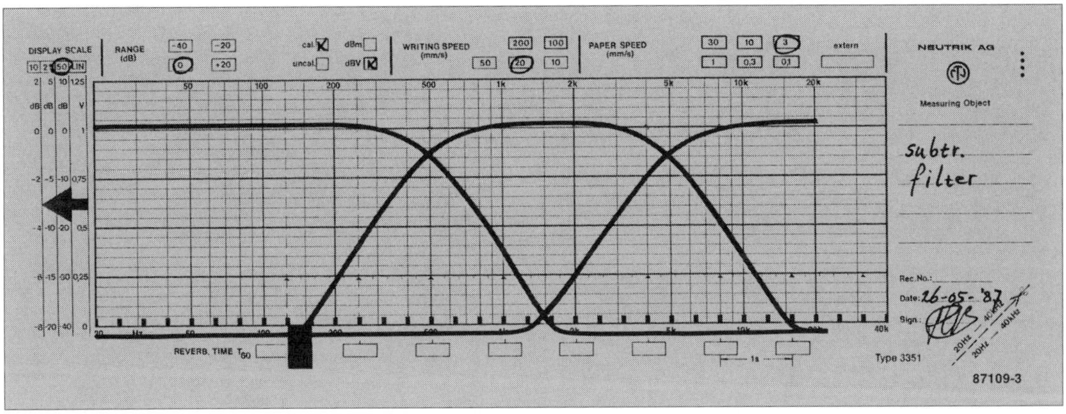

Fig. 14.3. Amplitude vs frequency characteristic of a 3-way phase-linear network.

since these give a sufficiently steep roll-off and do not unnecessarily complicate the circuit.

Like all fourth-order networks, the present one consists of two cascaded second-order sections. For the present purposes, these should be identical to ensure that the phase behaviour of the all-pass network will be the same as that of the filter.

The Linkwitz-Riley (Squared Butterworth) filter is eminently suitable for the present network, because it allows a fairly simple all-pass section to be designed with only two op amps. The resulting circuit has exactly the same phase behaviour as a fourth-order Linkwitz low-pass filter. Note that the cross-over frequencies are –6 dB points (as in all Linkwitz filters), since there is no phase shift between the two channels. The relative amplitude vs frequency characteristic is given in Fig. 14.3, while the three photographs in Fig. 14.6 illustrate the typical performance of the network. The photographs show the output voltage at the low- and middle-frequency terminals (a) slightly below the cross-over point, (b) at the cross-over point, and (c) slightly above the cross-over point. There are no discernible phase differences between the two signals anywhere.

Circuit description

In the circuit diagram, Fig. 14.4, op amp A_1 is used as a buffer between the input signal and the filter. If necessary, the input signal may be attenuated with P_1; the total amplification of the network is unity.

The low-pass filter is based on A_1 and A_2 and the associated all-pass filter on A_6 and A_7. The attenuation caused by bandpass filter A_6 is compensated by A_7. The low-pass section for the middle frequencies consists of A_8 and A_9. Here, two identical all-pass filters are required, and these are formed by A_4-A_5 in the low section and by A_{11}-A_{12} in the high section. This completes the low-pass filter.

For the middle-frequency section, the output signal of A_5 must be subtracted from that of A_9, which is effected by A_{10}.

Finally, the output signal of A_9 is subtracted once more from that of A_{12}, which is done by A_{13}. This completes the high-pass function.

The three outputs of the network are taken to preset potentiometers that enable matching each of them to the efficiency of the associated loudspeaker.

The quality of the power supply matches that of the cross-over network itself. Circuit IC_1 is a voltage regulator that, in conjunction with two external series transistors, provides symmetrical output voltage. Diodes D_5 and D_6 ensure that the regulator is not damaged at

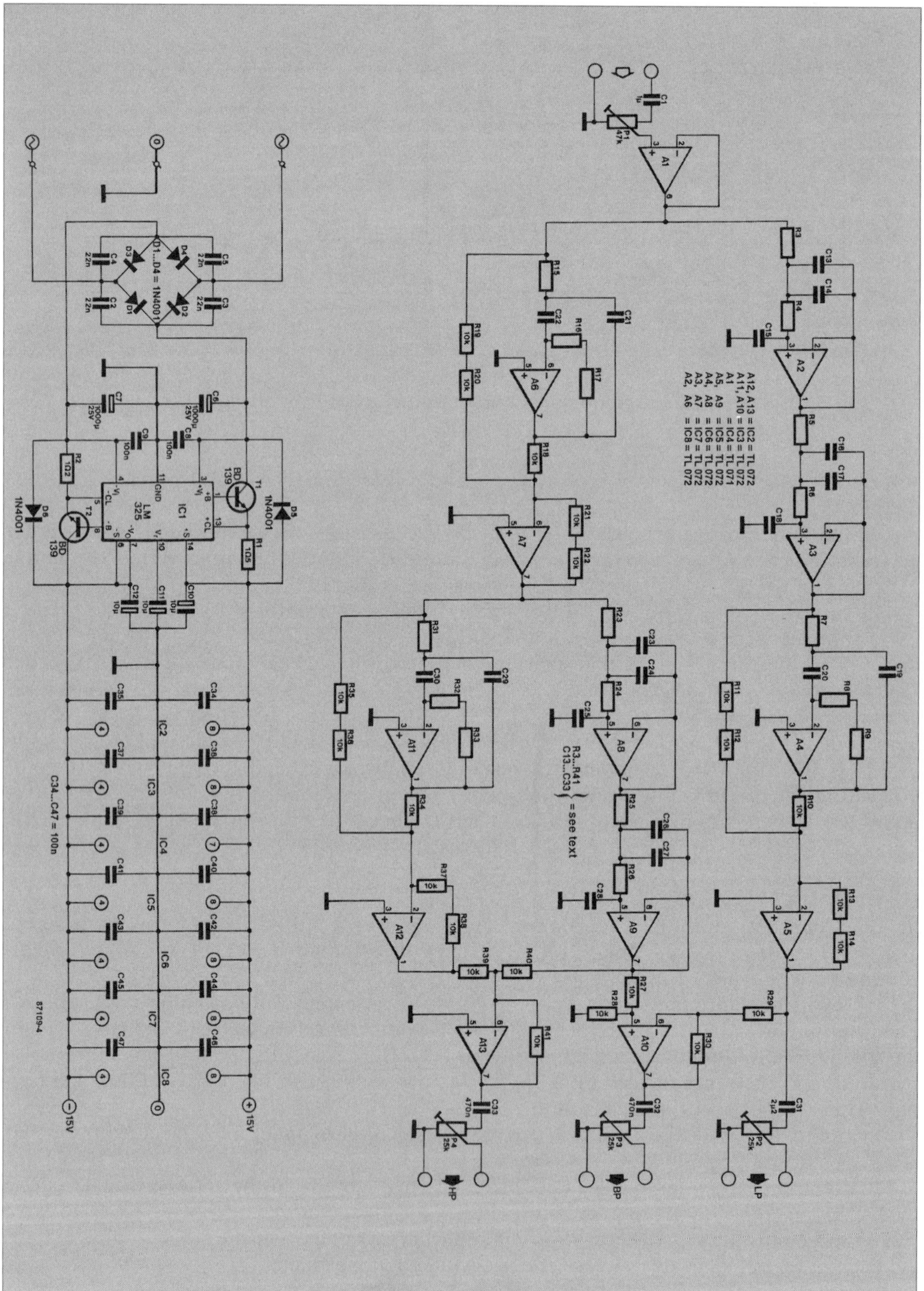

Fig. 14.4. Circuit diagram of the phase-linear network.

switch-off.

Construction

The network is best constructed on a printed-circuit board as shown in Fig. 14.5. The values of the components specified in the parts list pertain to crossover frequencies of 500 Hz and 5000 Hz. Different frequencies may be calculated with the Linkwitz formulas in Ref. 4.

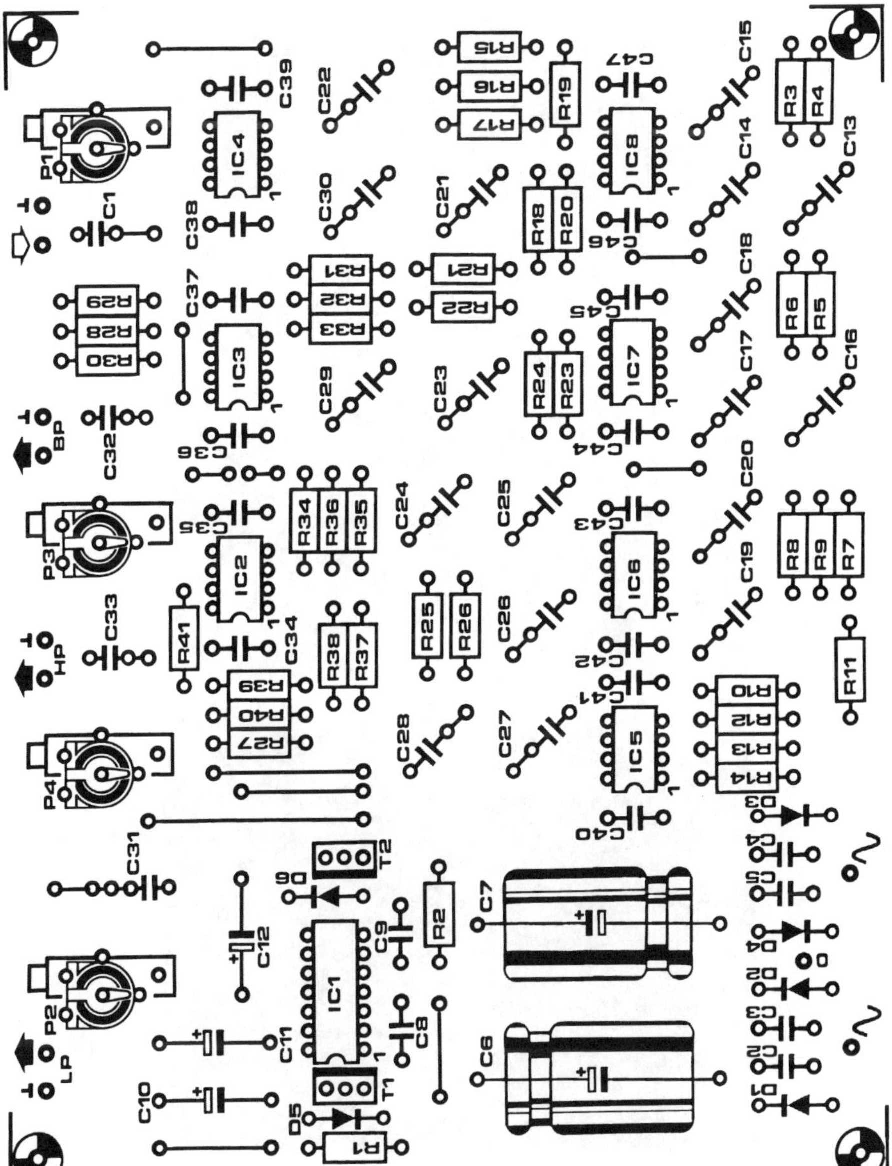

Fig. 14.5. Component layout of the PCB for the phase-linear network. The track layout is given on page 257.

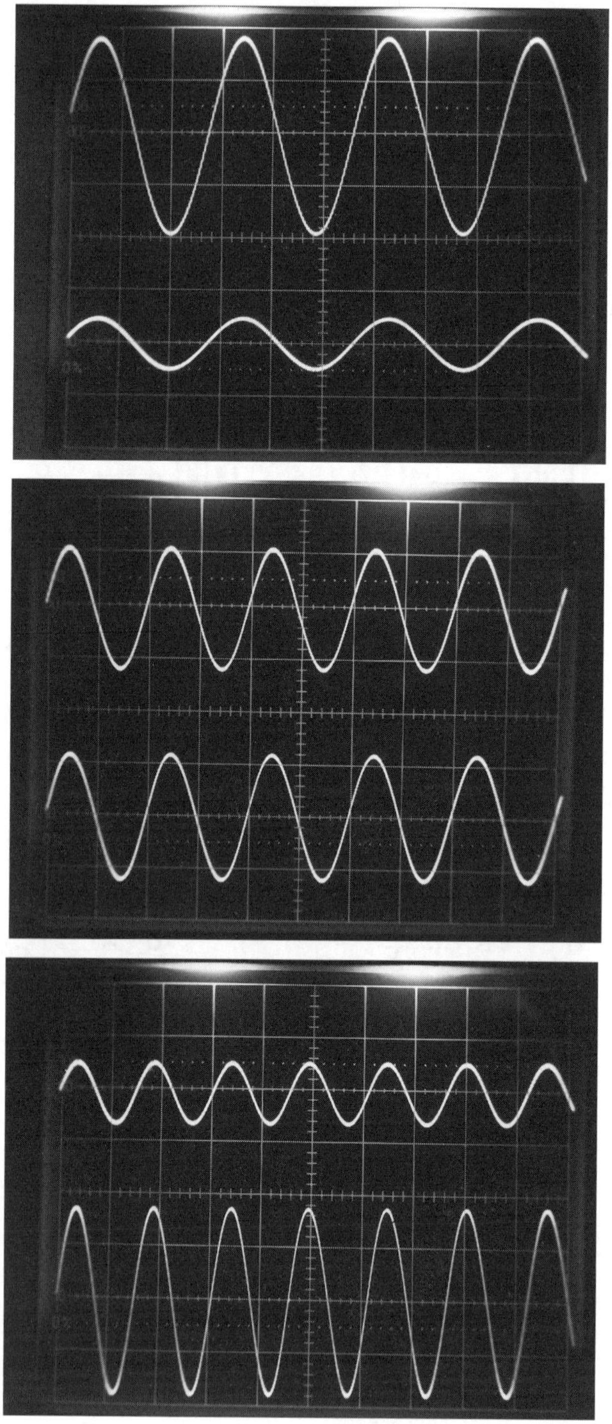

a

b

c

Fig. 14.6. The output voltage at the low- and middle- frequency terminals: (a) slightly below the cross-over point, (b) at the cross-over point, and (c) slightly above the cross-over point.

In several places, capacitors are shown in parallel and resistors in series: this is done to facilitate the use of as many components of the same value as possible. As usual, the choice of capacitors is determined mainly by their loss factor and cost, which normally results in plastic film types.

It should be noted that each board needs its own regulator IC; this is convenient where the network is fitted in the loudspeaker enclosure.

The impedance at the network outputs, depending on the positions of the presets, has a maximum value of 12 kΩ. Since this may be on the high side for certain output stages, the value of the presets may be reduced to 5 kΩ, which results in a maximum output impedance of of 2.5 kΩ. If this is done, the value of C_{31} should be increased to 4.7 μF. A useful rule of thumb is that the input impedance of the output stage must be at least ten times as large as the output impedance of the network.

The board may also be used to construct a two-way network in which the following components must be omitted: IC_2, IC_5, IC_6, R_7–R_{14}, R_{23}–R_{26}, R_{31}–R_{41}, C_{19}, C_{20}, C_{23}–C_{30}, C_{33}, P_4. Furthermore, a wire link must be fitted between pin 1 of A_3 and C_{31} and another one between pin 7 of A_7 and C_{32}.

Parts list

Resistors:

R_1, R_2 = 1.5 kΩ, 5%
R_3–R_9; R_{15}–R_{17}; R_{23}–R_{25}; R_{31}–R_{33} = 22.5 kΩ, 1%
R_{10}–R_{14}; R_{18}–R_{22}; R_{27}–R_{30}; R_{34}–R_{41} = 10 kΩ, 1%
P_1 = 47 kΩ, 5%, cermet
P_2–P_4 = 25 kΩ, 5%, cermet preset

Capacitors:

C_1 = 1.0 μF, plastic film
C_2–C_5 = 22 nF, ceramic
C_6, C_7 = 1000 μF, 25 V, electrolytic
C_8, C_9, C_{34}–C_{47} = 100 nF, ceramic
C_{10}–C_{12} = 10 μF, 25 V, electrolytic
C_{13}–C_{18}, C_{21}, C_{22} = 10 nF, 2.5%, polypropylene
C_{31} = 2.2 μF, plastic film
C_{32}, C_{33} = 470 nF, plastic film

Semiconductors:

D_1–D_6 = 1N4001
T_1 = BD139
T_2 = BD140

Integrated circuits:

IC_1 = LM325
IC_2, IC_3, IC_5–IC_8 = TL072; NE5532; LF353; LM833; OP215
IC_4 = TL071; NE5534; LF356; OP27; OP15

References:

1. S. Lipshitz & J. Vanderkooy, 'A Family of Linear-phase Cross-over Networks of High Slope Derived by Time Delay' – *Journal of the Audio Engineering Society*, Jan/Febr 1983.

2. S. Lipshitz & J. Vanderkooy, 'Is Phase Linearization of Loudspeaker Cross-over Networks Possible by Time Off-set and Equalization?' – *Journal of the Audio Engineering Society*, December 1984.

3. S. Lipshitz & J. Vanderkooy, 'Use of Frequency Overlap and Equalization to Produce High-slope Linear-phase Loudspeaker Cross-over Networks' – *Journal of the Audio Engineering Society*, March 1985.

4. 'Linkwitz Filters', *Elektor Electronics*, April 1987.

Chapter 15

Active mini sub-woofer

It is a basic fact of nature that the faithful reproduction of bass frequencies by a loudspeaker requires the displacement of a large volume of air. This in turn requires the cone of the drive unit to have a large area and a large linear movement. Since a drive unit in a box displaces large volumes of air at frequencies below the lower –3 dB cut-off point, which is, of course, very inefficient, it is necessary for good efficiency to design the enclosure in a way which ensures that the cut-off point lies well below 30 Hz.

These requirements are difficult to combine if the dimensions of the box are to be kept small. As often in life, it is therefore necessary to arrive at a compromise. However, if it is assumed that the user of the subwoofer is not going to need hundreds of watts of output power, that compromise works out very well.

Drive unit

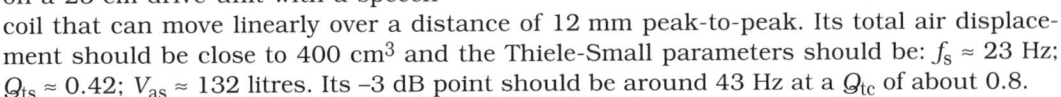

The design of the enclosure is based on a 25 cm drive unit with a speech coil that can move linearly over a distance of 12 mm peak-to-peak. Its total air displacement should be close to 400 cm^3 and the Thiele-Small parameters should be: $f_s \approx 23$ Hz; $Q_{ts} \approx 0.42$; $V_{as} \approx 132$ litres. Its –3 dB point should be around 43 Hz at a Q_{tc} of about 0.8.

Electronic correction

To render the lowest tones of a compact disk or vinyl record well audible. a simple electronic network was used in the prototype to straighten the lower part of the response of the subwoofer to just under 30 Hz. The diagram of the network, which was originally designed by Linkwitz, is shown in Fig. 15.2, while the simple formulas for calculating the values of the components are given on page 187.

The wanted correction is computed on the basis of the actual Thiele-Small parameters. First, Q_{tc} and f_c of the drive unit fitted in an enclosure are measured or calculated; then the required new Q_{tc} and f_c are chosen, after which the component values can be computed. It should be borne in mind that corrections must remain within certain limits, since the drive unit must be able to handle the additional large displacements. In the design as described, the maximum correction is just over 6 dB, which lowers the cut-off point by about 10 Hz.

A further point in connection with the calculation of the network is that in practice the results will be different from the theoretical overall response curve. Note, for instance, that the values of a number of components in Fig. 15.3 are quite different from the calculated values. This is because, after the values had been calculated, they were entered, together with the measured drive unit performance data, into a simulation program, on the results of which the values were adapted to give an optimum response curve (the components included C_1 and C_2).

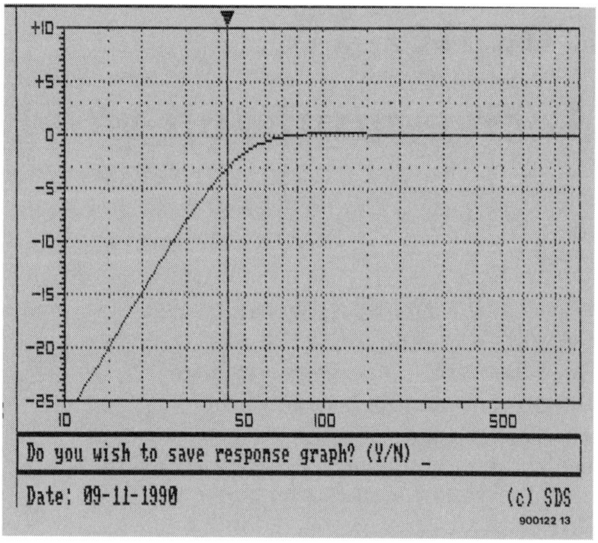

Fig. 15.1. Computed response curve of the prototype drive unit in a closed box.

Circuit IC_{1a} is a summing amplifier. The input signals from the left- and right-hand channels, which are at line level, enter via R_1-C_1 and R_2-C_2 respectively. Depending on the position of P_1, the sum of the signals is amplified to some degree and then applied to the correction network, which is based on IC_{1c}.

There are also two high-level inputs to which the signal from the integrated output amplifier may be connected. These signals are brought back to line level by R_3 and R_4.

Switch S_1 enables either the inverted or the normal signal to be selected. The inverted (by IC_{1d}) signal provides a phase-correct coupling with the existing loudspeakers.

The correction network, in conjunction with C_1 and C_2, provides a peak of just over 6 dB at 35 Hz, which results in the low –3 dB cut-off point of the subwoofer shifting down to about 28 Hz.

The correction network is followed by a third-order low-pass filter based on IC_{1b} with Butterworth characteristic. Four different cut-off points can be selected with switch S_2: 75 Hz, 100 Hz, 125 Hz and 150 Hz, to enable optimum coupling between the subwoofer and the existing loudspeakers to be obtained. One branch of the filter is shown in Table 15-1, which also gives formulas for calculating different cut-off points.

The output signal of IC_{1b} is applied to the power amplifier, which is described later.

Power for the correction network is derived from the power amplifier. Two voltage regulators, IC_2 and IC_3, reduce the line voltage to ±15 V.

Constructing the enclosure

The construction diagram for the enclosure is given in Fig. 15.6. The box

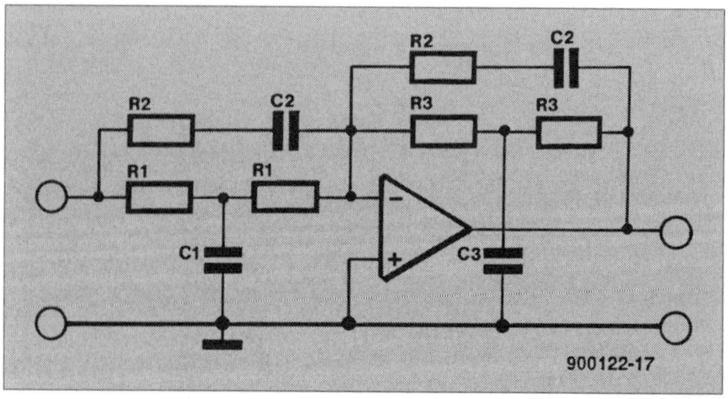

Fig. 15.2. The Linkwitz correction network.

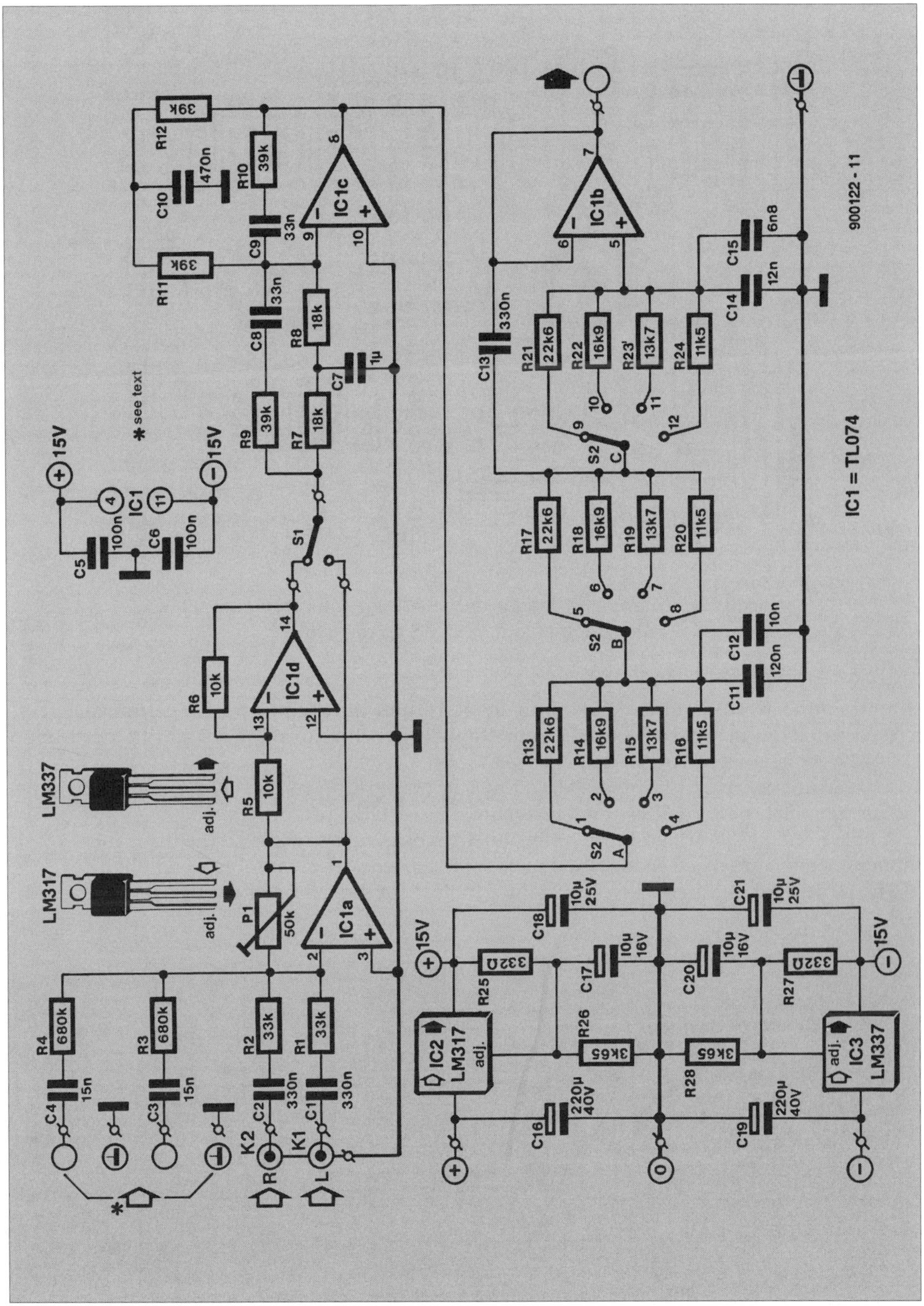

Fig. 15.3. Circuit diagram of the correction filter.

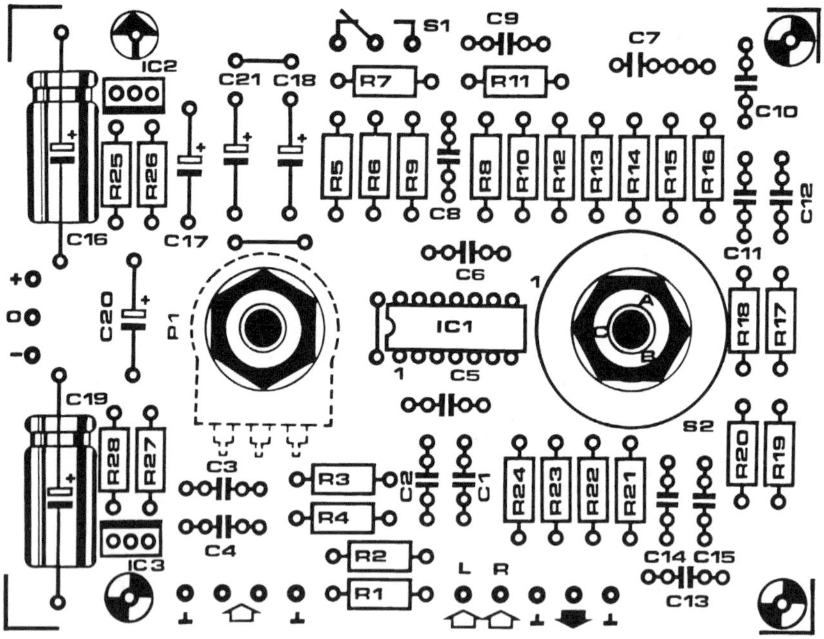

Fig. 15.4. Component layout of the PCB for the correction filter.
The track layout is given on page 259.

is a straightforward rectangular type made of 18 mm thick chipboard. The internal reinforcement struts may also be made from chipboard. All required parts can be cut from a 122×122 cm board—see opposite page.

A separate compartment is reserved at the rear of the box to house the electronics (power amplifier, power supply and correction network board).

A hole needs to be drilled in the rear panel for passing the cable to the drive unit. When that connection has been made, the hole should closed with a suitable wood filler.

The drive unit is screwed to the front panel; if the board used is sufficiently smooth,

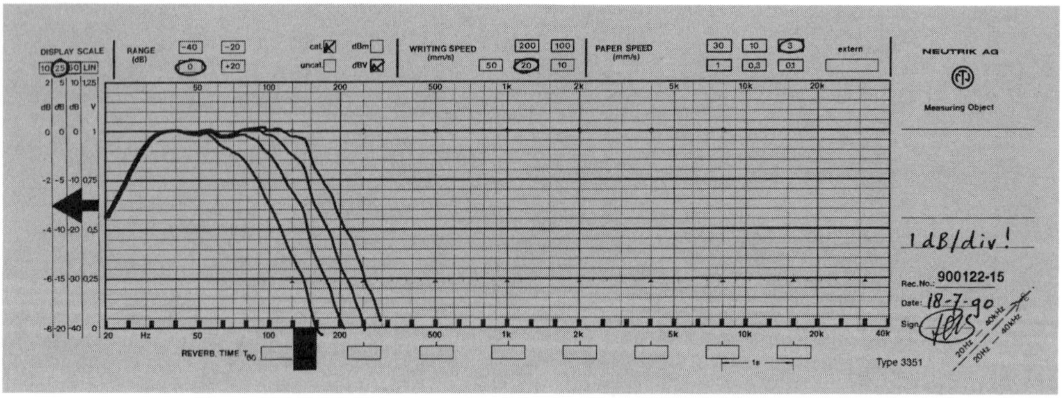

Fig. 15.5. Frequency response curves with four different high cut-off points.

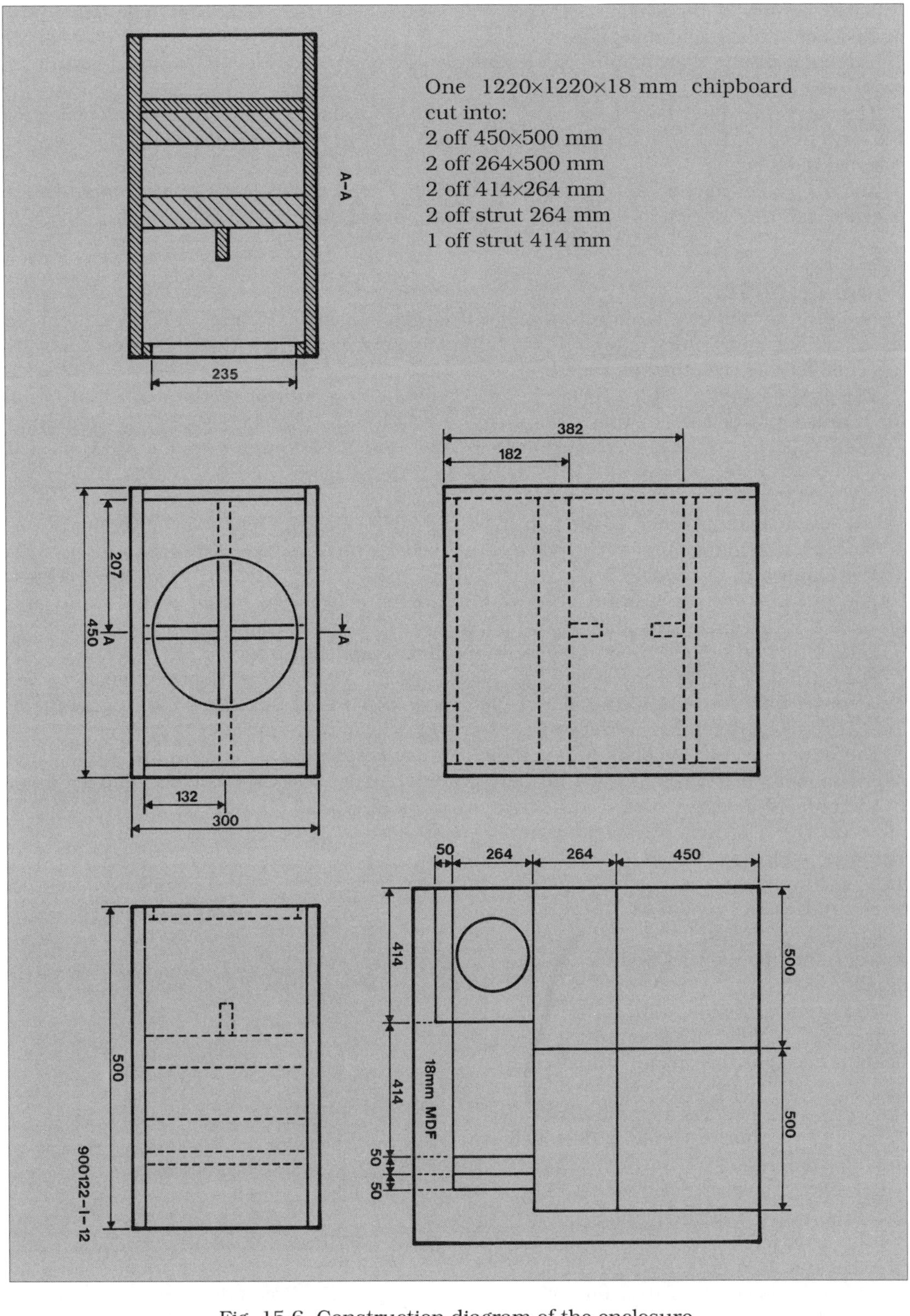

Fig. 15.6. Construction diagram of the enclosure.

this may be done without a gasket. The cable is connected to it with a car-type bullet plug and socket or blade and receptacle.

The enclosure is then filled with suitable wadding. Its exterior may be finished to personal taste.

The correction filter board—see Fig. 15.4—can now also be built. It provides space for the switches, so that only the phono connectors need to be fitted separately. This will be reverted to later.

The subwoofer may be tested at this stage by connecting a suitable power amplifier between it and the correction board. The correction board should be supplied with a voltage of 20–30 V.

Power amplifier

Although in principle any output amplifier that can deliver 50 W into an 8 Ω load may be used, many constructors may prefer to build the power amplifier that has been specially designed for use with the subwoofer.

The power amplifier is a hybrid circuit consisting of a control section based on an op amp, and a power section that uses discrete transistors—see Fig. 15.8. The op amp, a Type OP16 from PMI, is a precision type with JFET inputs and a slew rate of 25 V μs^{-1}. It has its own power supply of ±15 V, which is derived from the 30 V mains supply via R_{15}-D_4 and R_{16}-D_5.

The input signal is taken to the non-inverting input of the op amp via C_1. The input impedance is determined almost entirely by R_1 (since the op amp has JFET inputs).

The bandwidth of the OP16 is restricted to some extent by a 2.2 nF capacitor between the output and inverting input, and a 100 Ω resistor between the inverting input and ground. This arrangement may be compared to the compensation capacitor between the outputs of the first differential amplifier in a conventional output stage.

The output of the op amp drives the power section via a current source based on T_1. This source ensures a stable setting of the quiescent current through the output transistors. The voltage reference in the source is provided by a high-efficiency LED (D_1).

The power section consists of a complementary compound configuration, T_3–T_6. Normally, an emitter follower is used in the output to ensure adequate current amplification. In the present design, current amplification alone (a typical characteristic of an emitter follower) is not sufficient, because the signal excursion at the output of the op amp is limited to about ±12 V. Some additional amplification is, therefore, needed. A compound circuit provides current amplification as well as voltage amplification.

The voltage amplification in the present circuit is determined by the amplification factor of the output transistors and potential divider R_9-R_{10} between the output transistors and the drivers. To make sure that the op amp does not provide too high an output voltage, which would limit the output current, the amplification of the compound circuit is restricted to ×4.

Notable in this output stage configuration is the location of the emitter resistors of the output transistors, which

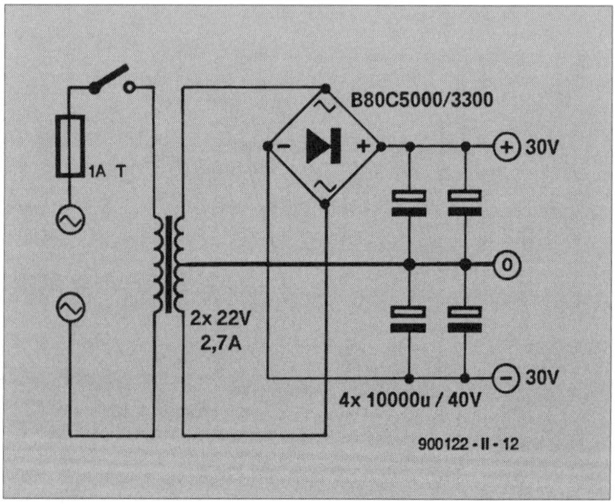

Fig. 15.7. Circuit diagram of the power supply for the power amplifier.

are connected to the power rails.

The quiescent current level is set with variable 'zener diode' T_2-P_1-R_4. Transistor T_2 is clamped to the heat sink between the output transistors to ensure good thermal coupling. Capacitors C_7 and C_{13} provide a.c. decoupling of the 'zener diode'.

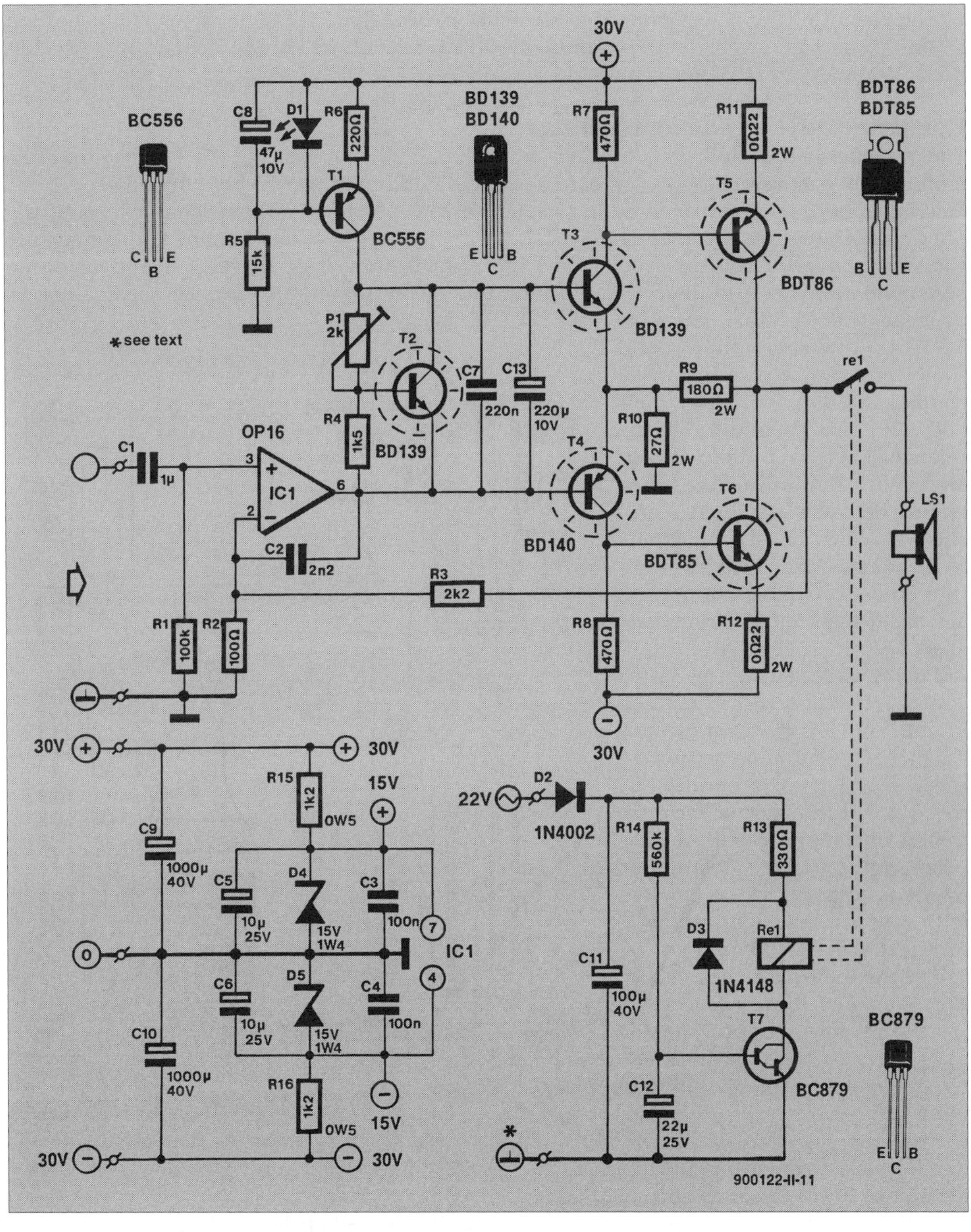

Fig. 15.8. Circuit diagram of the dedicated output amplifier.

The feedback loop of the overall amplifier consists of resistors R_2 and R_3, which set the overall amplification to ×23 (27 dB).

The circuit around T_7 and Re_1 provides a delay of a few seconds between power on and the connection between the loudspeaker and output stage being established. It derives power from the main power supply via D_2: this ensures that the relay is deenergized as soon as the power is switched off.

The circuit of the power supply is straightforward—see Fig. 16.9. Apart from the four 10 000 μF capacitors shown, two more 1 000 μF capacitors on the board provide additional decoupling of the power lines.

Constructing the power amplifier

The amplifier is best built on a printed-circuit board as shown in Fig. 15.9. Apart from the mounting of transistors T_2–T_6, the construction should not present any difficulties.

Transistors T_2–T_6 may be fitted in various ways, depending on the mechanical construction. If an aluminium L-section is used, they can be fitted above the board and fastened to the L-section, which in turn is screwed to the heat sink. It is, however, also feasible to screw the amplifier and filters to an aluminium sheet of suitable size, which then serves as heat sink. In that case, fit T_2–T_6 to the sheet first, bend their terminal wires upward a couple of millimetres above their body and pass them through the relevant holes in the board. Make sure that sufficient space is left between the board and sheet to allow solder connections to be made. Also, bear in mind that the transistors must be electrically isolated from the sheet.

For clarity's sake, the latter construction, on a 3 mm thick aluminium sheet, is shown in Fig. 15.10. The dimensions of the sheet allow it to be fitted in the space at the back of the subwoofer enclosure. For that purpose, glue four triangular wooden supports in the corners of that space to which the built-up sheet is screwed later on.

Fit the boards to the sheet with the aid of 10 mm spacers.

Fit the power supply as far away from the boards as possible to avoid any possibility of hum.

Note the separate earth connection for the delay circuit (indicated in Fig. 15.9 by ⊥ and an *) to the central earthing point. Do not make a direct connection

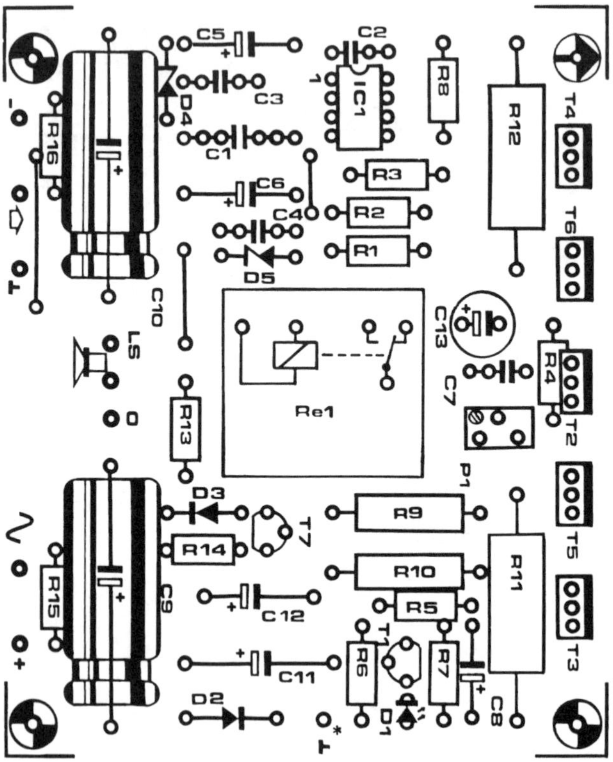

Fig. 15.9. Component layout of the PCB for the output amplifier. The track layout is given on page 259.

Fig. 15.10. (Opposite page). Wiring diagram of the output amplifier and its power supply.

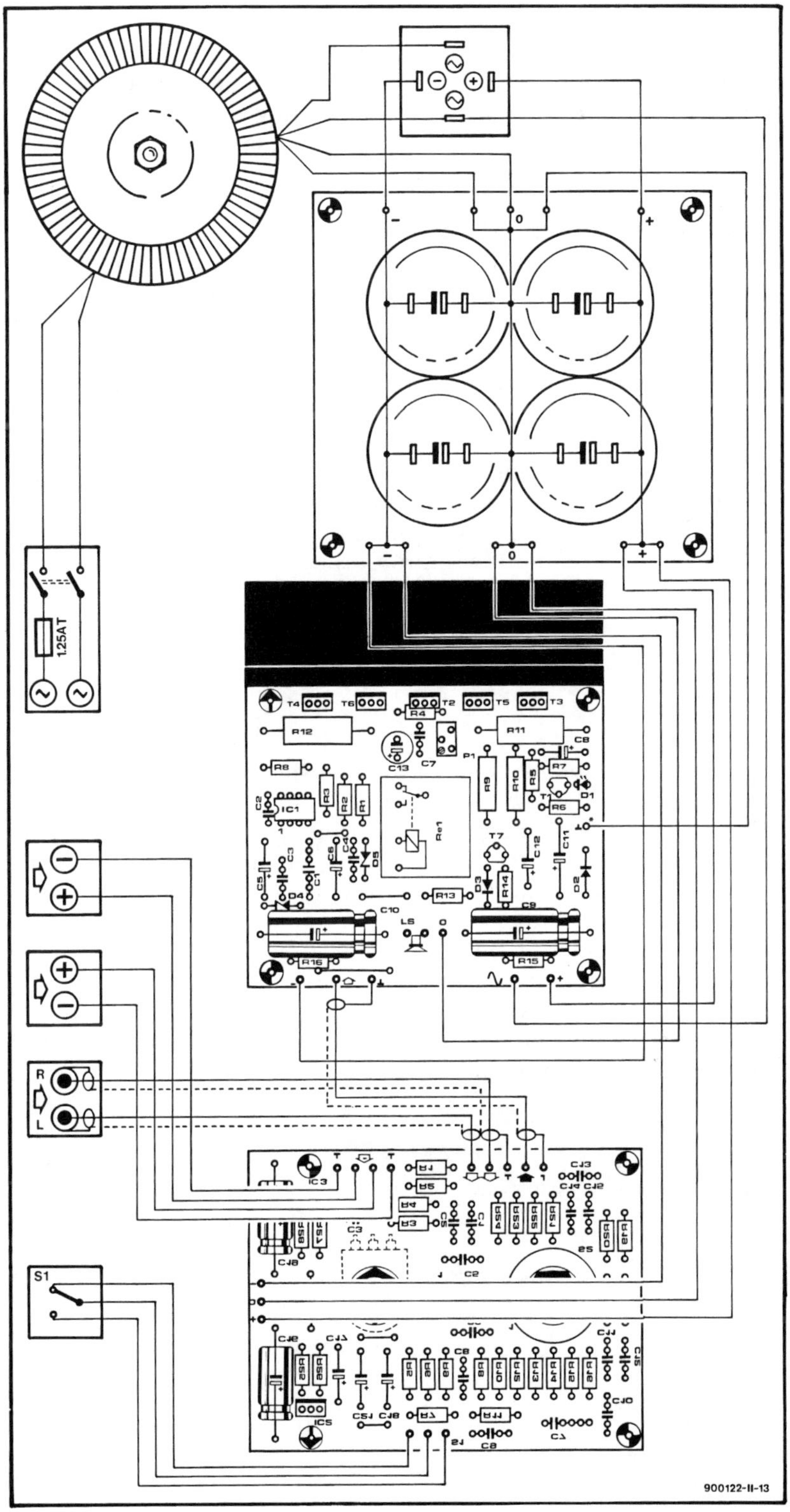
1.25AT
S1
R
L
900122-II-13

between the two earthing points on the amplifier board.

Do not yet connect the loudspeaker to the amplifier.

When everything is ready, set P_1 for minimum resistance and switch on the mains. Next, adjust P_1 for a quiescent current through the output amplifier of 100 mA: this is measured with a millivoltmeter across R_{11} or R_{12}, where the reading should be 22 mV.

Finally, switch off the mains, connect the loudspeaker to the amplifier and close the loudspeaker enclosure.

Connecting the subwoofers into the system

There are two ways in which the subwoofers may be connected to an existing audio system. If the system has discrete preamplifiers and output amplifiers, or an external connection between these units when integrated, the best way is to feed the output of the preamplifier to the subwoofers via a screened audio cable. If that is not possible, connect the (second pair of) loudspeaker terminals of the system to the banana sockets on the subwoofer.

When the connections between the audio system and its loudspeaker boxes are long, it is possible to extend them from those boxes to the subwoofers, since the latter should in any case be near the loudspeakers for optimum performance.

The low cut-off point of the existing system and the subwoofers may be matched in several ways. When separate preamplifiers and output amplifiers are used, a simple first-order high-pass filter may be provided by altering the value of the input capacitor of the power amplifier. If the input impedance, Z, of the power amplifier is known, the value of the capacitor for a cut-off frequency, f, is given by

$$C = 1/2\pi fZ.$$

Another way is adapting the cross-over network in the loudspeaker boxes. This is not so simple, however, because in the low-frequency range the resonance peak of the subwoofer will have an effect, so that the filter can not be terminated into a pure resistance.

A third possibility is leaving everything as it is. Particularly with small loudspeaker boxes whose cut-off frequency is fairly high—normally 75–100 Ω—it is perfectly all right to just connect the subwoofers into the system.

The location of the subwoofers is not important, but they should preferably be not too far from the loudspeakers. Critical listeners may like them between the loudspeakers.

The required sound level is set with the potentiometer at the back of the subwoofers.

Finally, if needed, the input signals may be inverted with a phase switch. This will probably require some experimentation.

Parts list (1 loudspeaker)

CORRECTION NETWORK & ENCLOSURE

Resistors:

R_1, R_2 = 33 kΩ
R_3, R_4 = 680 kΩ
R_5, R_6 = 10 kΩ
R_7, R_8 = 18 kΩ
R_9–R_{12} = 39 kΩ
R_{13}, R_{17}, R_{21} = 22.6 kΩ, 1%
R_{14}, R_{18}, R_{22} = 16.9 kΩ, 1%
R_{15}, R_{19}, R_{23} = 13.7 kΩ, 1%
R_{16}, R_{20}, R_{24} = 11.5 kΩ, 1%
R_{25}, R_{27} = 332 Ω, 1%

Linkwitz correction network formulas

Needed data:
Q_{tc} (= quality factor of drive unit in closed box)
f_c (= resonance frequency of drive unit)

Wanted new data:
$Q_{tc'}$ (= new quality factor with correction network)
$f_{c'}$ (= new resonance frequency with correction network).

Conditions for chosen factors:

$$k = \frac{f_c / f_{c'} - Q_{tc} / Q_{tc'}}{Q_{tc} / Q_{tc'} - f_{c'} / f_c} > 0$$

where k is the pole-shifting factor.

Calculation:

$$R_2 = 2kR_1$$

$$R_3 = R_1(f_c / f_{c'})^2$$

$$C_1 = \frac{2Q_{tc}(1+k)}{2\pi f_c R_1}$$

$$C_2 = \frac{1}{4\pi f_c Q_{tc} R_1 (1+k)}$$

$$C_3 = C_1(f_{c'} / f_c)^2$$

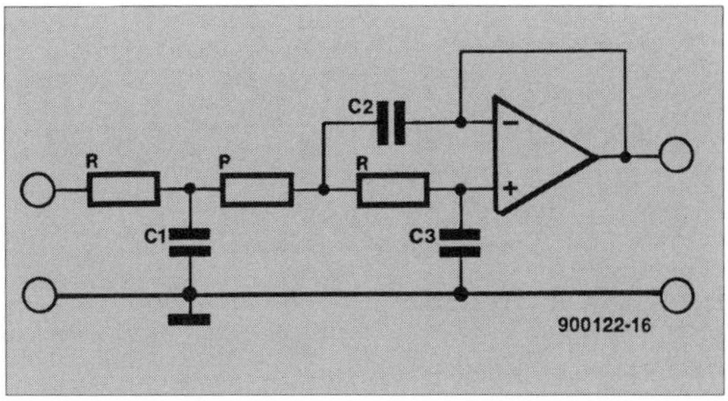

Butterworth	Bessel
$C_1 = 0.2215/fR$	$C_1 = 0.1572/fR$
$C_2 = 0.5644/fR$	$C_2 = 0.2265/fR$
$C_3 = 0.03221/fR$	$C_3 = 0.04039/fR$

Table 15-1.

R_{26}, R_{28} = 3.65 kΩ, 1%
P_1 = 50 kΩ (47 kΩ) preset

Capacitors:
C_1, C_2, C_{13} = 330 nF
C_3, C_4 = 15 nF
C_5, C_6 = 100 nF
C_7 = 1 µF
C_8, C_9 = 33 nF
C_{10} = 470 nF
C_{11} = 120 nF

C_{12} = 10 nF
C_{14} = 12 nF
C_{15} = 6.8 nF
C_{16}, C_{19} = 220 µF, 40 V
C_{17}, C_{20} = 10 µF, 16 V
C_{18}, C_{21} = 10 µF, 25 V

Integrated circuits:
IC_1 = TL074
IC_2 = LM317
IC_3 = LM337

Miscellaneous:
S_1 = toggle switch, single change-over
S_2 = rotary switch, 3-pole, 4-position
2 off phono sockets
4 off banana sockets
1 off drive unit as specified in text
Cabinet wadding as required
18 mm thick chipboard 122×122 cm (for cutting diagram, see Fig. 15.6)

POWER AMPLIFIER
Resistors:
R_1 = 100 kΩ
R_2 = 100 Ω
R_3 = 2.2 kΩ
R_4 = 1.5 kΩ
R_6 = 220 Ω
R_7, R_8 = 470 Ω
R_9 = 180 Ω, 2.5 W
R_{10} = 27 Ω, 2.5 W
R_{11}, R_{12} = 0.22 Ω, 5 W
R_{13} = 330 Ω, 1 W
R_{14} = 560 Ω
R_{15}, R_{16} = 1.2 kΩ, 0.5 W
P_1 = 2 kΩ, multi-turn preset with top adjustment

Capacitors:
C_1 = 1 µF
C_2 = 2.2 nF
C_3, C_4 = 100 nF
C_5, C_6 = 10 µF, 25 V
C_7 = 220 nF
C_8 = 47 µF, 10 V
C_9, C_{10} = 1000 µF, 40 V
C_{11} = 100 µF, 40 V
C_{12} = 22 µF, 25 V
C_{13} = 220 µF, 10 V, radial

Semiconductors:
D_1 = LED, 3 mm, red, high efficiency
D_2 = 1N4002

D_3 = 1N4148
D_4, D_5 = zener diode, 15 V, 1.4 W
T_1 = BC556
T_2, T_3 = BD139
T_4 = BD140
T_5 = BDT86 or BD912
T_6 = BDT85 or BD911
T_7 = BC879

Integrated circuits:
IC_1 = OP16

Miscellaneous:
Re_1 = relay, 24 V, single change-over
Mains transformer, secondary 2×22 V, 2.7 A
4 off electrolytic capacitors, 10 000 μF, 40 V
Bridge rectifier B80C5000/3300

Part 6

Tests

Chapter 16

Phase check unit for audio systems

Reversed phase connections in an audio equipment system give strange and unpredictable effects, such as the unwanted attenuation or boosting of a particular frequency range, jet-plane effects, whistling noises, or amplifier output power that does not seem to produce any usable sound level. The phase check unit helps to avoid these problems. Consisting of a simple transmitter, receiver and a good/fault indicator, it enables the system to be checked from the input (microphone or line input) right through to the output (loudspeaker or line output).

The transmitter supplies positive or negative needle pulses, which are fed either electrically to an equipment input via the line output socket, or acoustically to a microphone via the built-in loudspeaker. Accordingly, the receiver has an electrical input (line) and an acoustic input (microphone).

The drawings in Fig. 16.1. illustrate two ways of using the transmitter and receiver for phase tests on audio equipment. Figure 16.1a shows the set-up used to check the polarity of a microphone, and Fig. 16.1b that used to ensure a loudspeaker is connected the right way. The LEDs on the receiver provide a quick indication whether or not the received pulses have the same polarity as the transmitted pulses. If the receiver indicates the opposite polarity of the transmitter, there is almost certainly a reversed signal connection somewhere in the system.

Transmitter

The needle pulses are generated by oscillator IC_{1a} (see Fig. 16.2.), which is built from a NAND gate with two Schmitt trigger inputs. When the supply voltage is on, these inputs as-

sume complementary logic levels, that is, one is high and the other is low. Consequently, the output of the gate is high. Capacitor C_2 is charged via resistor R_1 until the voltage across it reaches the high threshold voltage of about 5.5 V, whereupon the gate changes state (output low = 0). Capacitor C_2 is discharged via D_1 and R_2 until the low threshold voltage of about 3 V is reached, whereupon the gate changes state again, which causes C_2 to be charged anew.

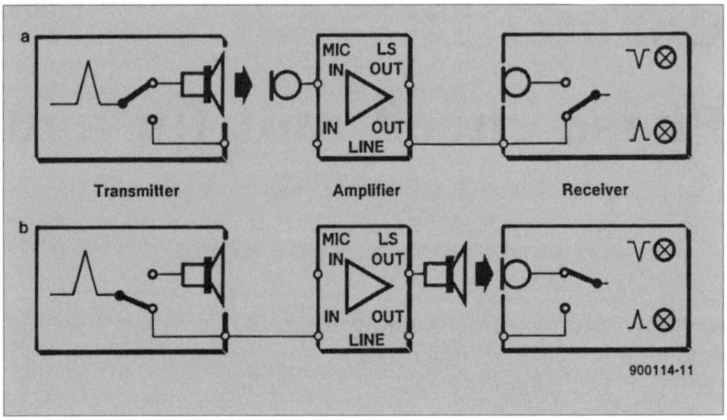

Fig. 16.1. Application examples of the phase-check system.

This process is cyclical and results in a self-oscillating circuit. Since R_2 is much smaller than R_1, the discharge time of C_2 is much shorter than its charging time. As a result, the

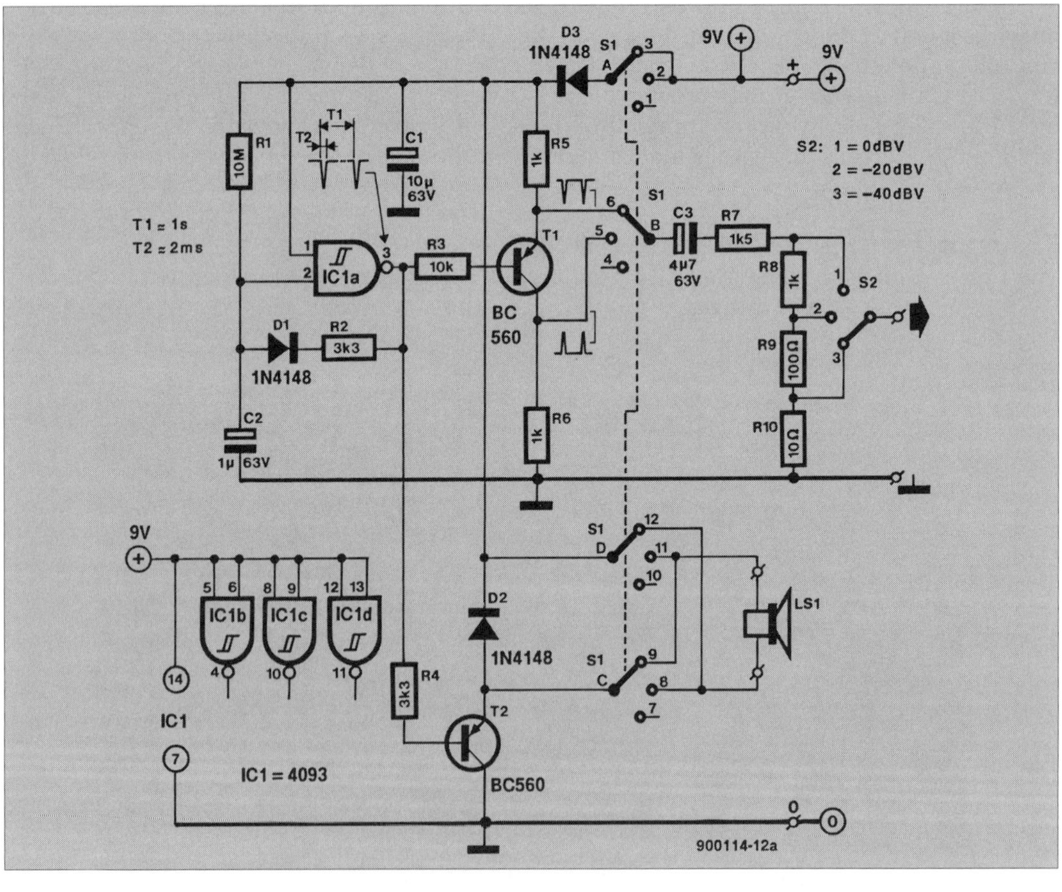

Fig. 16.2. Circuit diagram of the pulse transmitter.

on/off (mark/space) ratio of the output signal is about 0.002. Note that 'off' in this context means 'logic high' since the gate is a NAND type.

The oscillator output signal is fed to two sub-circuits. One is a small loudspeaker driver, based on emitter follower T_2. The loudspeaker connections can be swapped with switch S_1 (contacts c and d). When an oscilloscope is connected to the loudspeaker, it indicates negative-going needle pulses when the switch is at its centre position, and positive-going pulses when the switch is at its upper position (in the diagram). Likewise, in the other signal branch, the polarity is changed by switching transistor T_1 from a common-emitter circuit (S_1 in centre position) to a common-collector circuit (S_1 in upper position). Coupling capacitor C_3 applies the test signal to an attenuator that supplies output levels of 1 V_{pp} (0 dBV), –20 dBV, and –40 dBV.

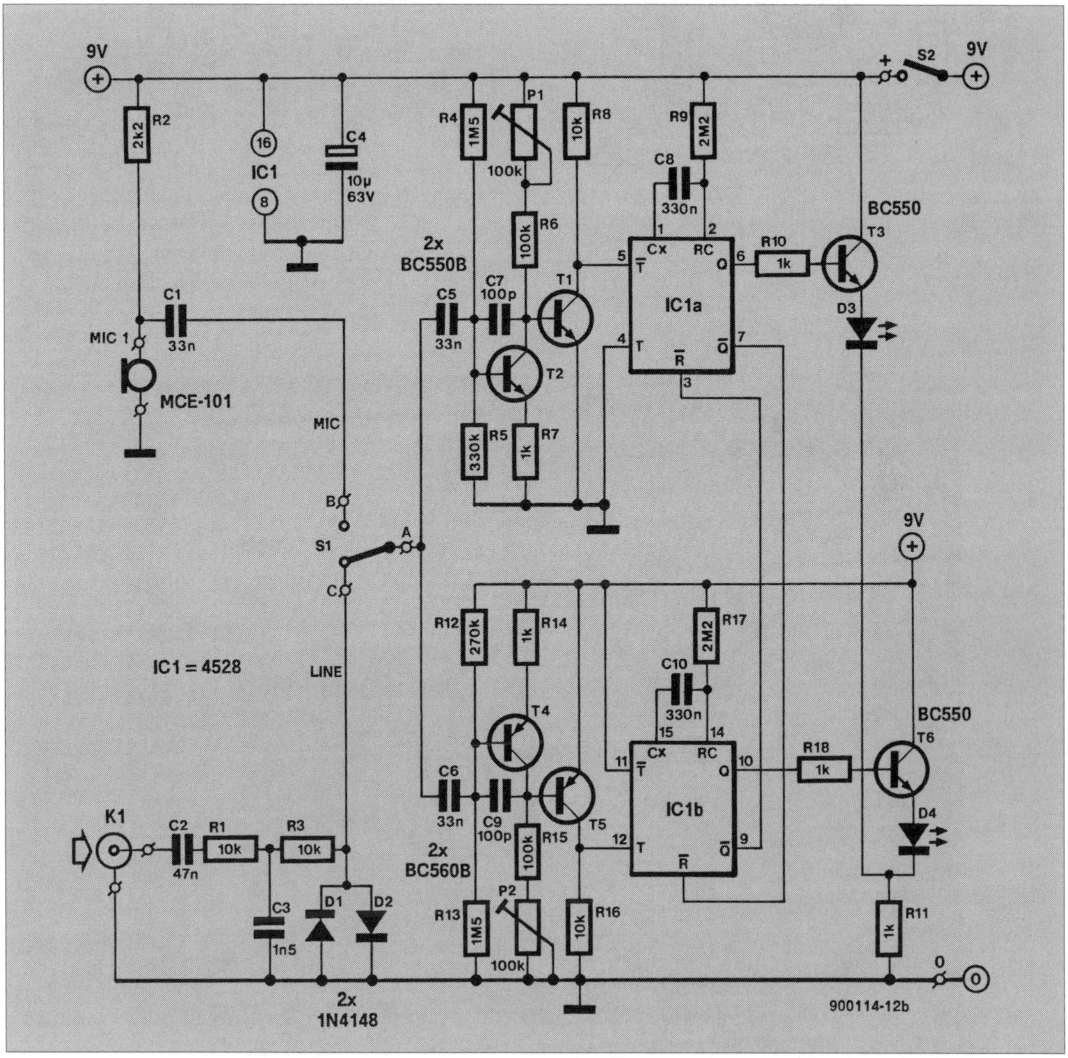

Fig. 16.3. Circuit diagram of the pulse receiver.

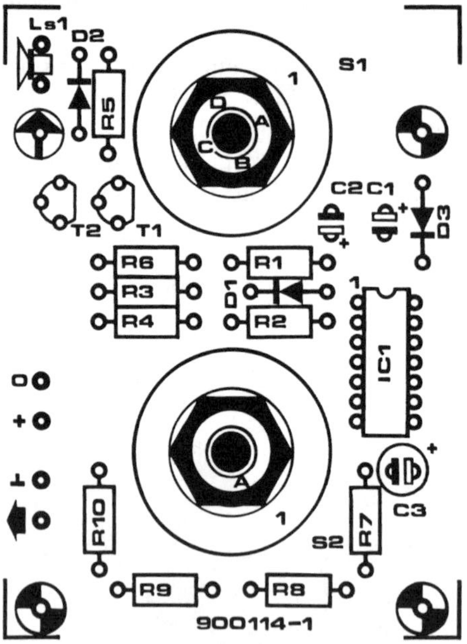

Fig. 16.4. Component layout of the PCB for the pulse transmitter. The track layout is given on page 261.

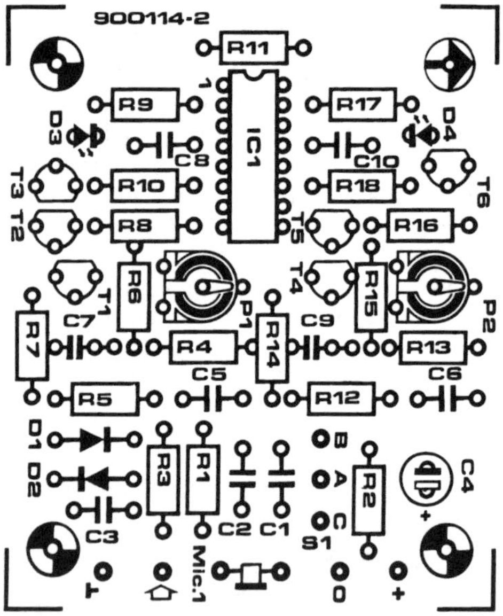

Fig. 16.5. Component layout of the PCB for the pulse receiver. The track layout is given on page 261.

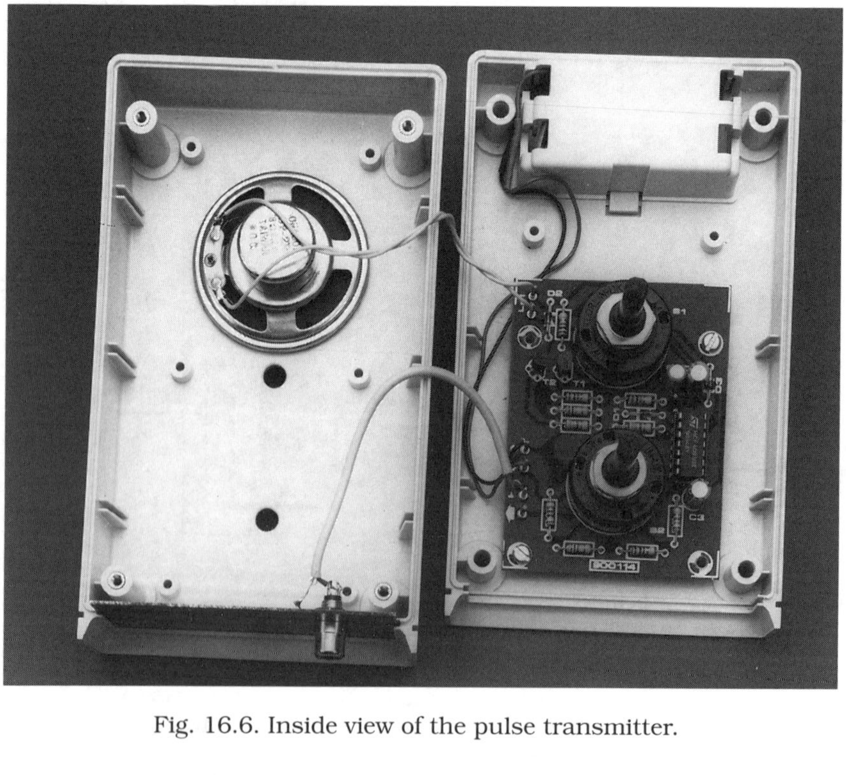

Fig. 16.6. Inside view of the pulse transmitter.

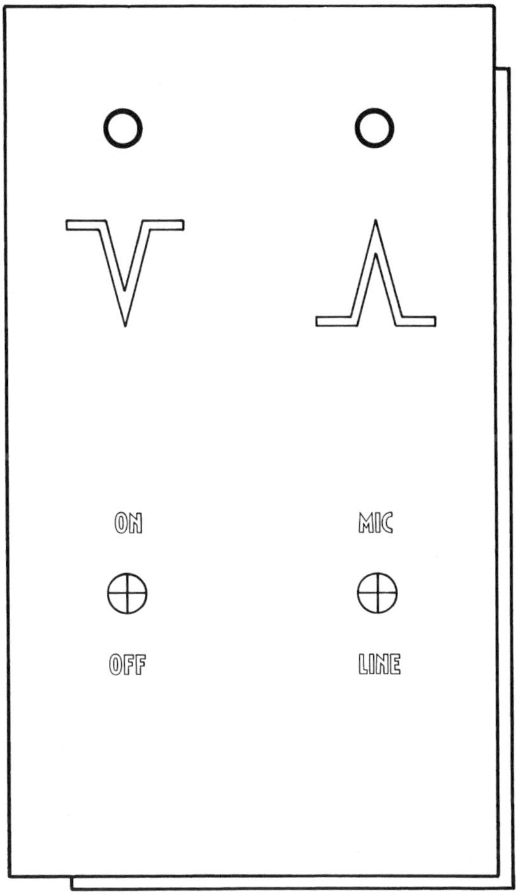

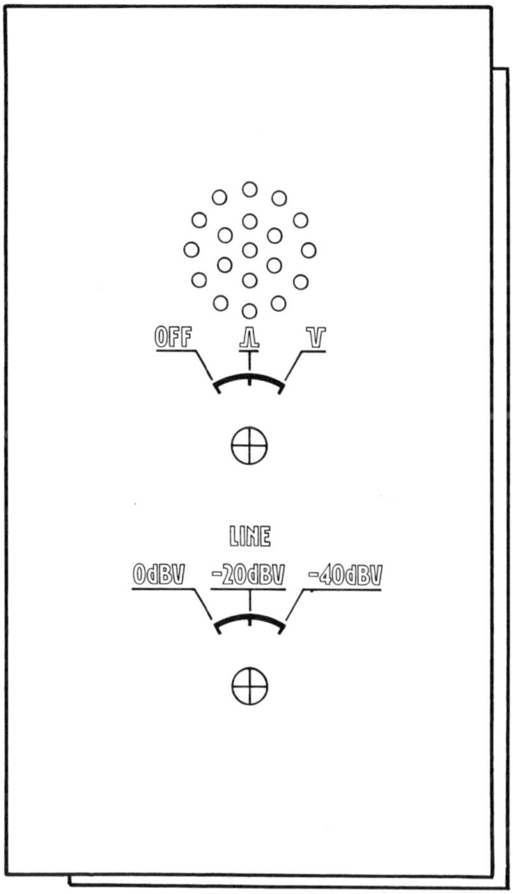

Fig. 16.7. Suggested front panel layout for the receiver (left) and transmitter (right).

Receiver

The receiver contains two almost identical detectors—see Fig. 16.3. The test signal is applied to voltage amplifiers T_1-T_2 and T_4-T_5 by either the electret microphone or the signal source connected to K_1. In the latter case, the signal is applied via high-pass filter R_1-C_3 to limiter D_1-D_2. The input source, microphone or line, is selected with switch S_1. The voltage amplifiers are complementary circuits: T_1-T_2 amplifies the negative pulses and T_4-T_5 the positive pulses.

The two monostables in IC_1 have different networks at their trigger inputs to enable them to respond to the edges of negative pulses (IC_{1a}) or positive pulses (IC_{1b}). To prevent the trailing edge of a pulse triggering the wrong monostable, IC_{1a} and IC_{1b} disable one another when one of them is actuated. The monostables thus allow the circuit to determine whether a pulse starts with a positive (rising) or a negative (falling) edge. The two LEDs, D_3 and D_4, indicate the respective polarities. The monostable times are set at about 0.5 s with R_9-C_8 and R_{17}-C_{10}. This causes the active LED to flicker.

Building and testing

The receiver and transmitter are best built on the printed-circuit boards shown in

Fig. 16.4 and 16.5. Be sure to fit all polarized components (electrolytic capacitors, ICs, transistors and diodes) the correct way around. Also, make sure that the two rotary switches on the transmitter board are fitted as shown on the component layout (note the '1' mark, and the letters that indicate the poles). On completion of the boards, each should be fitted in a dedicated case, the front panel of which may be finished as shown in Fig. 16.7. Figure 16.6. shows the inside of the pulse transmitter.

Interconnect the transmitter and receiver via their line sockets, and check that the LED indication of the receiver accords with the polarity set on the transmitter. If the LEDs remain off, IC_1 in the receiver may not have sufficient gain. In that case, adjust P_1 and P_2 until the receiver does trigger correctly.

Parts list

TRANSMITTER

Resistors:

R_1 = 10 MΩ
R_2, R_4 = 3.3 kΩ
R_3 = 10 kΩ
R_5, R_6, R_8 = 1 kΩ
R_7 = 1.5 kΩ
R_9 = 100 Ω
R_{10} = 10 Ω

Capacitors:

C_1 = 10 µF, 63 V, radial
C_2 = 1 µF, 63 V, radial
C_3 = 4.7 µF, 63 V, radial

Semiconductors:

D_1–D_3 = 1N4148
T_1, T_2 = ~BC560

Integrated Circuits:

IC_1 = 4093

Miscellaneous:

S_1 = rotary switch, 4-pole, 3-position, for board mounting
S_2 = rotary switch, 1-pole, 12-position, for board mounting
LS_1 = loudspeaker, 8 Ω, 50 mm dia.
1 off ABS enclosure
1 off clip for 9 V (PP3) battery
1 off phono socket

RECEIVER

Resistors:

R_1, R_3, R_8, R_{16} = 10 kΩ
R_2 = 2.2 kΩ
R_4, R_{13} = 1.5 MΩ
R_5 = 330 kΩ
R_6, R_{15} = 100 kΩ
R_7, R_{10}, R_{11}, R_{14}, R_{18} = 1 kΩ
R_9, R_{17} = 2.2 MΩ
R_{12} = 270 kΩ

P_1, P_2 = 100 kΩ horizontal preset

Capacitors:
C_1, C_5, C_6 = 33 nF
C_2 = 47 nF
C_3 = 1.5 nF
C_4 = 10 µF, 63 V, radial
C_7, C_9 = 100 pF
C_8, C_{10} = 330 nF

Semiconductors:
D_1, D_2 = 1N4148
D_3 = LED, red
D_4 = LED, green
T_1–T_3, T_6 = BC550B
T_4, T_5 = BC560B

Integrated circuits:
IC_1 = 4528

Miscellaneous:
Mic_1 = electret microphone
K_1 = phono socket
S_1 = miniature SPDT switch
S_2 = miniature SPST switch
1 off clip for 9 V (PP3) battery
1 off ABS enclosure

Chapter 17

PC-controlled a.f. measurement systems

An overview

The evolution of the personal computer over the past ten years has led to the present (1995) situation where nearly every one seems to have or use a computer, be it for personal or business use. Today, an investment of less than £1,000 ($ 1600) buys a PC with computing power and storage capacity which was thought impossible or at least 'futuristic' only a few years ago. Coupled with its graphics capacities, that makes the PC eminently suited to more tasks than run-of-the-mill word processing or twiddling with spreadsheets.

Developers of measurement equipment have been long aware of this potential, and a variety of PC-controlled measurement equipment and plug-in cards is currently available. Incidentally, many 'ordinary' measurement systems also contain a full-blown microprocessor, often with a keyboard to control the relevant measurement equipment, and a display to visualize data. The catalogues of companies like Hewlett Packard and Tektronics list beautiful high-end test equipment with a plethora of features and control options, all by virtue of built-in computing power. However, this type of equipment is not covered by this chapter,

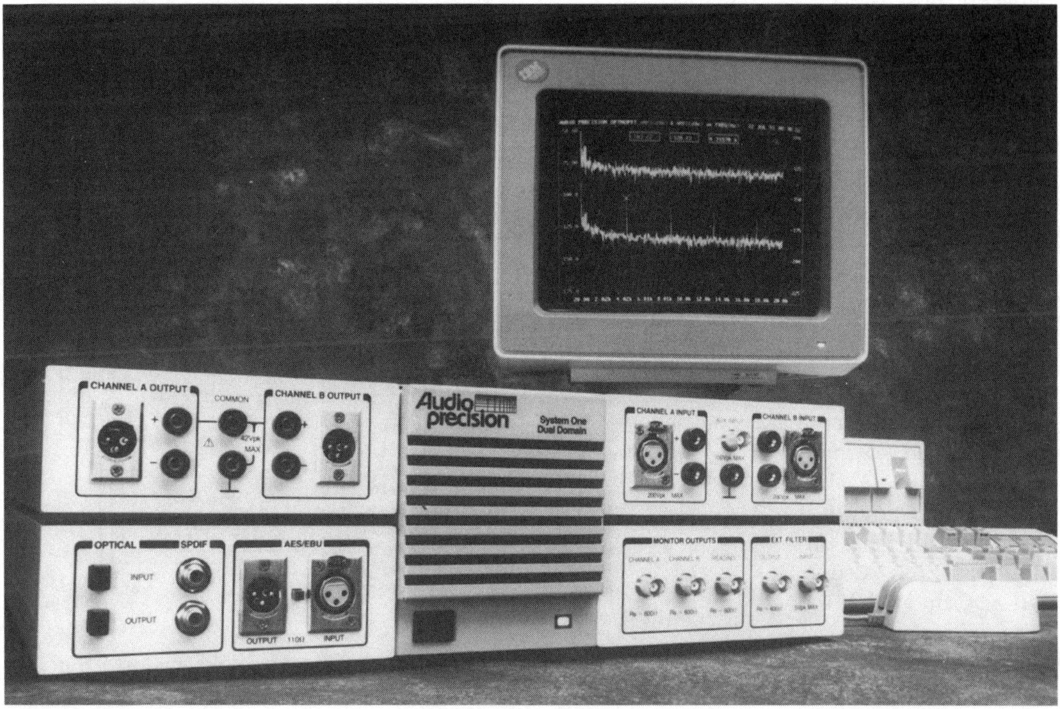

which is restricted to audio measurement systems that can be linked to a PC, via either a cable or an insertion card. The bulk of this type of equipment is intended for measurements in the audio field. Consequently, cards functioning as a voltmeter and/or oscilloscope are not covered here. Summarizing, this chapter concentrates on instruments for extensive measurements on audio equipment, including frequency response and distortion measurements.

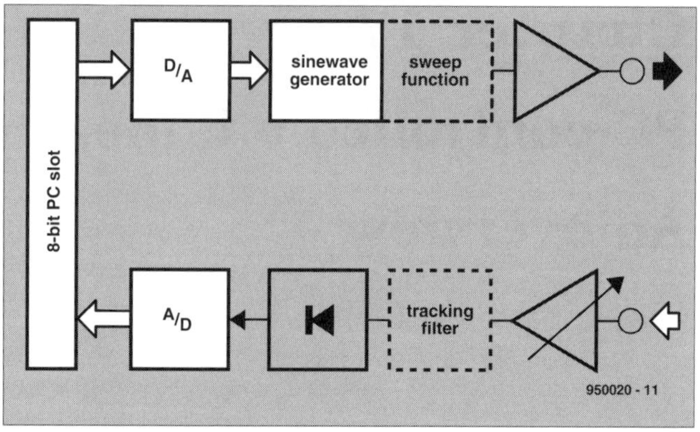

Fig. 17.1. Block schematic of a computer-controlled frequency-response plotter.

Classification

From a point of view of their construction, audio measurement systems may be divided into two classes:

- external measurement systems, where the function of the PC is restricted to effecting the settings on the instrument, and visualizing (and, possibly, processing also) the measured data;
- internal measurement systems which are plugged into the PC and share part of the pc hardware (for instance, an insertion card which copies measured data directly to the PC's memory).

Initially, the differences between these two categories will appear to be marginal. It is even possible for certain measurement systems to come either as an internal or an external version. The main advantage of an external system is that it is easy to relocate to another computer system.

The major difference between an internal and an external PC-controlled measurement system lies in the accuracy that can be achieved. Although a high-resolution ADC may be fitted on an insertion card, it will be hard to achieve the specified accuracy

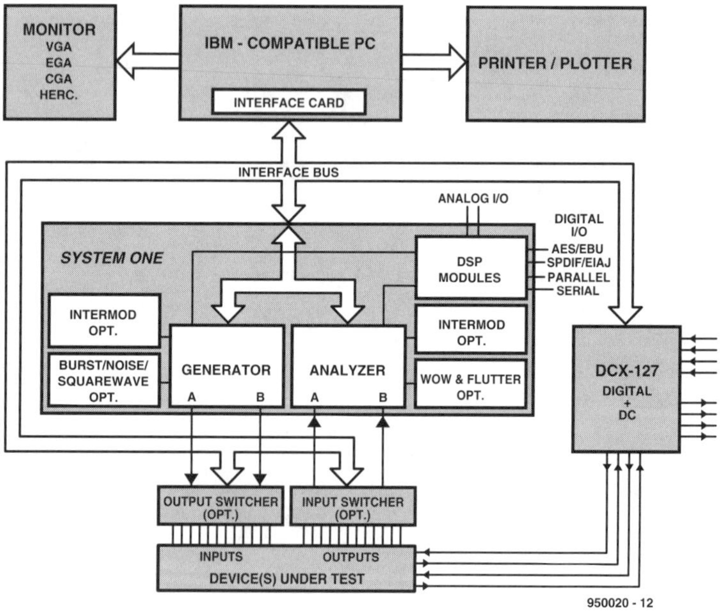

Fig. 17.2. Block diagram of the Audio Precision System One analogue measurement system.

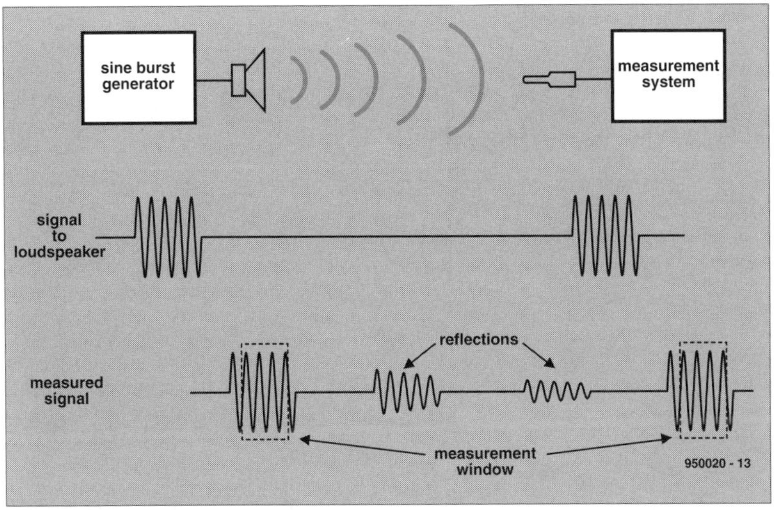

Fig. 17.3. Gated sine-wave measurements are based on sending bursts of sine waves to the test microphone. The receiver is disabled outside the gate period, so that sound reflections caused by the test room are ignored.

in the face of the high noise level that exists inside a PC. This noise is caused by the switch-mode power supply, as well as by switching and data signals with fast edges. The sum total of these effects severely curtails measurements of small signals. This is where the external measurement system has the edge because it allows sensitive parts like an ADC to be better screened. Only a handful of specialized manufacturers are capable of producing insertion cards which afford enough screening and power line decoupling to ensure high measurement accuracy. Therefore, do not expect too much from an audio measurement system in the form of an insertion card — though it may have a 16-bit A-D converter, this will probably achieve an accuracy of 12 or 13 bits at the most. If you are after really accurate audio measurements, for instance, distortion levels produced by hi-fi amplifiers, there is no alternative but to use an external measurement system. For measurements with 'relaxed' requirements, for instance, sound pressure level (SPL) or frequency response measurements on loudspeaker systems, a plug-in card is good enough, and often much cheaper than an external system.

Evidently, you have to take into account that the performance of any plug-in card depends largely on its internal design. By the same token, do not expect miracles from a measurement system just because it is external. In both cases, the quality strongly depends on the price.

Types of audio measurement system

There is such a bewildering variety of measurements that can be performed on audio equipment that it is impossible to list them all in this article. Only look at a company like Bruel & Kjaer, which supplies dozens of different instruments for audio measurements only. Fortunately, the diversity is much smaller if we focus on PC-controlled systems.

Frequency response plotters

The simplest type of audio measurement card contains a sweep generator and a voltmeter. A sinusoidal signal is applied to the equipment under examination (for instance, a loudspeaker or an amplifier), and the voltmeter is used to measure signal level levels at a number of points during the sweep operation. The frequency response is displayed on the PC screen (see block diagram in Fig. 17.1). Functionally, this corresponds to the well-known mechanical frequency response plotter. Unfortunately, a sine-wave sweep gives unreliable test results in a 'normal', i.e., non-anechoic, test room because that produces reflections. This can be compensated partly by the use of narrow-band tracking filters, or by a wobbulator, which produces rapid variations of the instantaneous frequency within a third or an octave.

Analogue measurement systems

Distortion measurements require special equipment with steep-skirted filters to separate the base frequency from the higher harmonics. The best known test instrument in this field is with-

External instruments

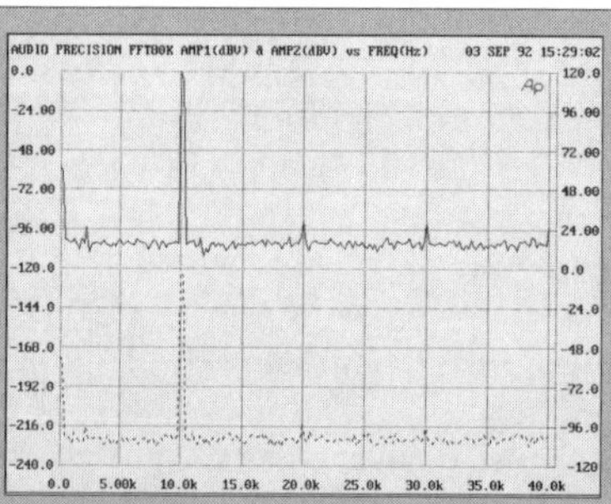

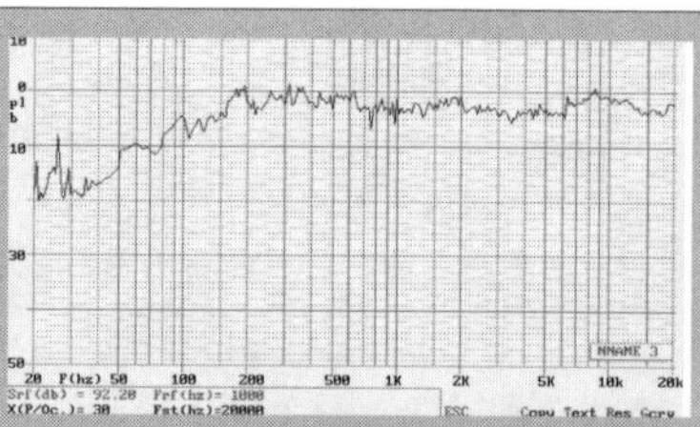

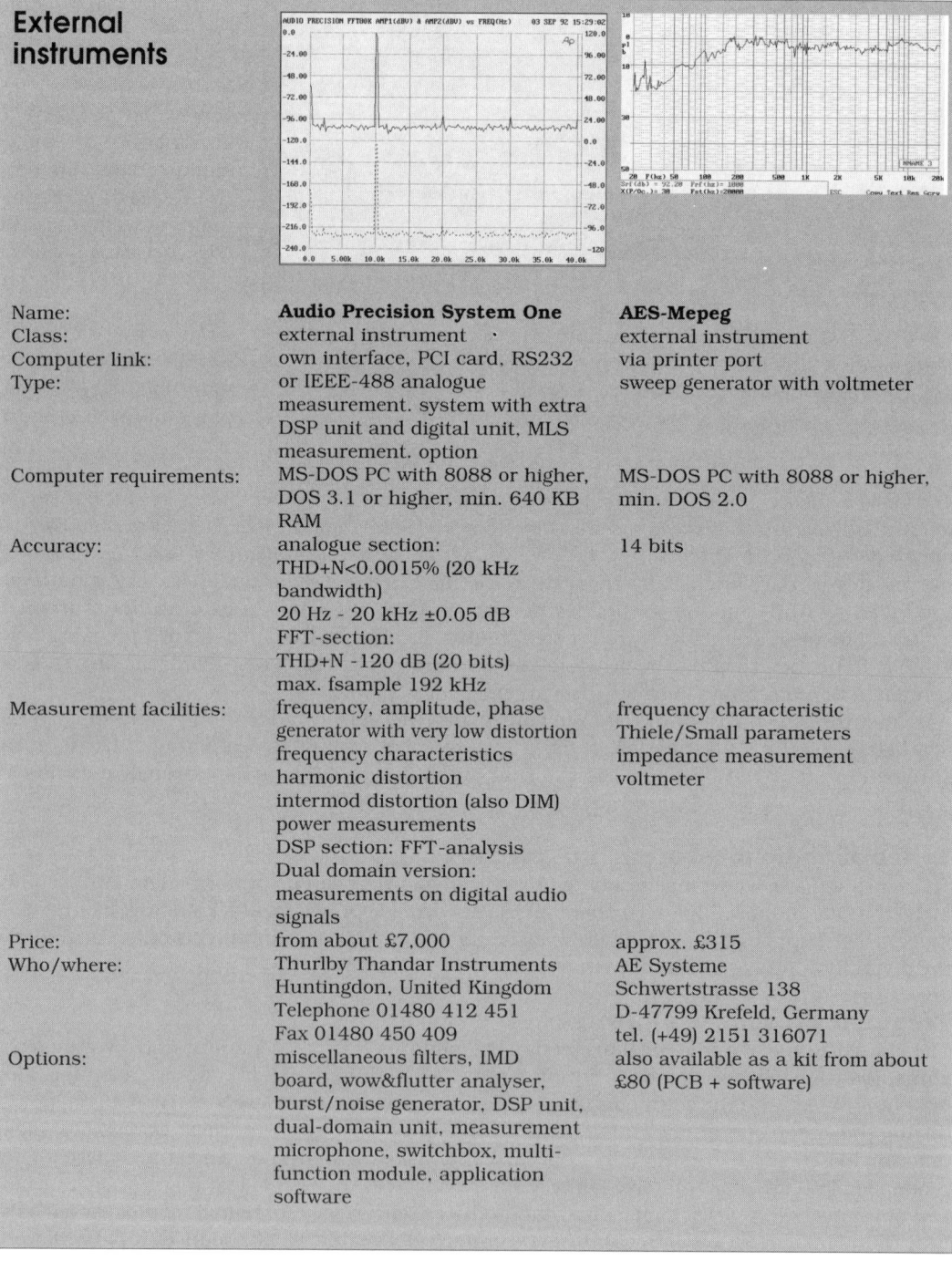

Name:	**Audio Precision System One**	**AES-Mepeg**
Class:	external instrument	external instrument
Computer link:	own interface, PCI card, RS232 or IEEE-488	via printer port
Type:	analogue measurement. system with extra DSP unit and digital unit, MLS measurement. option	sweep generator with voltmeter
Computer requirements:	MS-DOS PC with 8088 or higher, DOS 3.1 or higher, min. 640 KB RAM	MS-DOS PC with 8088 or higher, min. DOS 2.0
Accuracy:	analogue section: THD+N<0.0015% (20 kHz bandwidth) 20 Hz - 20 kHz ±0.05 dB FFT-section: THD+N -120 dB (20 bits) max. fsample 192 kHz	14 bits
Measurement facilities:	frequency, amplitude, phase generator with very low distortion frequency characteristics harmonic distortion intermod distortion (also DIM) power measurements DSP section: FFT-analysis Dual domain version: measurements on digital audio signals	frequency characteristic Thiele/Small parameters impedance measurement voltmeter
Price:	from about £7,000	approx. £315
Who/where:	Thurlby Thandar Instruments Huntingdon, United Kingdom Telephone 01480 412 451 Fax 01480 450 409	AE Systeme Schwertstrasse 138 D-47799 Krefeld, Germany tel. (+49) 2151 316071
Options:	miscellaneous filters, IMD board, wow&flutter analyser, burst/noise generator, DSP unit, dual-domain unit, measurement microphone, switchbox, multi-function module, application software	also available as a kit from about £80 (PCB + software)

out doubt the Audio Precision System One (see page 204), which has been around for more than ten years, and has become a kind of standard for measurements on hi-fi audio equipment. The instrument is, obviously, external, and contains all sub-circuits for frequency response plotting, distortion measurements, and more. A version is available with a built-in FFT analyser, as well as an optional extension for measurements on digital audio signals. Prices of the APS One start at about £7,000. That may seem a lot, but considering the impressive performance and possibilities, this sum buys one of the best measurement systems in its class.

The block diagram of the APS One is reproduced in Fig. 17.2. The ingredients are those of an ordinary measurement system: a low-distortion oscillator, input amplifiers with measurement filters, a frequency meter and a voltmeter. The only non-standard block is the interface between the PC and the measurement system. It allows you, for instance, to set the

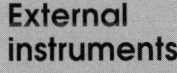

External instruments

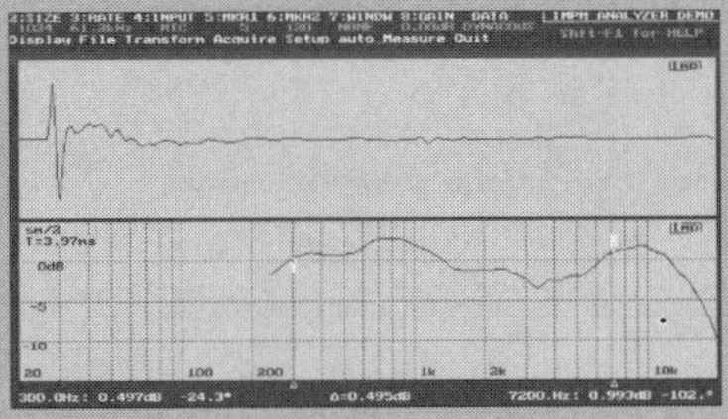

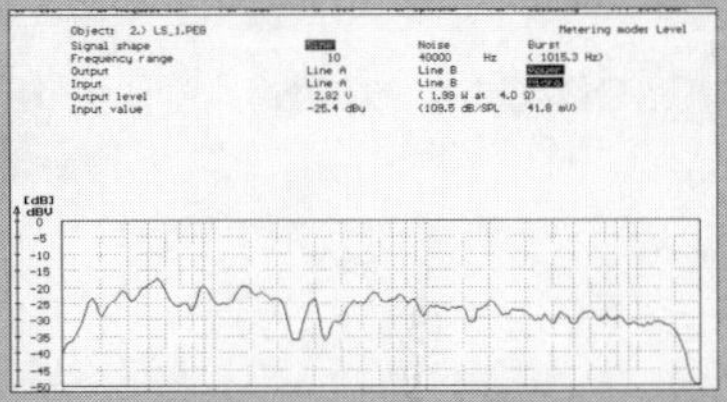

Name:	**IMP 2.0**	**Kemsonic AMS PC/ST type 1656**
Class:	external instrument	external instrument
Computer link:	via Centronics port	via RS232 interface
Type:	generator + FFT-analyser, MLS measurement option	sweep generator with voltmeter (and optional tracking filter)
Computer requirements:	MS-DOS-PC with 8088 or higher, min. MS-DOS 3.1, min. 640 KB RAM	MS-DOS-PC, min. MS-DOS 3.1, min. 640 KB RAM (Atari version also available)
Accuracy:	12 bits, f(sample) max. 61.4 kHz	not stated
Measurement facilities:	frequency characteristic phase characteristic impedance measurement Thiele/Small parameters waterfall spectrum	frequency characteristics impedance measurements Thiele/Small parameters room acoustics
Price:	approx. £395	from approx. £425
Who/where:	Clear Sound I. Brouwersteeg 4 NL-8911 BZ Leeuwarden, The Netherlands tel. (+31) 58 159927	Kemsonic Audio Measurement Systems GmbH Teutoburger Strasse 37 D-4800 Bielefeld, Germany tel. (+49) 521 175314, fax (+49) 521 176931
Options:	price incl. measurement microphone	1/3-octave tracking filter, phase measurement card, misc. microphones, vibration transducer, in-/output-module, software for quality checking

frequency and output level of the oscillator, while the measurement results are returned to the computer in digitized form. Next, the PC converts the data into, for example, a frequency curve which may be viewed on the monitor. Despite (or, perhaps, thanks to) the fairly conventional setup of this system, the accuracy is very high, and sets a standard for all other PC-controlled measurement systems. The FFT section has more facilities, including breaking down a signal into its frequency components, and performing MLS measurements. The (optional) dual domain unit enables the user to analyse digital audio signals.

Gated sine-wave measurements

Long before computer-controlled measurements had reached sufficient power to enable FFT analyses to be performed in a simple way, a method had been discovered (by, among others,

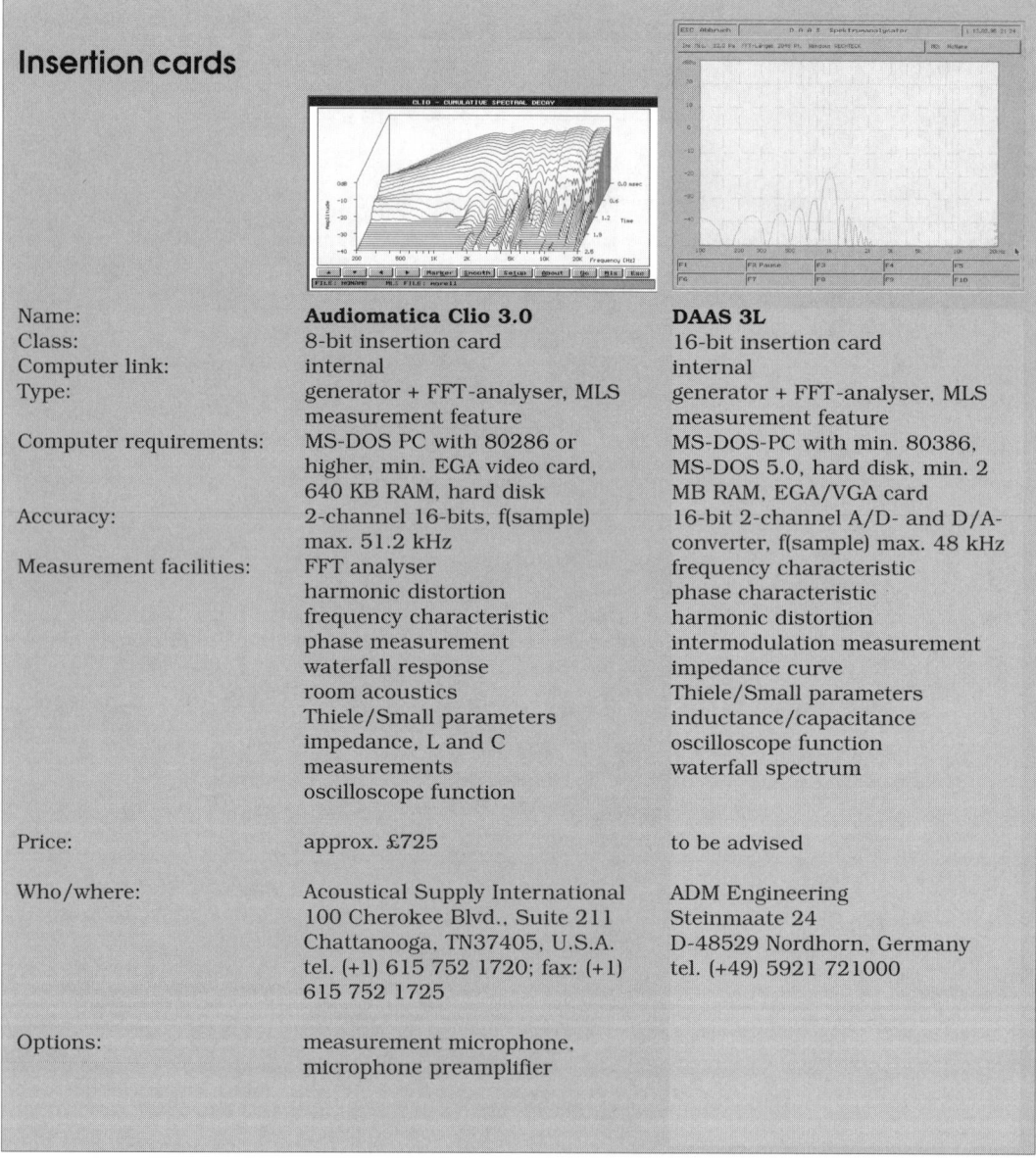

Insertion cards

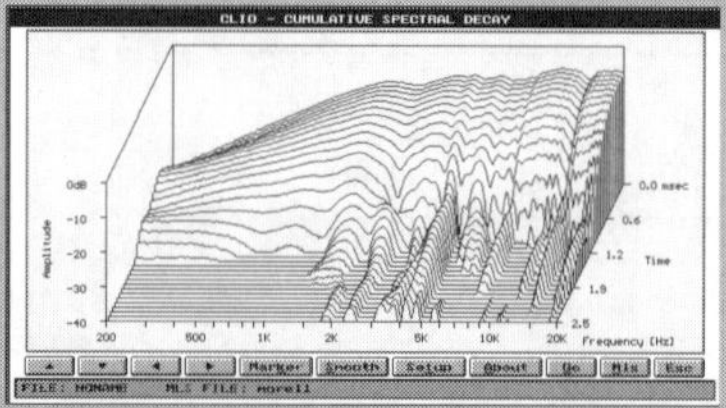

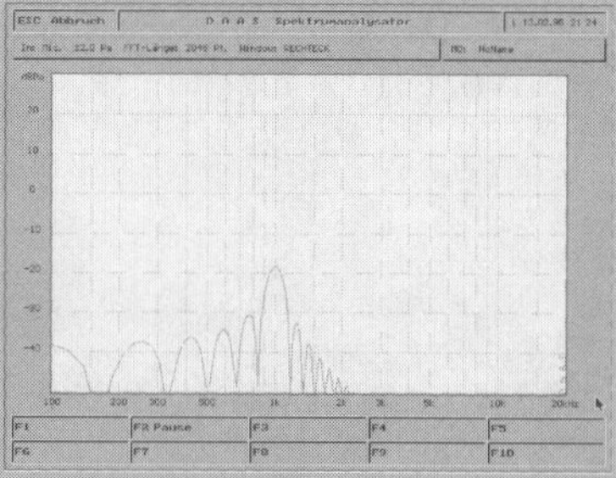

Name:	**Audiomatica Clio 3.0**	**DAAS 3L**
Class:	8-bit insertion card	16-bit insertion card
Computer link:	internal	internal
Type:	generator + FFT-analyser, MLS measurement feature	generator + FFT-analyser, MLS measurement feature
Computer requirements:	MS-DOS PC with 80286 or higher, min. EGA video card, 640 KB RAM, hard disk	MS-DOS-PC with min. 80386, MS-DOS 5.0, hard disk, min. 2 MB RAM, EGA/VGA card
Accuracy:	2-channel 16-bits, f(sample) max. 51.2 kHz	16-bit 2-channel A/D- and D/A-converter, f(sample) max. 48 kHz
Measurement facilities:	FFT analyser harmonic distortion frequency characteristic phase measurement waterfall response room acoustics Thiele/Small parameters impedance, L and C measurements oscilloscope function	frequency characteristic phase characteristic harmonic distortion intermodulation measurement impedance curve Thiele/Small parameters inductance/capacitance oscilloscope function waterfall spectrum
Price:	approx. £725	to be advised
Who/where:	Acoustical Supply International 100 Cherokee Blvd., Suite 211 Chattanooga, TN37405, U.S.A. tel. (+1) 615 752 1720; fax: (+1) 615 752 1725	ADM Engineering Steinmaate 24 D-48529 Nordhorn, Germany tel. (+49) 5921 721000
Options:	measurement microphone, microphone preamplifier	

KEF), to run accurate measurements on loudspeakers. A serious problem with loudspeaker measurements is that the test microphone picks up room reflections when a frequency response plot is being performed. This can be avoided, but only by carrying out the test in an anechoic room which unfortunately is very expensive and very large.

The gated sine-wave measurement involves driving the loudspeaker under test with a burst-shaped signal at a particular frequency. The system measures the level of the received signal during a short period only, during which the burst reaches the microphone for the first time (Fig. 17.3). All subsequent reflections are thereby ignored and can not cause measurement errors. In this way, a fairly accurate frequency response measurement can be performed on loudspeakers in an ordinary room. An example of a modern measurement system based on this principle is LMS—see next page.

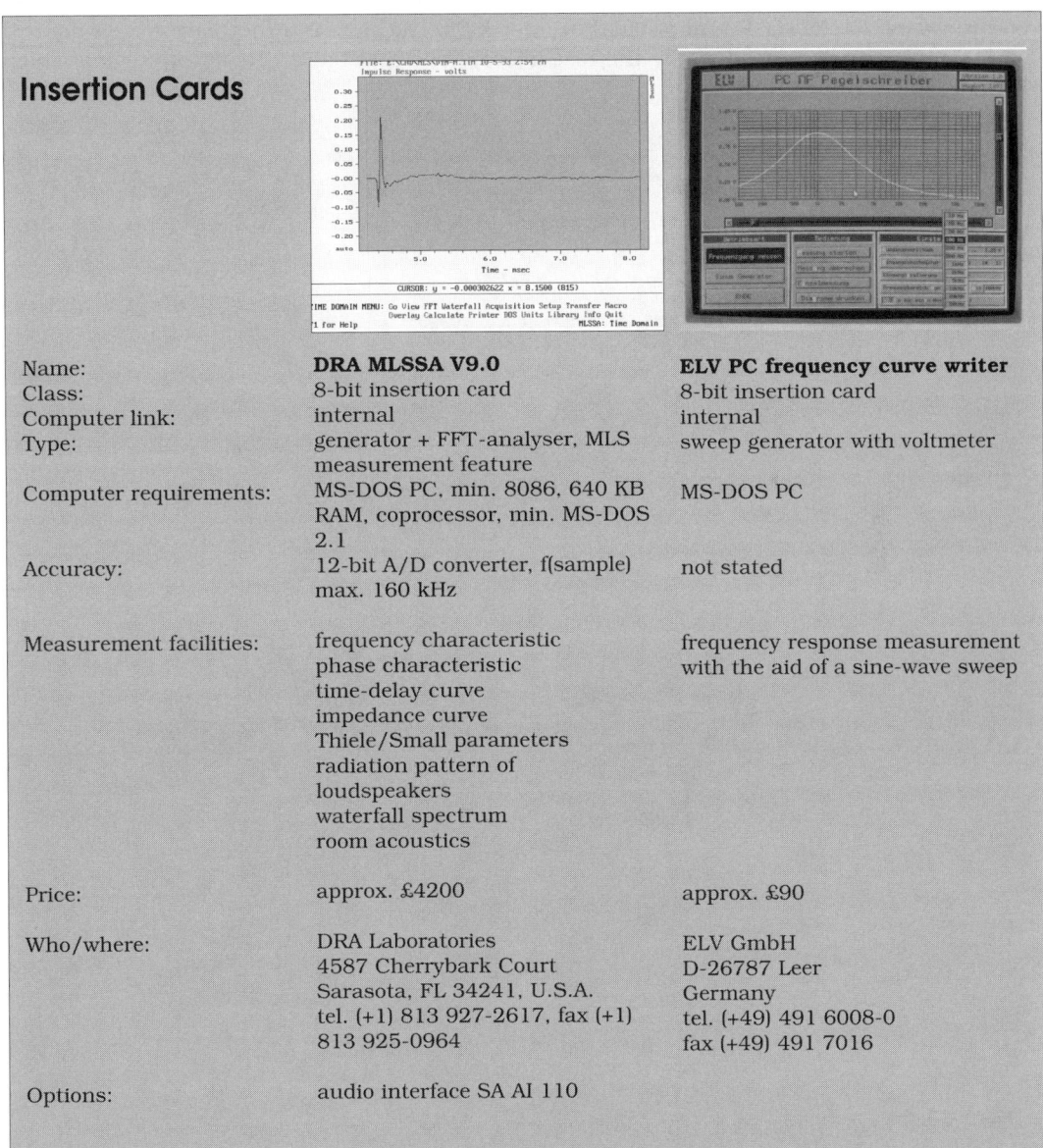

Insertion Cards

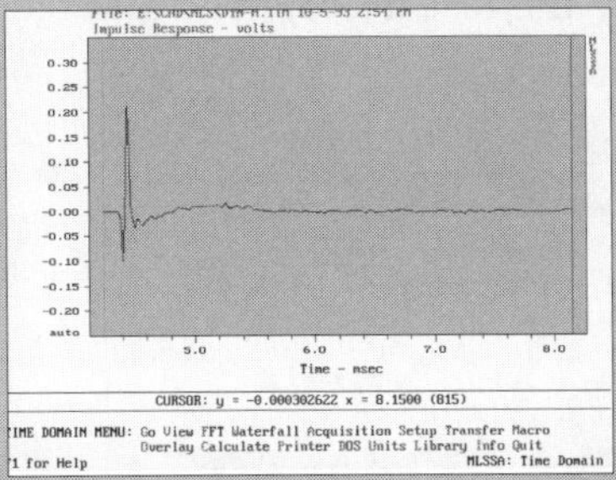

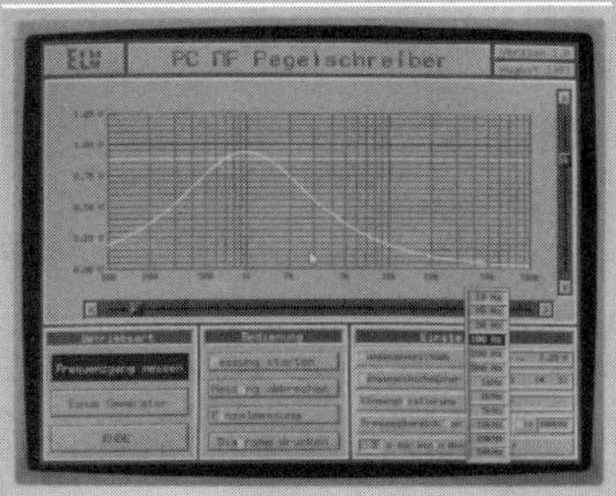

Name:	**DRA MLSSA V9.0**	**ELV PC frequency curve writer**
Class:	8-bit insertion card	8-bit insertion card
Computer link:	internal	internal
Type:	generator + FFT-analyser, MLS measurement feature	sweep generator with voltmeter
Computer requirements:	MS-DOS PC, min. 8086, 640 KB RAM, coprocessor, min. MS-DOS 2.1	MS-DOS PC
Accuracy:	12-bit A/D converter, f(sample) max. 160 kHz	not stated
Measurement facilities:	frequency characteristic phase characteristic time-delay curve impedance curve Thiele/Small parameters radiation pattern of loudspeakers waterfall spectrum room acoustics	frequency response measurement with the aid of a sine-wave sweep
Price:	approx. £4200	approx. £90
Who/where:	DRA Laboratories 4587 Cherrybark Court Sarasota, FL 34241, U.S.A. tel. (+1) 813 927-2617, fax (+1) 813 925-0964	ELV GmbH D-26787 Leer Germany tel. (+49) 491 6008-0 fax (+49) 491 7016
Options:	audio interface SA AI 110	

Pulse measurements

Another method which was frequently used in the past (in particular, by KEF), involved measuring the pulse response of a system. A very short pulse (called dirac) in principle contains an infinitely wide frequency spectrum. With the aid of an FFT analysis, the shape of the measured response to the pulse allows the transfer function of the system to be determined. In principle, this method allows the effects of the room to be virtually eliminated by an FFT analysis only on the first pulse only received by the test microphone. A number of measurement systems including Clio (page 206) and PC Audiolab (page 209) are capable of running pulse response tests alongside other measurement methods.

MLS systems

The acronym MLS stands for Maximum Length Sequence, and refers to a smart FFT method introduced by Douglas D. Rife about in the mid-1980s. The basis is a short, accurately defined noise signal. This is picked up by the test microphone and fed to the FFT analyser, which

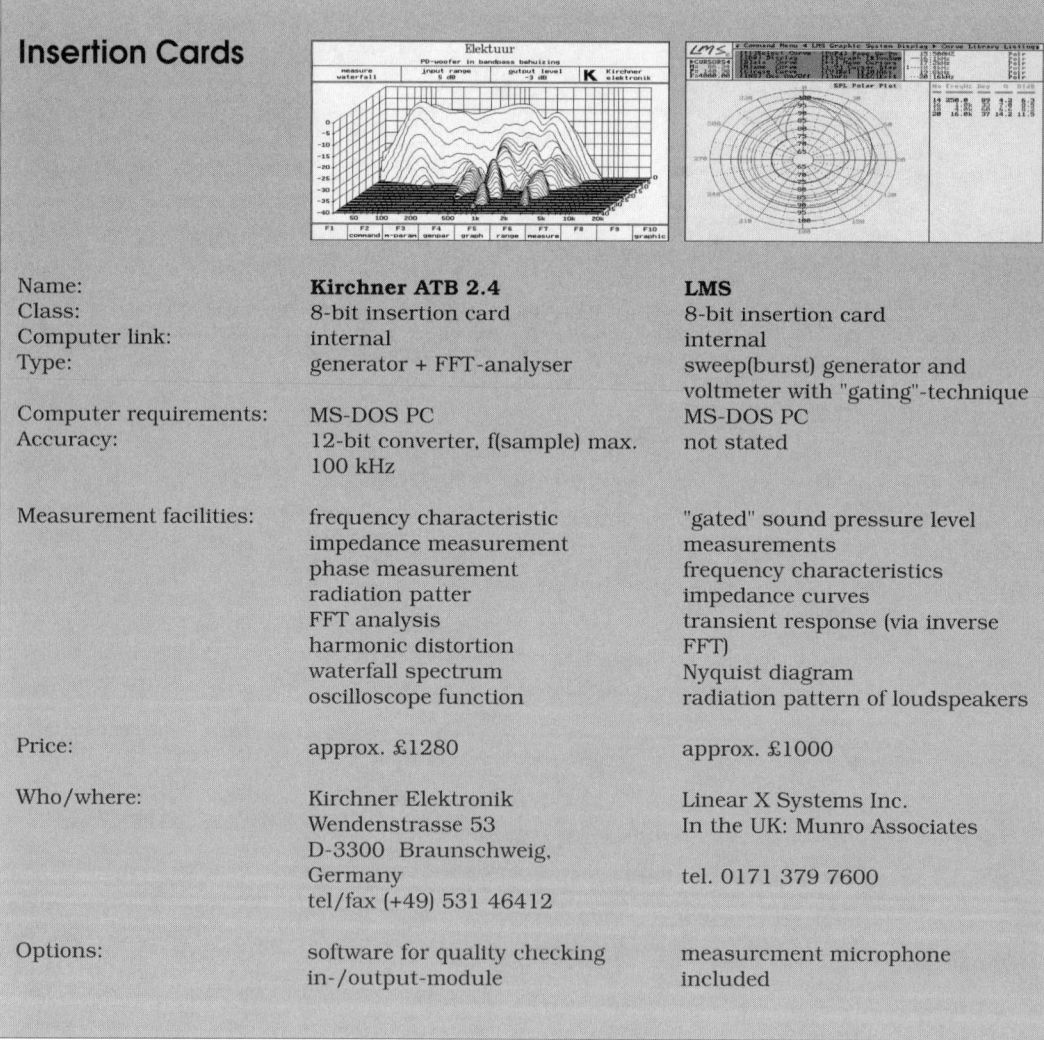

Insertion Cards

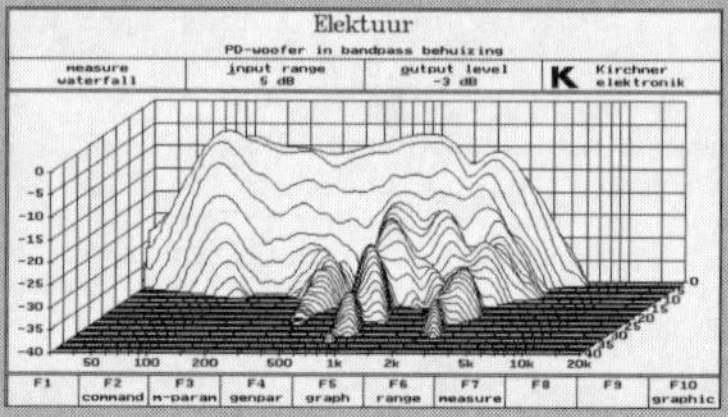

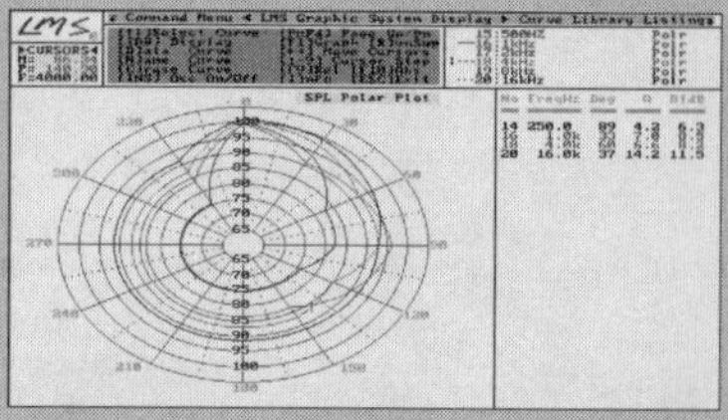

Name:	**Kirchner ATB 2.4**	**LMS**
Class:	8-bit insertion card	8-bit insertion card
Computer link:	internal	internal
Type:	generator + FFT-analyser	sweep(burst) generator and voltmeter with "gating"-technique
Computer requirements:	MS-DOS PC	MS-DOS PC
Accuracy:	12-bit converter, f(sample) max. 100 kHz	not stated
Measurement facilities:	frequency characteristic impedance measurement phase measurement radiation patter FFT analysis harmonic distortion waterfall spectrum oscilloscope function	"gated" sound pressure level measurements frequency characteristics impedance curves transient response (via inverse FFT) Nyquist diagram radiation pattern of loudspeakers
Price:	approx. £1280	approx. £1000
Who/where:	Kirchner Elektronik Wendenstrasse 53 D-3300 Braunschweig, Germany tel/fax (+49) 531 46412	Linear X Systems Inc. In the UK: Munro Associates tel. 0171 379 7600
Options:	software for quality checking in-/output-module	measurcment microphone included

'dissects' it, and computes the pulse response of the system. The short length of the noise 'burst' and the fact that the FFT window can be set for the calculation of, for example, the frequency response, it is possible to virtually eliminate the effect of the room. The de facto standard in this field is the MLSSA system (page 207). This plug-in card has a resolution of 12 bits which is quite sufficient for measurements on loudspeakers. Loudspeakers manufacturers consider MLSSA to be a standard.

Product overview

The four test methods described above cover most of the currently available audio measurement systems. Depending on individual requirements, there is a choice from many systems,

Software

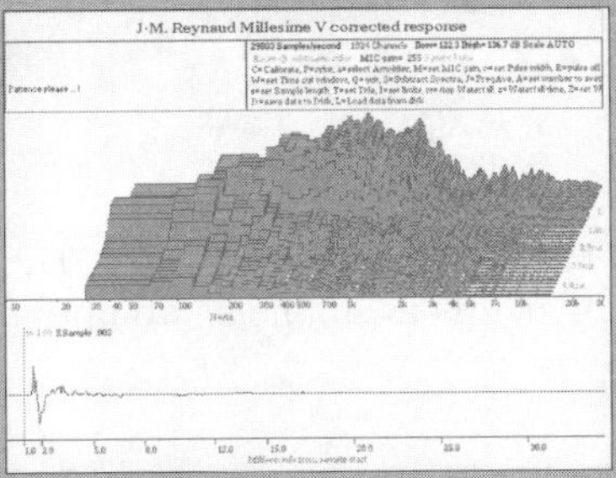

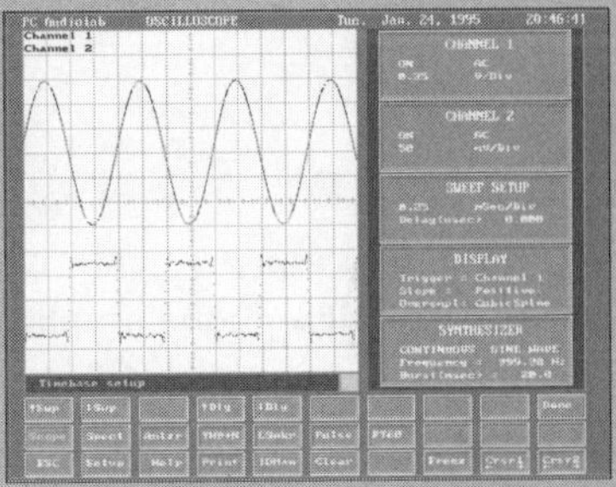

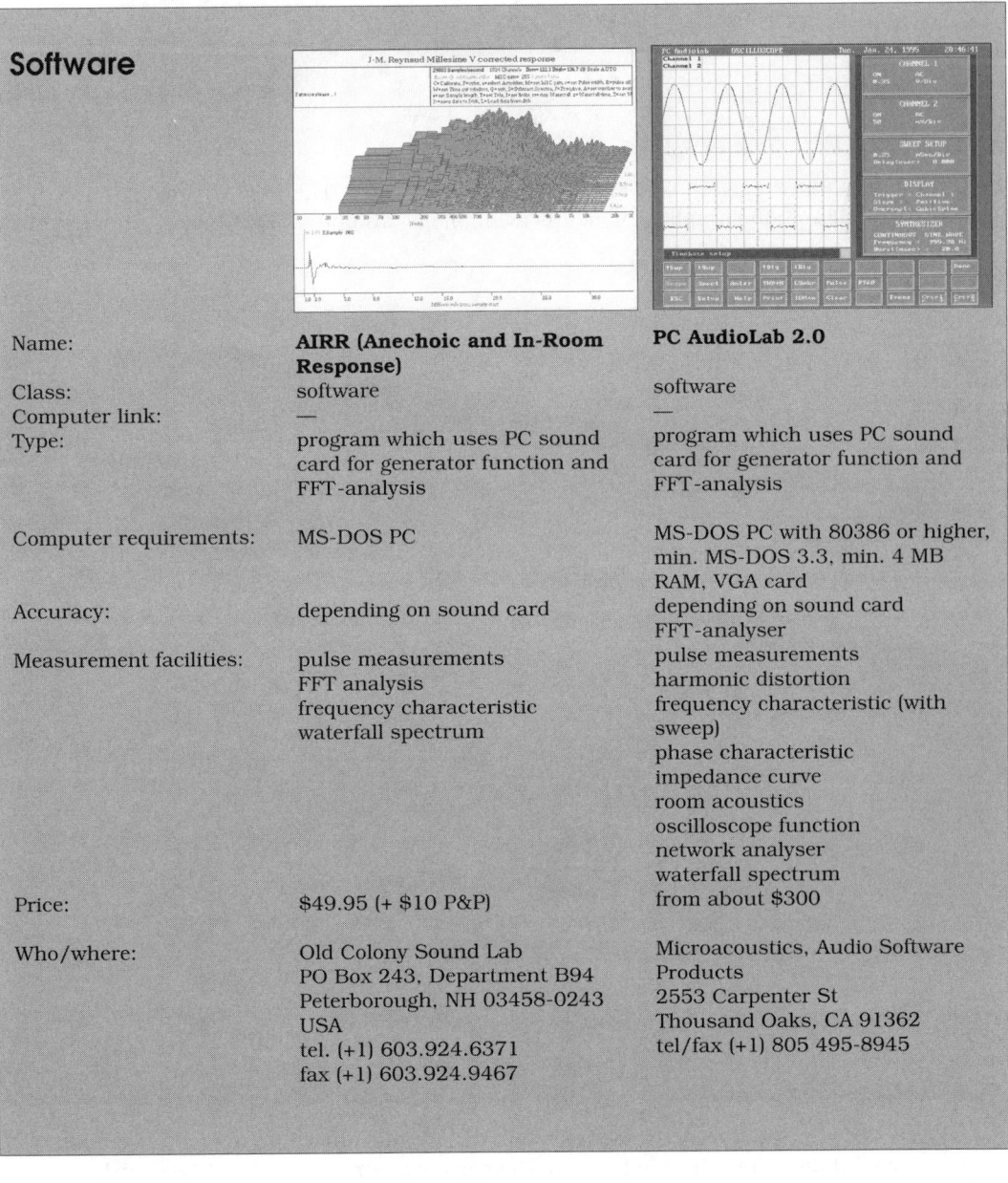

Name:	**AIRR (Anechoic and In-Room Response)**	**PC AudioLab 2.0**
Class:	software	software
Computer link:	—	—
Type:	program which uses PC sound card for generator function and FFT-analysis	program which uses PC sound card for generator function and FFT-analysis
Computer requirements:	MS-DOS PC	MS-DOS PC with 80386 or higher, min. MS-DOS 3.3, min. 4 MB RAM, VGA card
Accuracy:	depending on sound card	depending on sound card FFT-analyser
Measurement facilities:	pulse measurements FFT analysis frequency characteristic waterfall spectrum	pulse measurements harmonic distortion frequency characteristic (with sweep) phase characteristic impedance curve room acoustics oscilloscope function network analyser waterfall spectrum
Price:	$49.95 (+ $10 P&P)	from about $300
Who/where:	Old Colony Sound Lab PO Box 243, Department B94 Peterborough, NH 03458-0243 USA tel. (+1) 603.924.6371 fax (+1) 603.924.9467	Microacoustics, Audio Software Products 2553 Carpenter St Thousand Oaks, CA 91362 tel/fax (+1) 805 495-8945

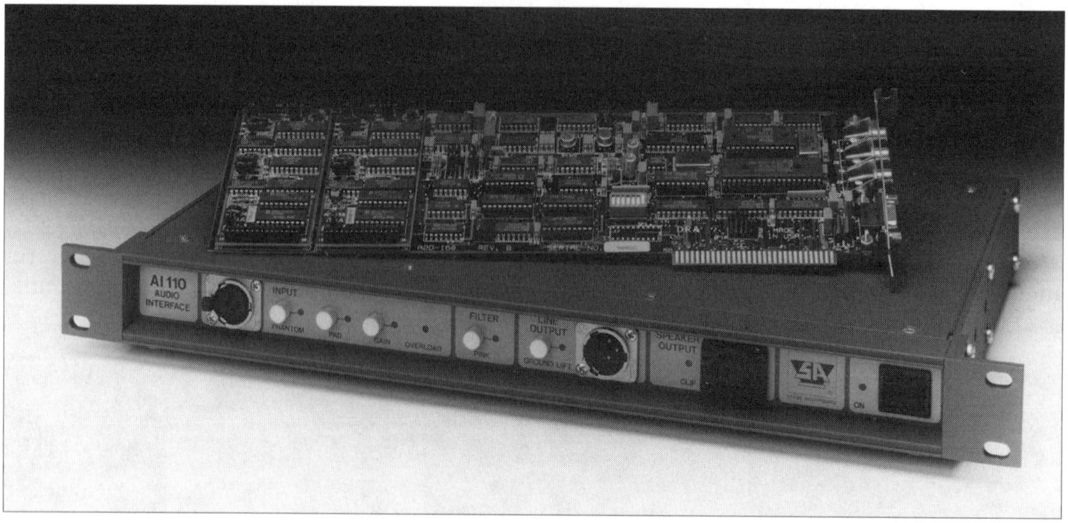

Fig. 17.4. The industry standard MLSSA measurement card. The card is shown on top of a power/microphone amplifier which is available as an option from Sage Company.

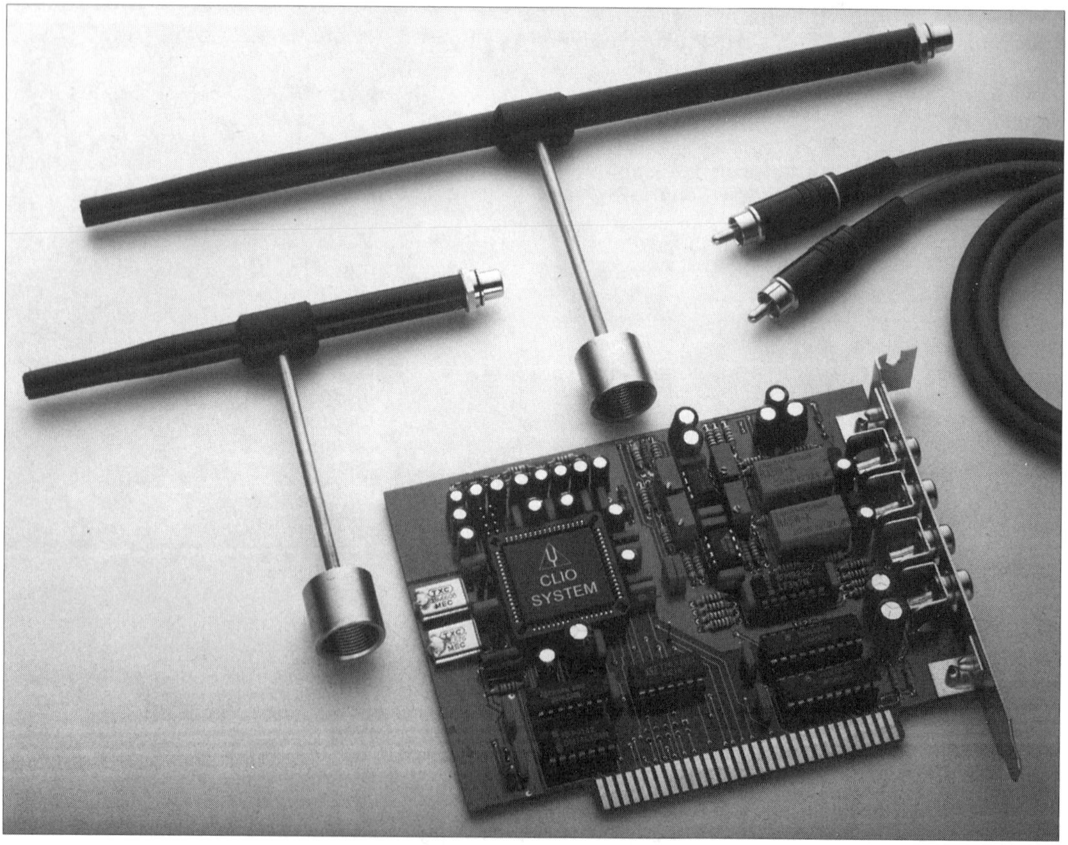

Fig. 17.5. A modern measurement system called Clio, shown with a couple of test microphone. Note the small size of the card compared with the MLSSA card.

12 of which are summarized on pages 204–209. Such a brief overview can, of course, not possibly cover all available products in the field. Prices start at about £85 ($ 130) for a simple frequency response plotter, while those of systems capable of performing more serious measurements, including FFT analysis and MLSS-based testing, start at about £350 ($ 550).

Two items from the overview, AIRR and PC Audiolab deserve special attention. Normally, a measurement system consists of a card, or a box with a connection cable to the computer. This makes the system fairly expensive. To reduce cost, AIRR and PC Audiolab make use of the sound card in the PC. Today's generation of Soundblaster and Adlib compatible cards offer 16-bit sound recording and playback facilities at prices just under £100 ($ 160). Backed up by the appropriate software, such cards offer measurement options comparable to those of a system like MLSSA, at much lower cost. However, these developments are in their infancy (1995) and it may take some time for affordable software to become available

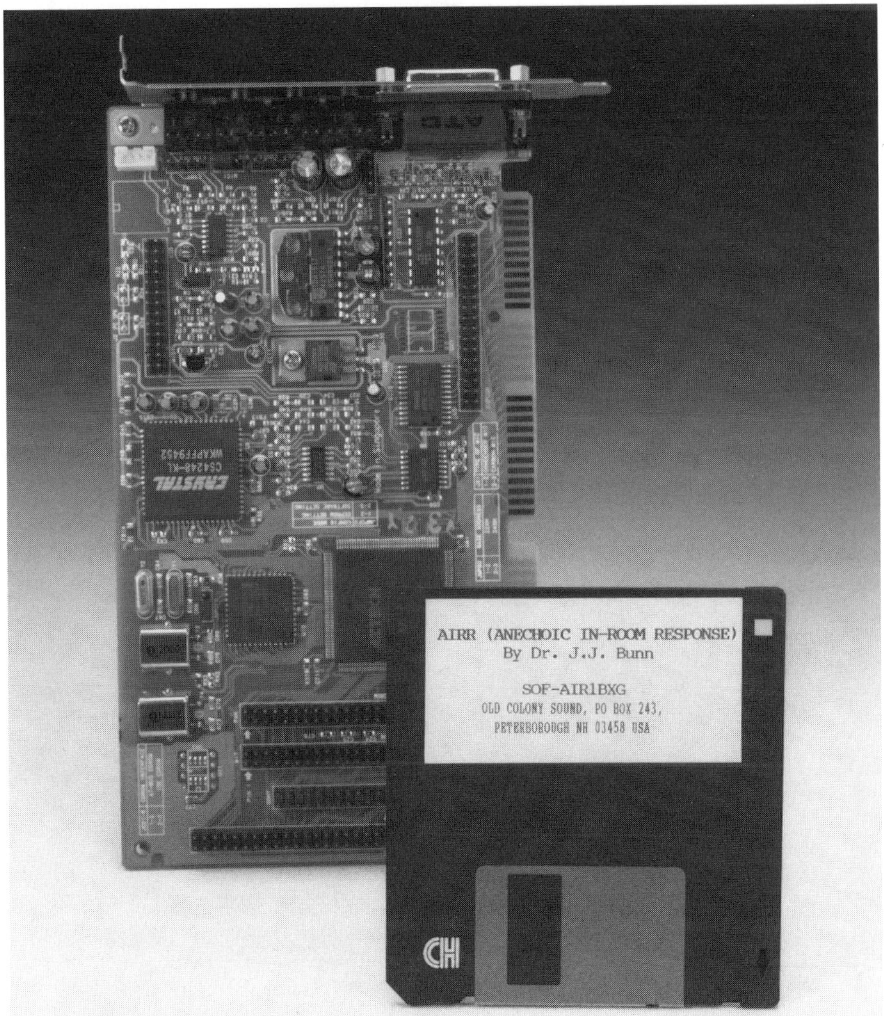

Fig. 17.6. Nowadays it is possible to carry out pulse response and frequency response plotting, as well as distortion measurements on a 'shoestring' with the aid of programs like AIRR and PC Audiolabs, which use a computer's internal sound card.

which offers the same facilities as MLSSA. Readers keen on having a first go at this challenging field could do worse than buying the inexpensive AIRR software from Old Colony Sound Lab (for address, see page 209).

APPENDIX

PCB track layouts

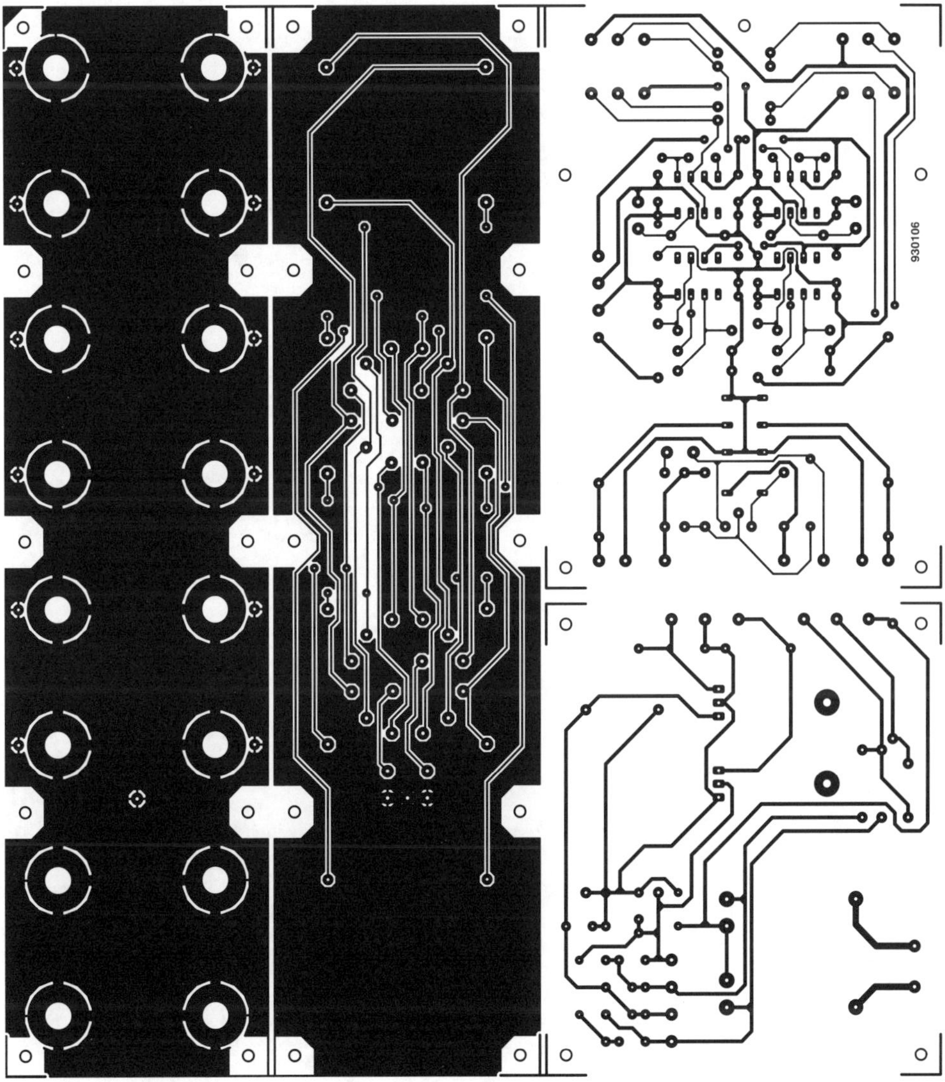

Fig. 1.4.(b). Track layout of the PCB for Preamplifier I (reduced to 70.7% = A3→A4).

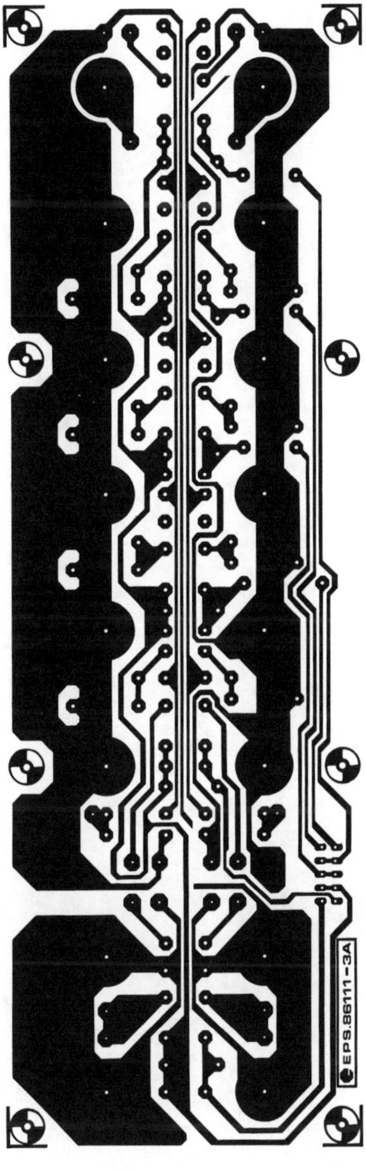

Fig. 2.3.(b). Track layout of the PCB for the busboard (reduced to 70.7% = A3→A4).

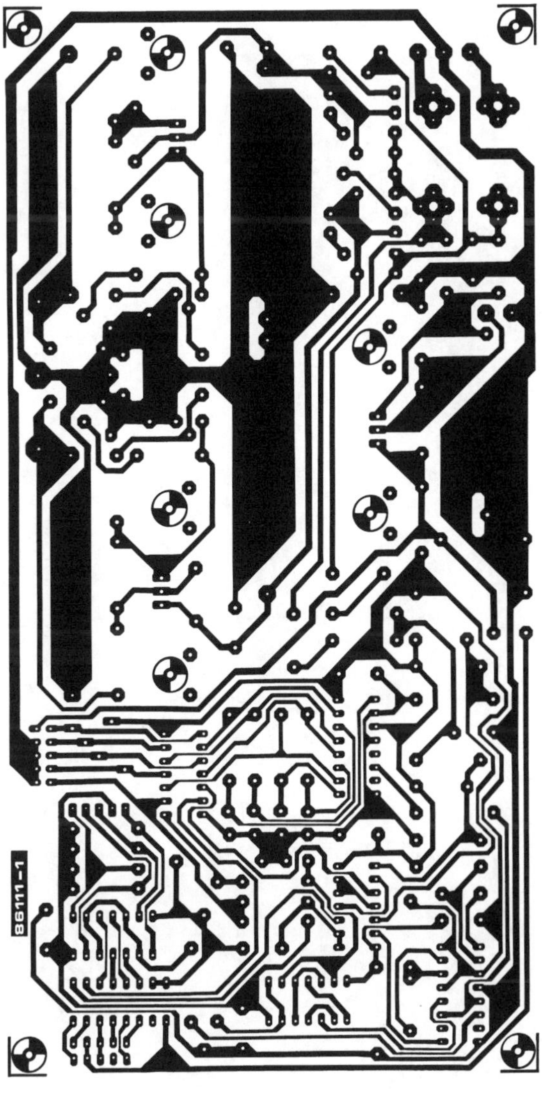

Fig. 2.8.(b). Track layout of the PCB for the power supplies and relay control circuits. (Reduced to 70.7% = A3→A4).

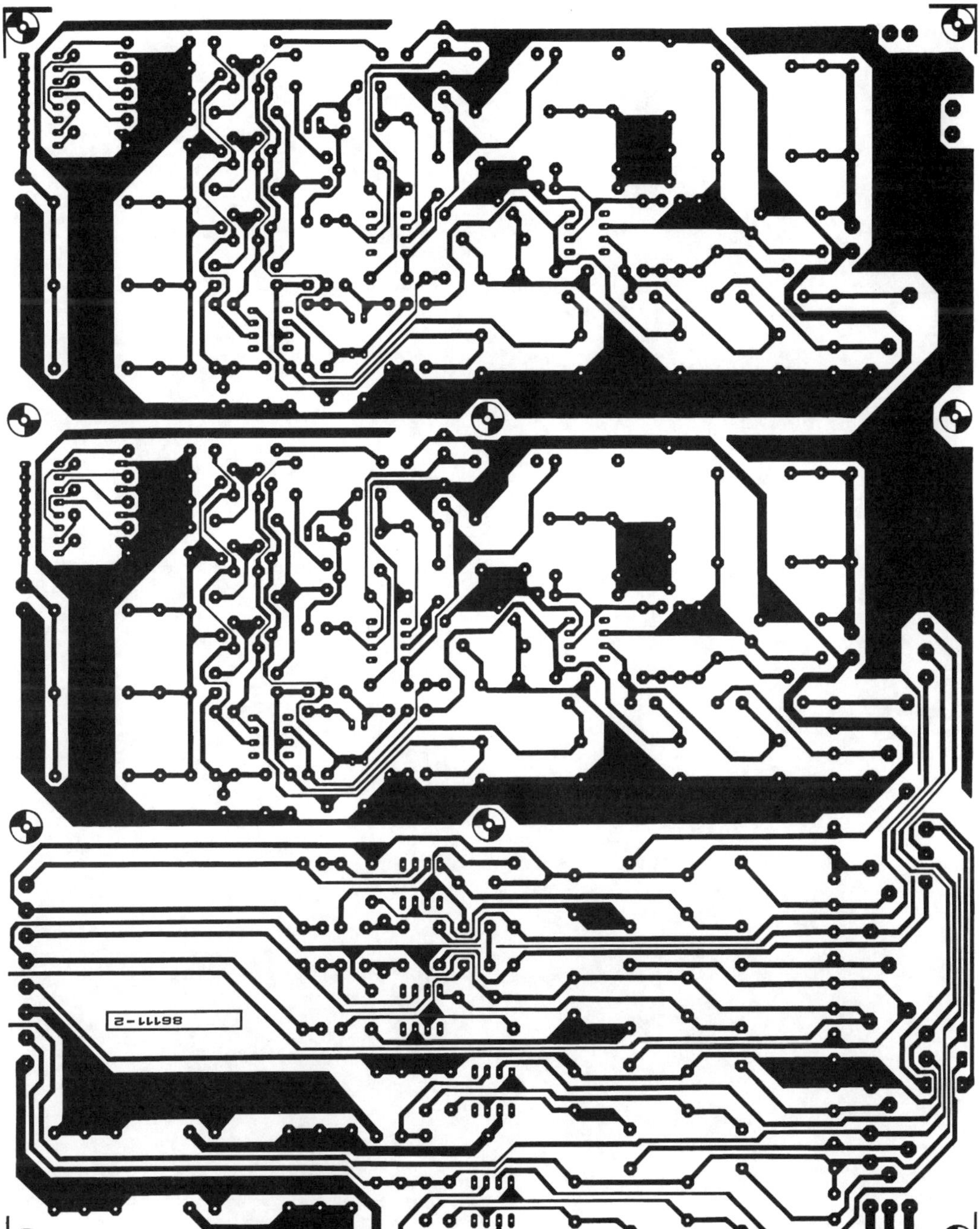

Fig. 2.14.(b). Track layout of the motherboard. (Reduced to 70.7% = A3→A4).

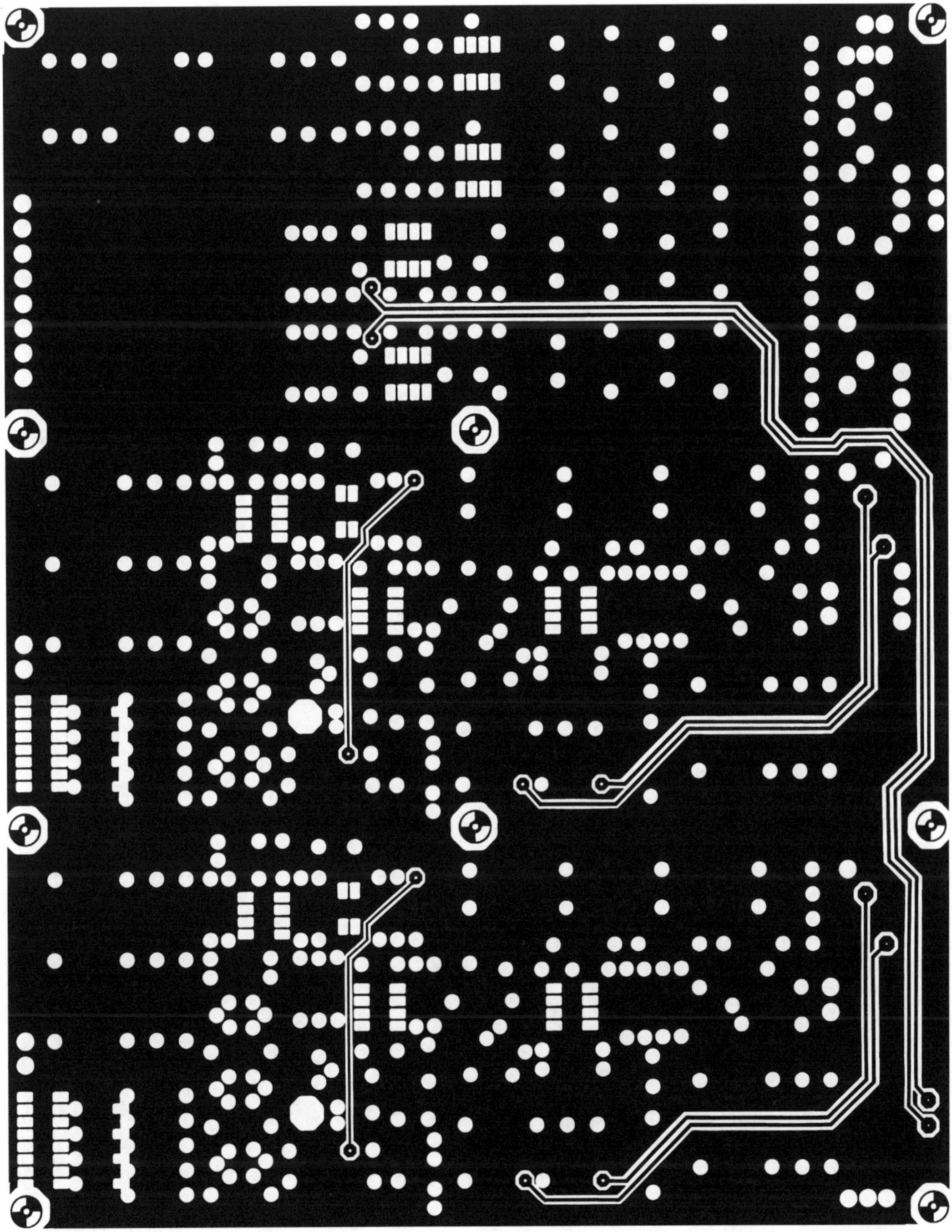

Fig. 2.14.(c). Component overlay for the motherboard. (Reduced to 70.7% = A3→A4).

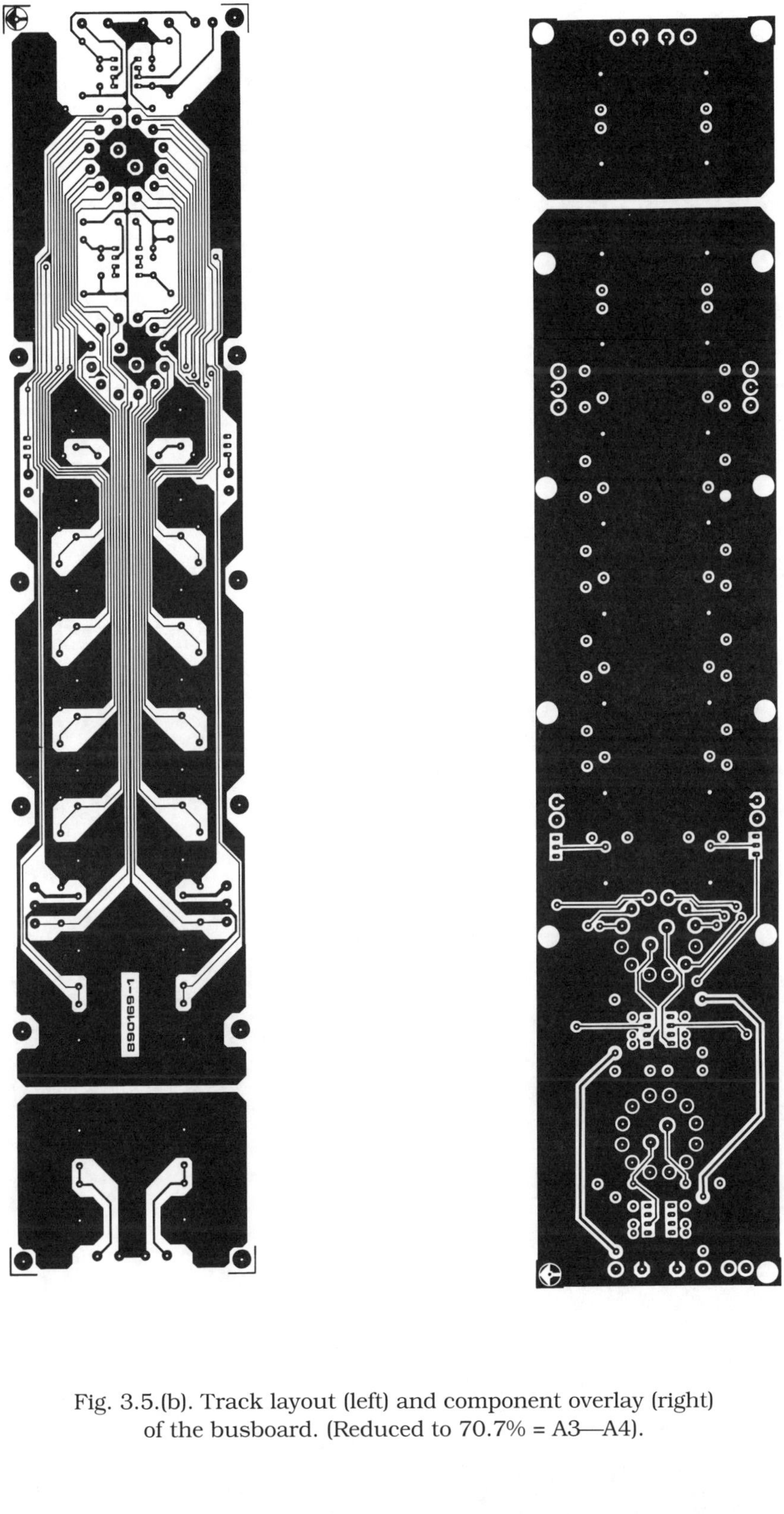

Fig. 3.5.(b). Track layout (left) and component overlay (right) of the busboard. (Reduced to 70.7% = A3—A4).

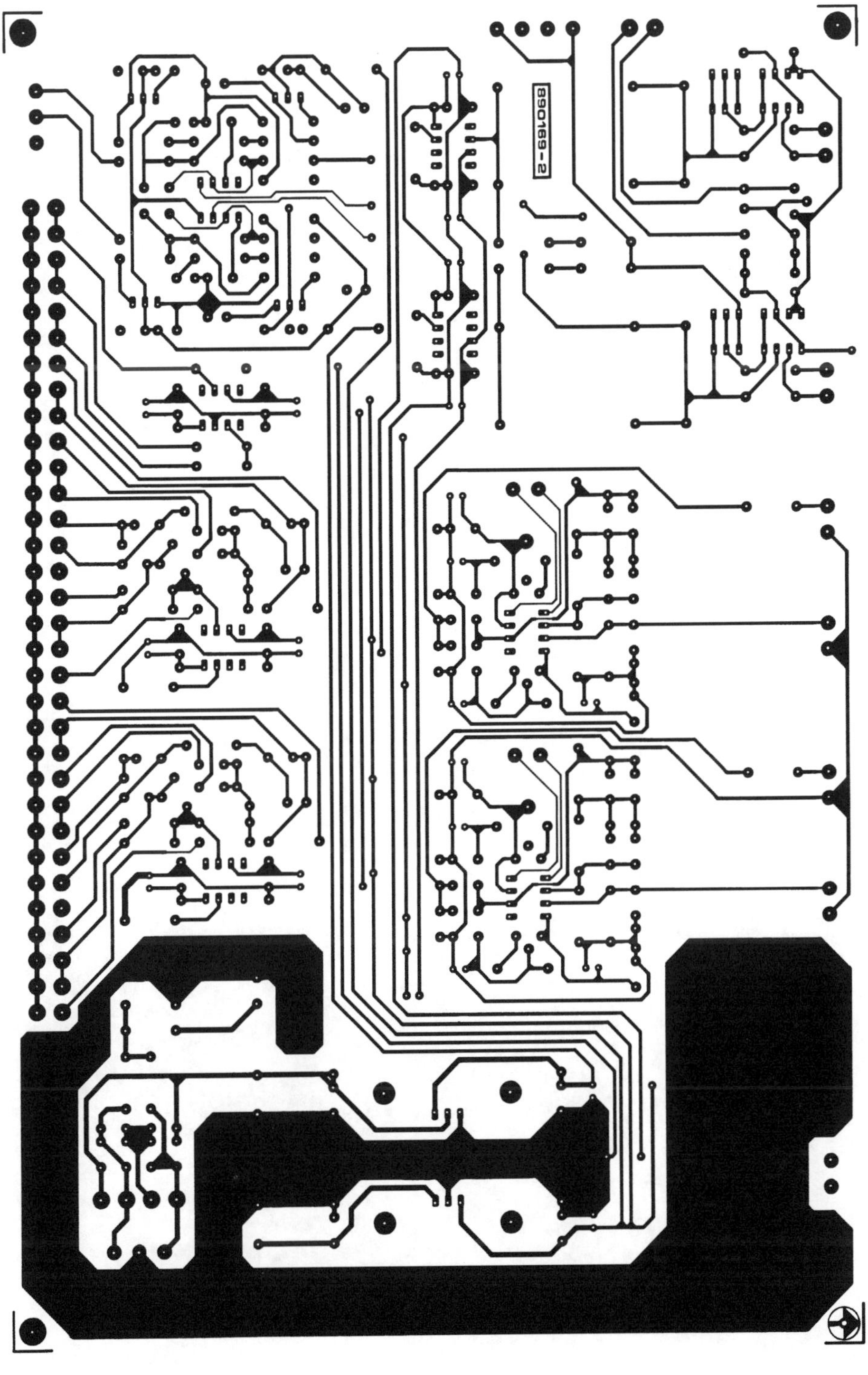

Fig. 3.6.(b). Track layout of the motherboard.

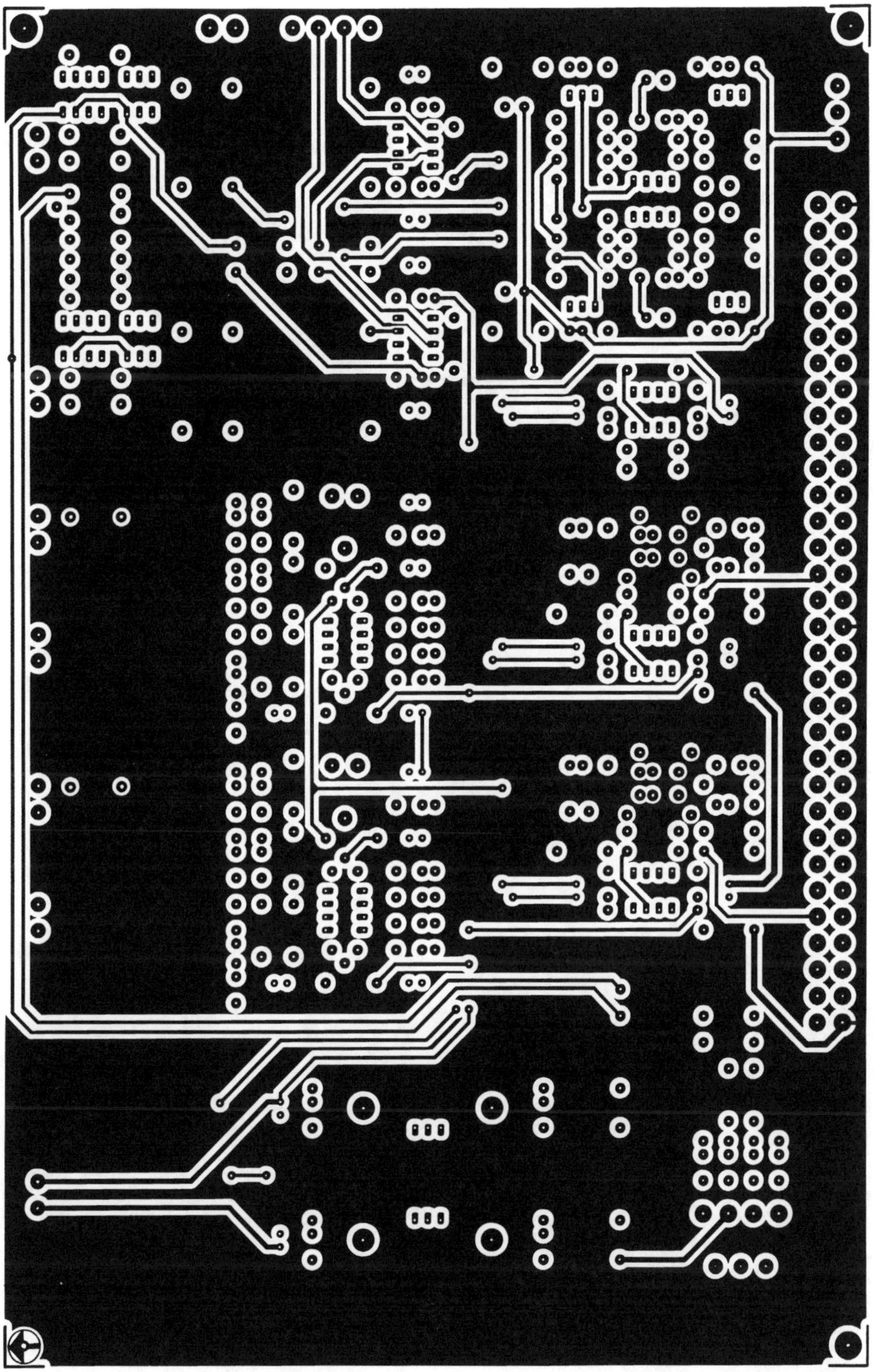

Fig. 3.6.(c). Component overlay for the motherboard.

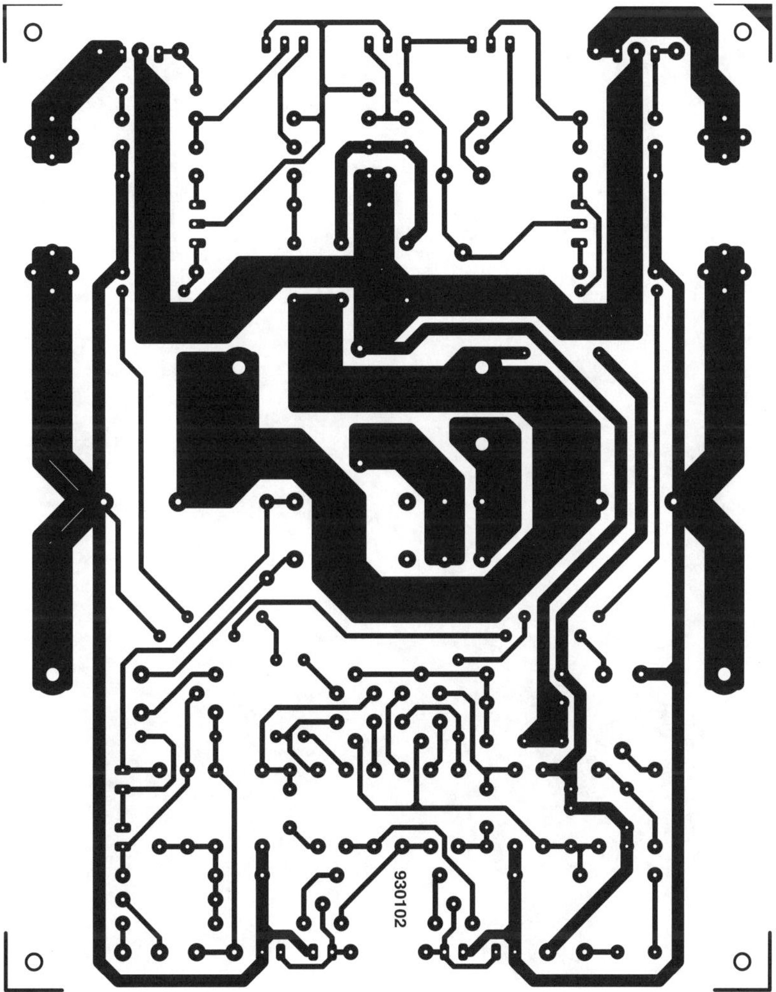

Fig. 4.3.(b). Track layout of the PCB for the 60 W power amplifier.

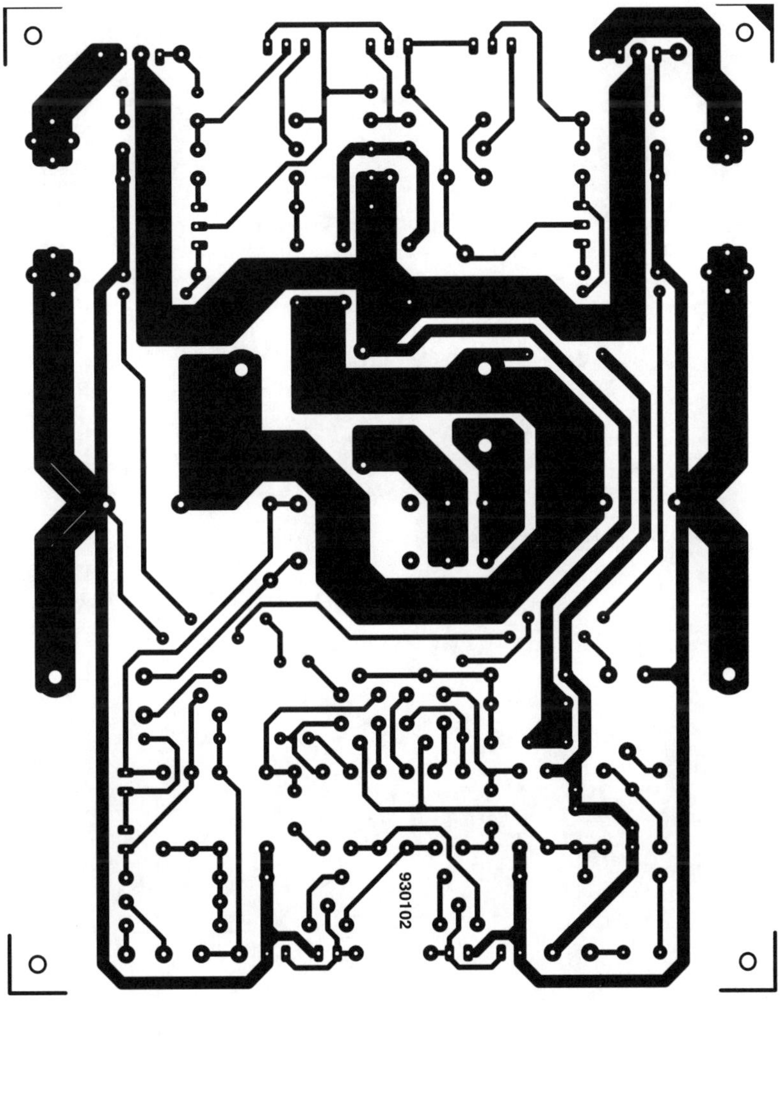

Fig. 5.3.(b). Track layout of the PCB for the 90 W power amplifier.

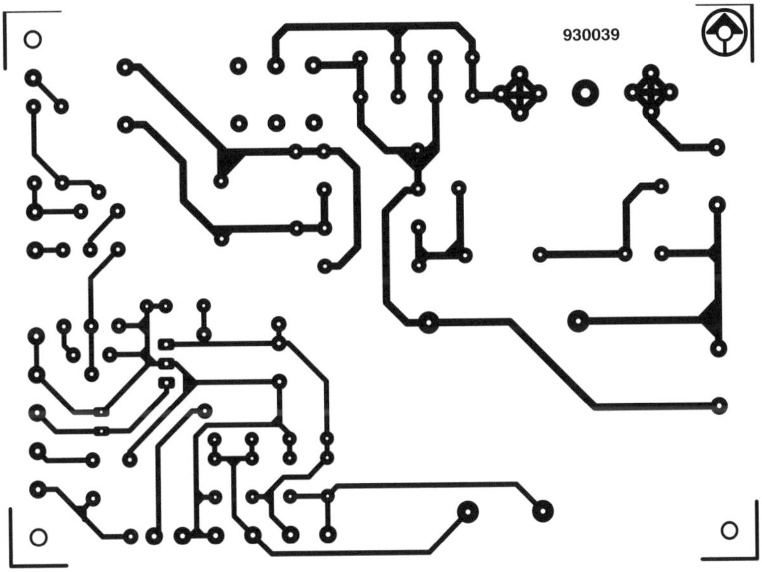

Fig. 6.6.(b). Track layout of the PCB for the auxiliary circuit.

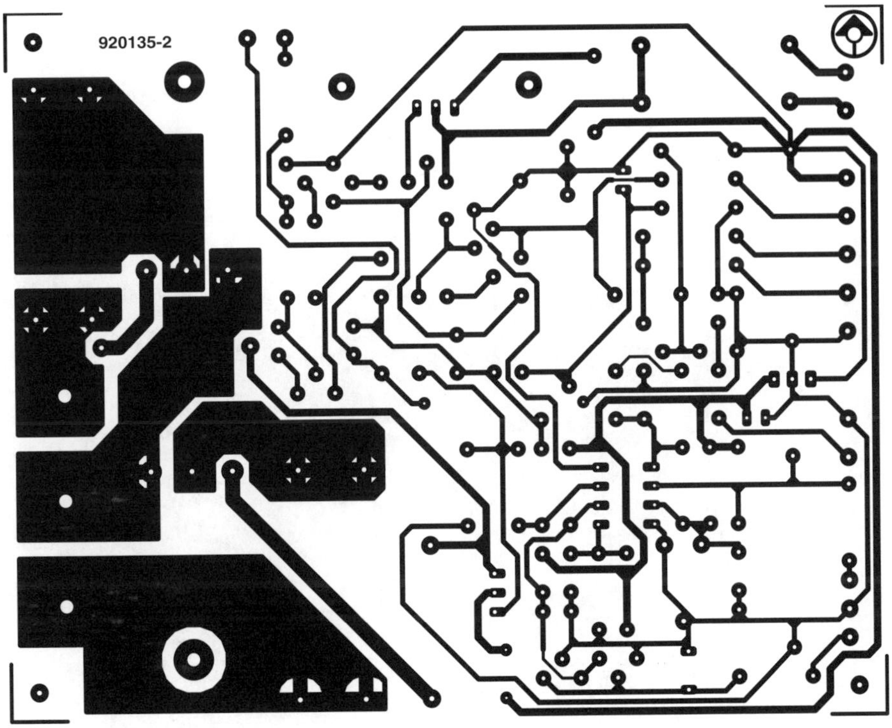

Fig. 6.7.(b). Track layout of the PCB for the 100 W amplifier.

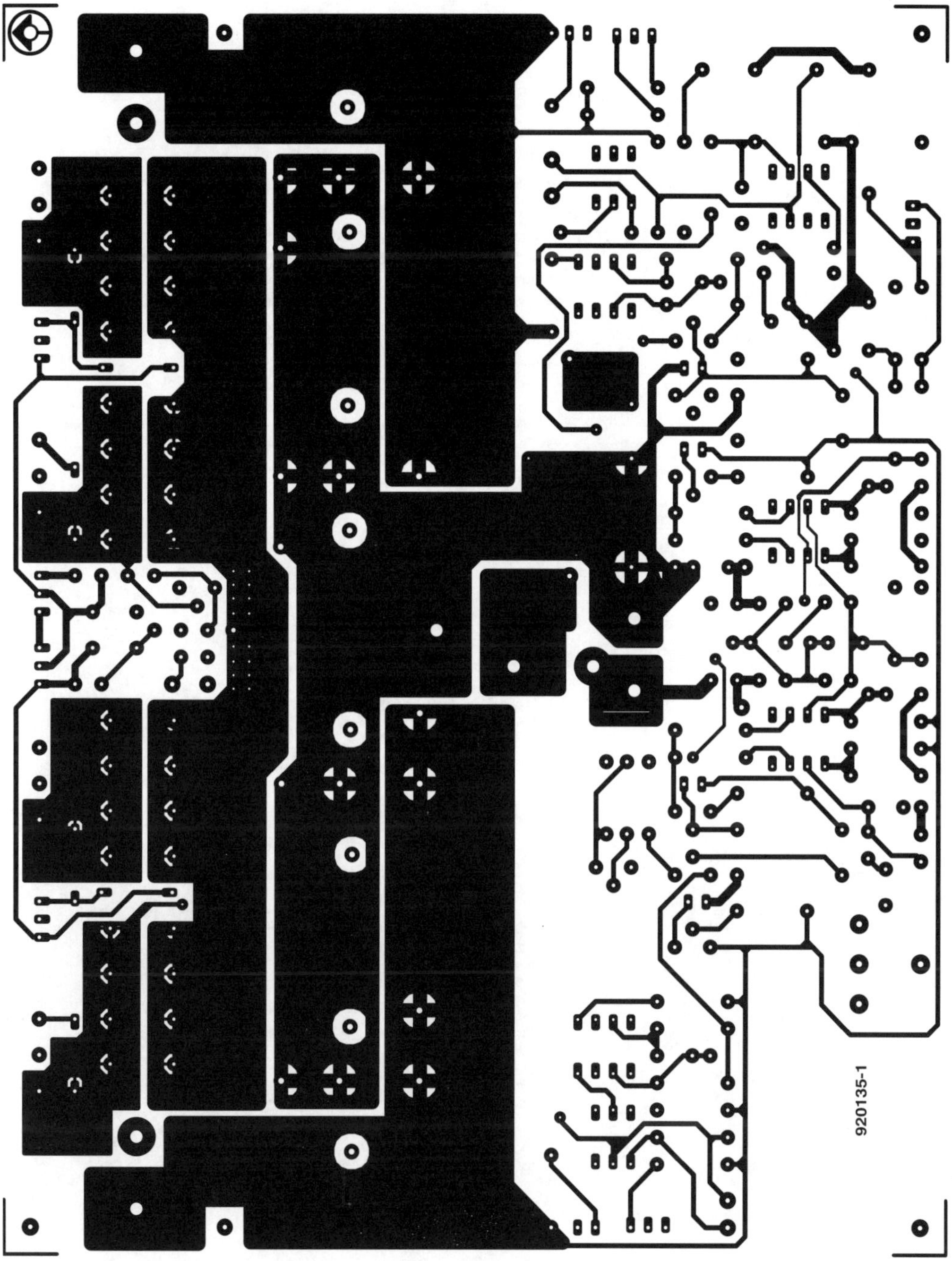

Fig. 6.8.(b). Track layout of the PCB for the protection circuit.

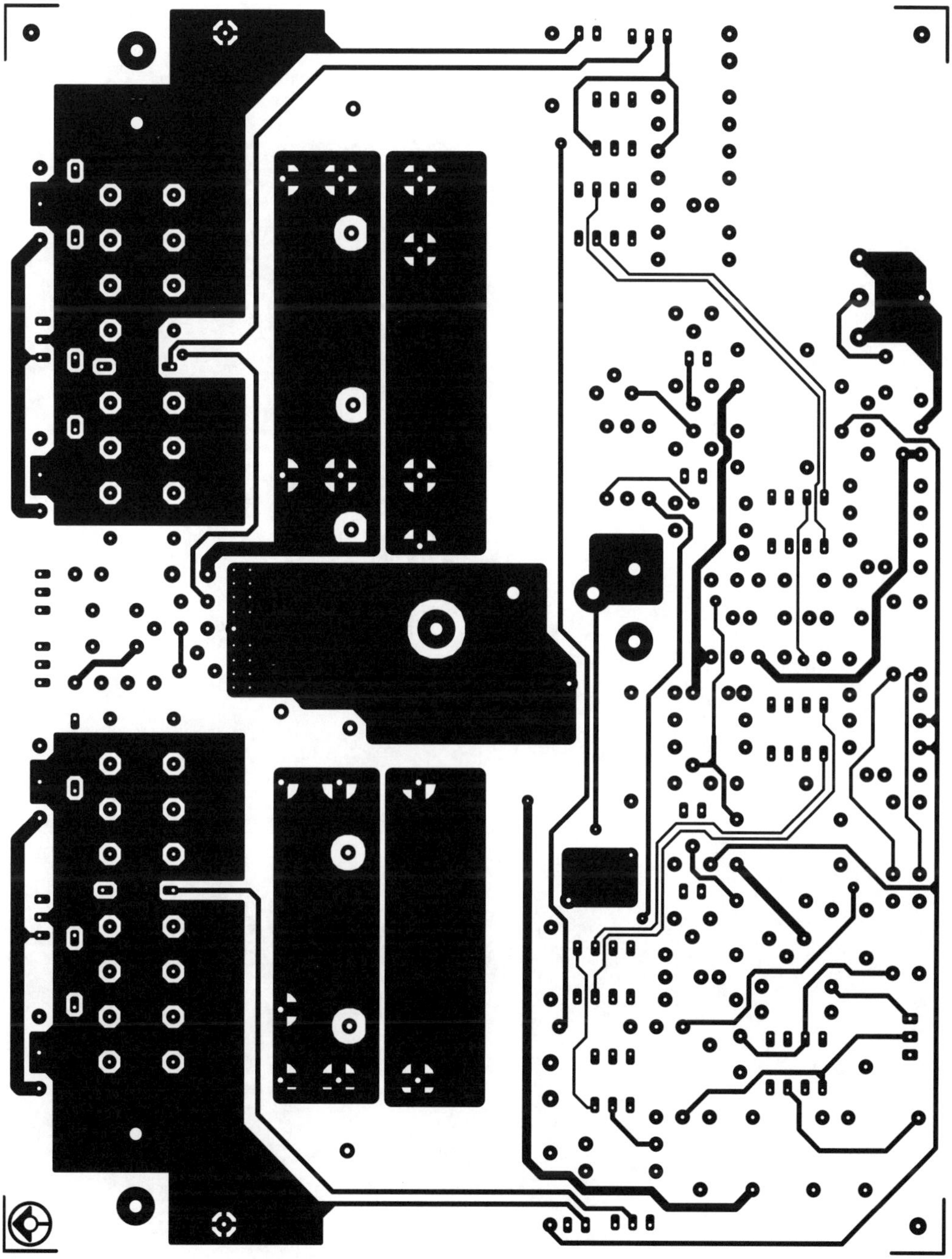

Fig. 6.8. (c). Component overlay of the PCB for the protection circuits.

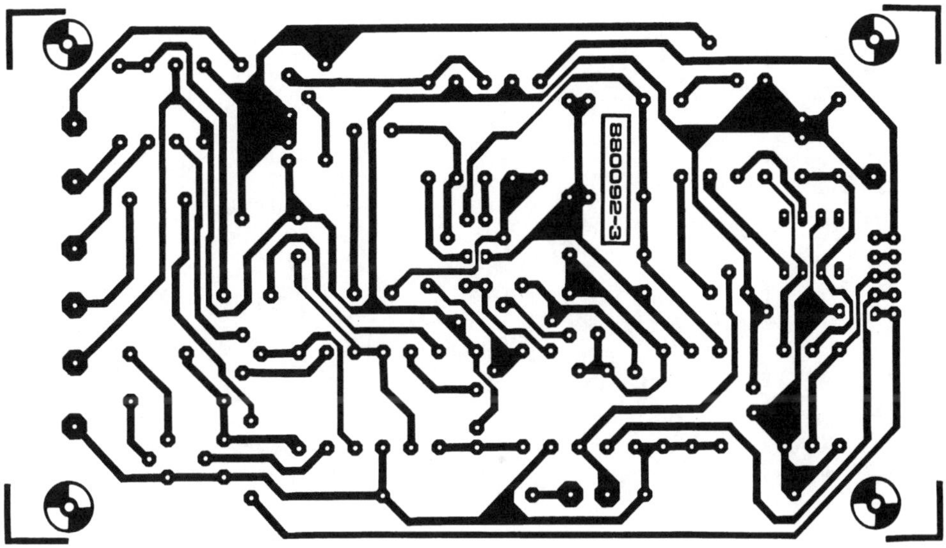

Fig. 7.5.(b). Track layout of the PCB for the protection circuits.

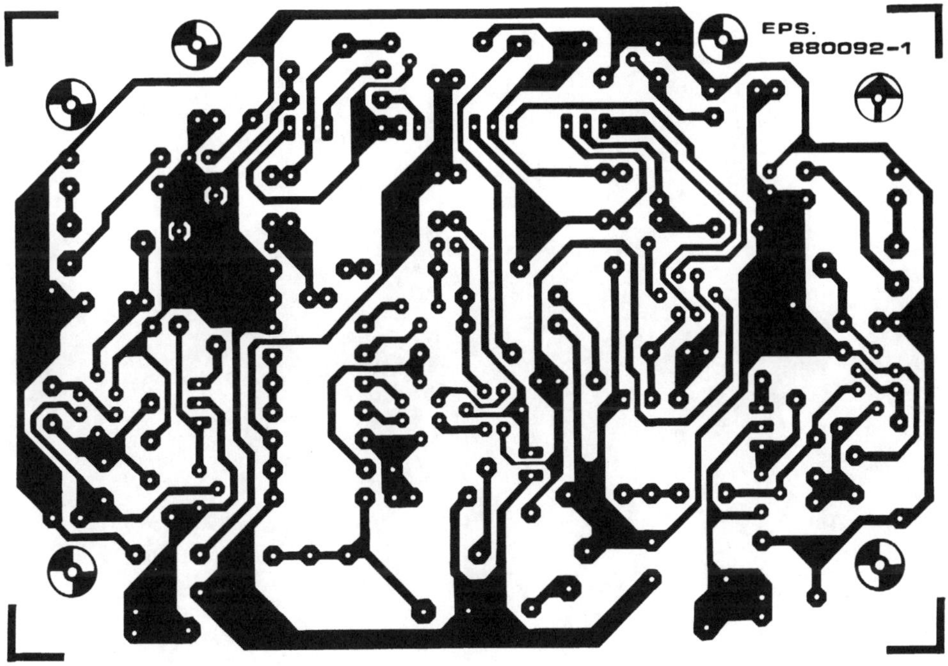

Fig. 7.6.(b). Track layout of the PCB for the voltage amplifier.

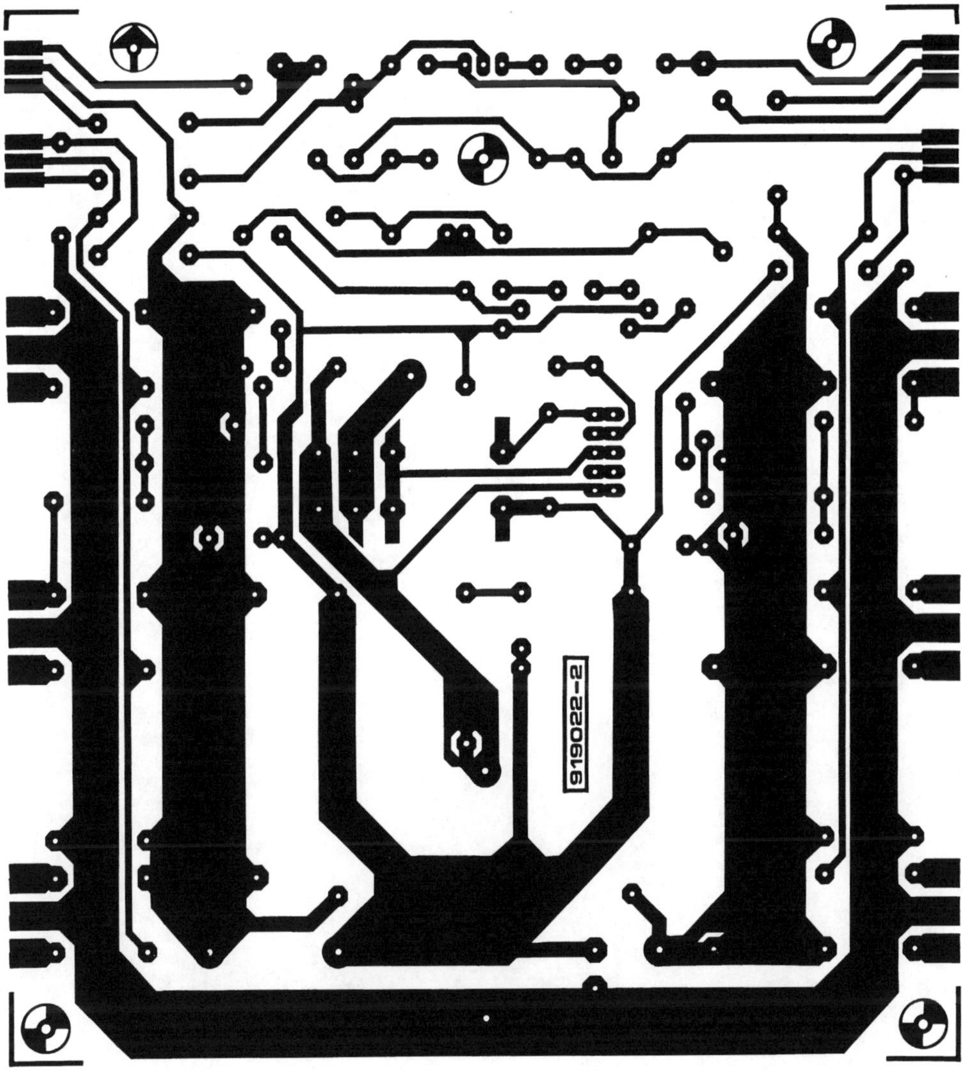

Fig. 7.7.(b). Track layout of the PCB for the current amplifier.

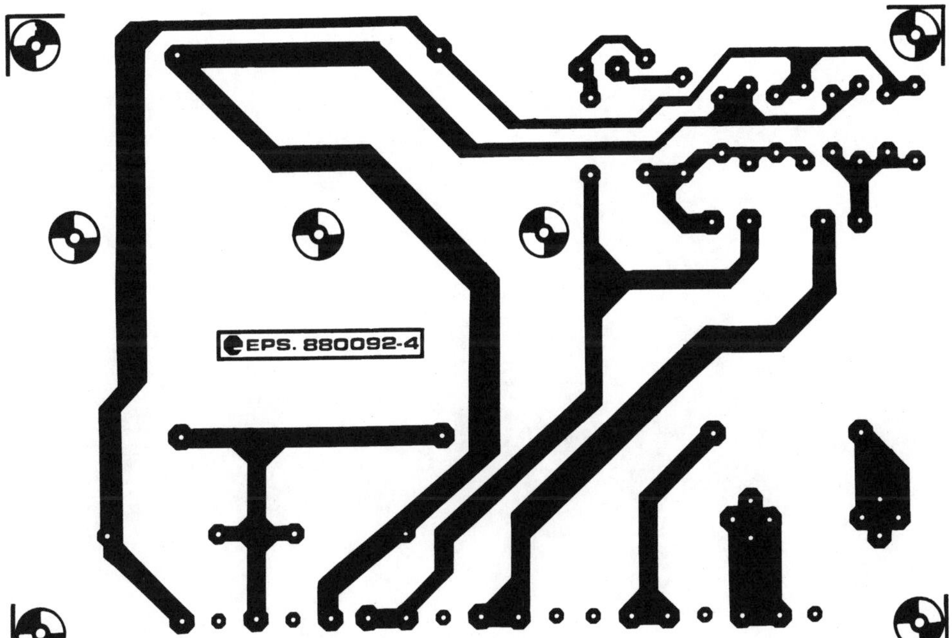

Fig. 7.8.(b). Track layout of the PCB for the power supply.

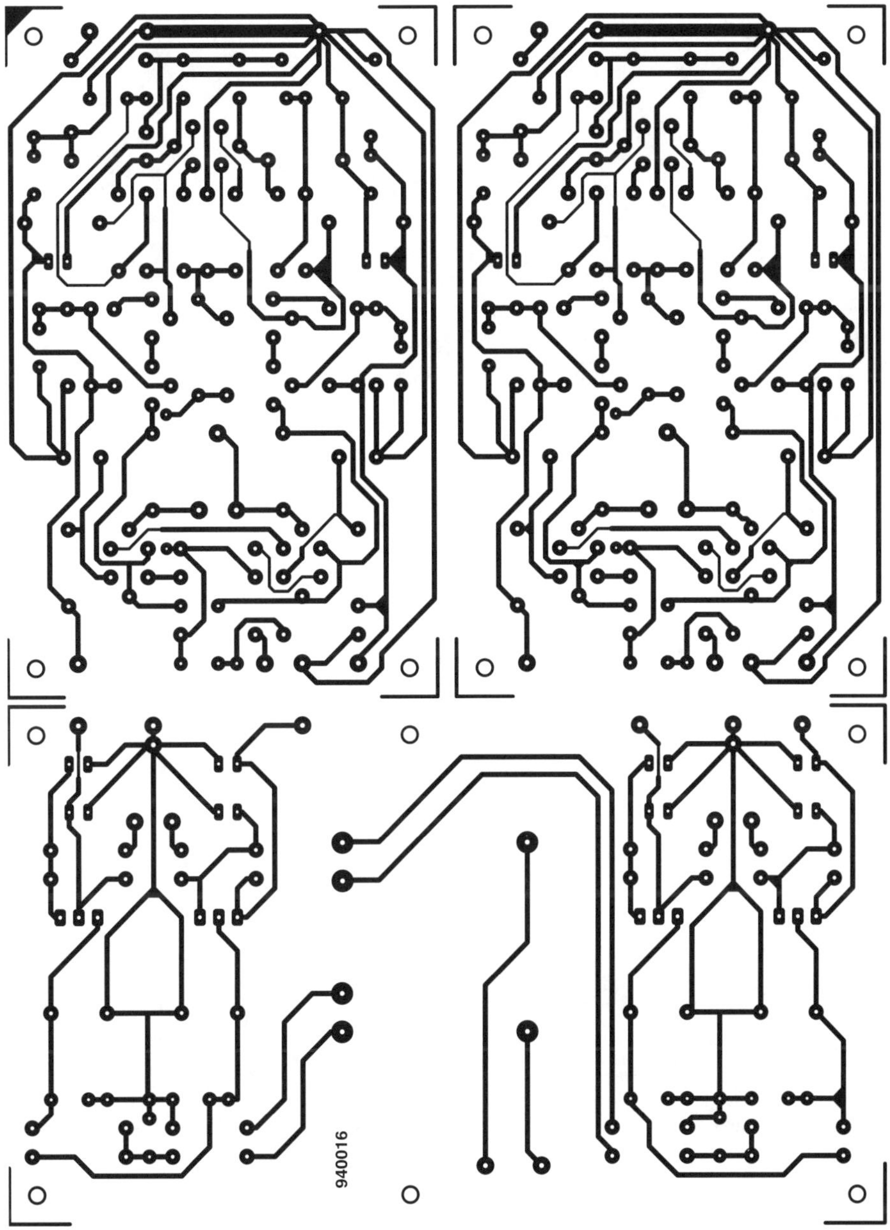

Fig. 9.4.(b). Track layout of the PCB for Headphone Amplifier II.

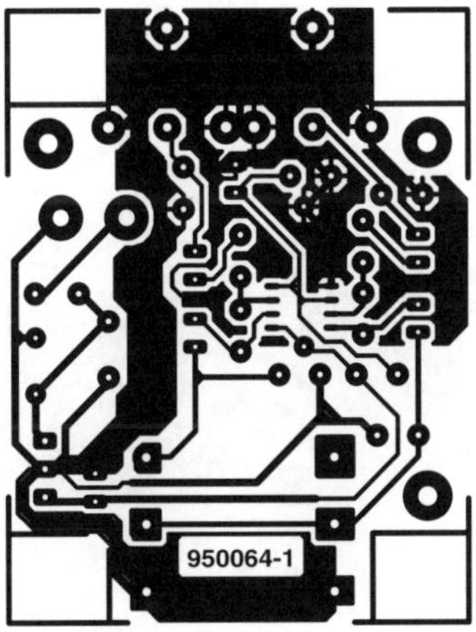

Fig. 10.4. (b). Track layout of the PCB for Headphone amplifier III.

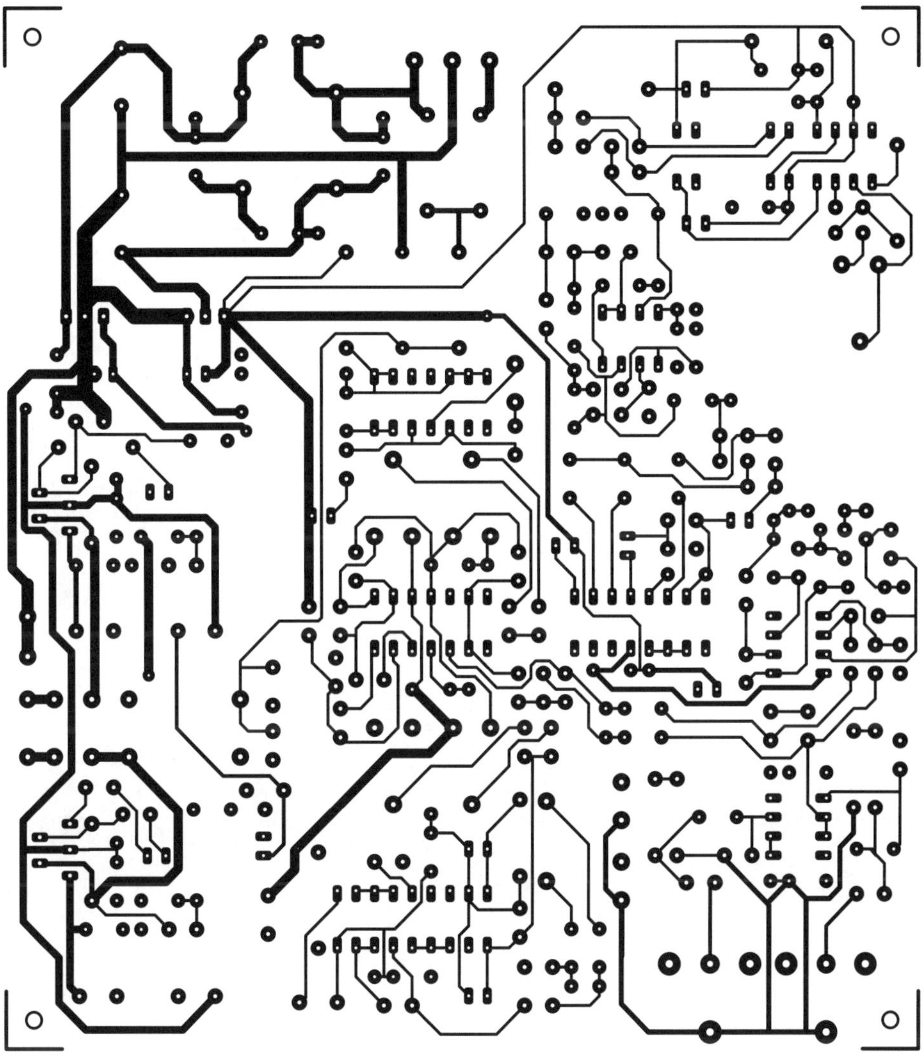

Fig. 12.5.(b). Track layout of the PCB for the surround sound processor.

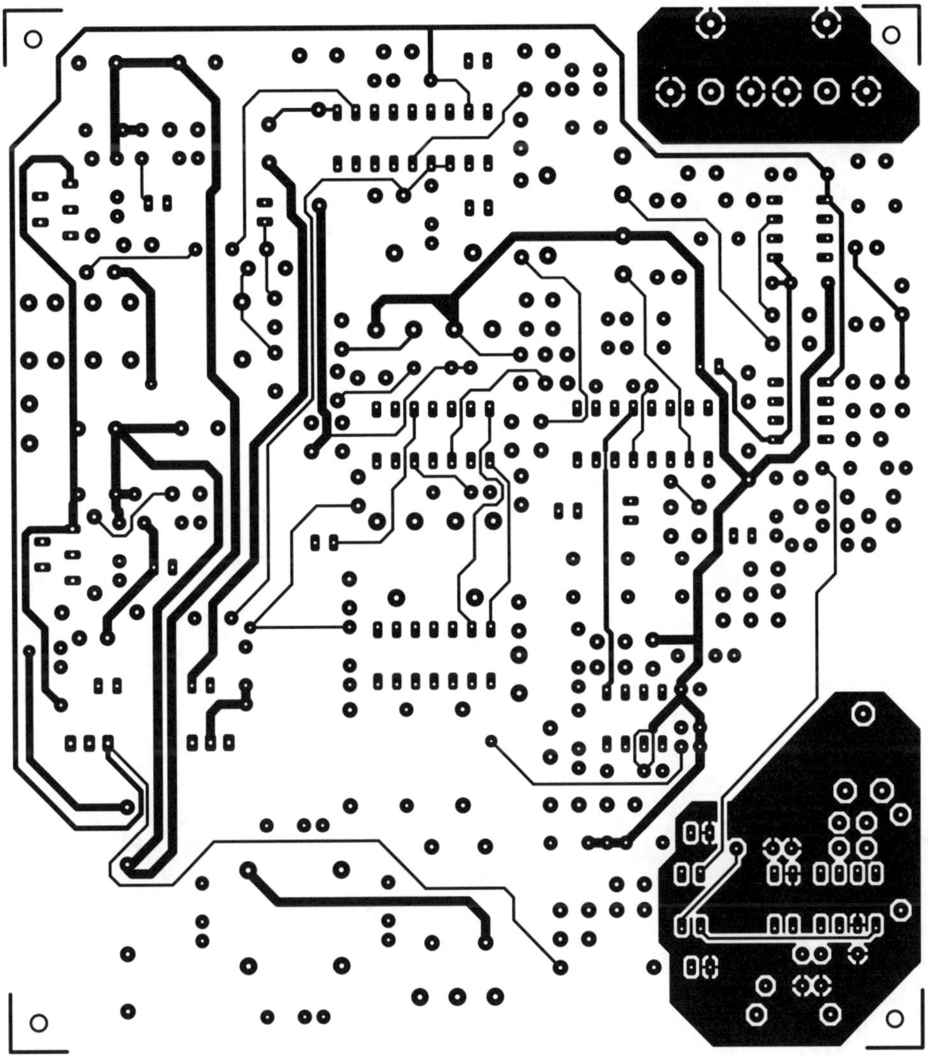

Fig. 12.5.(c). Component overlay of the PCB for the surround sound processor.

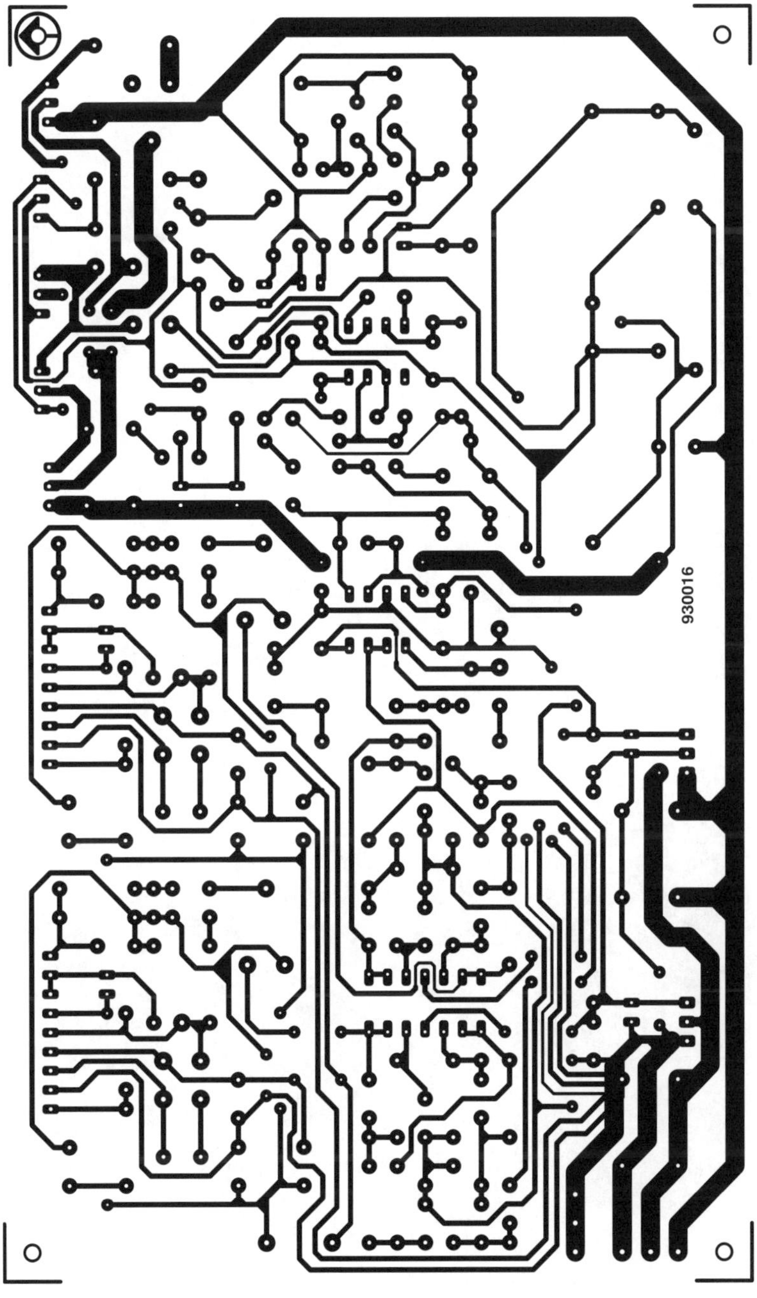

Fig. 13.4.(b). Track layout of the PCB for the 3-way loudspeaker system.

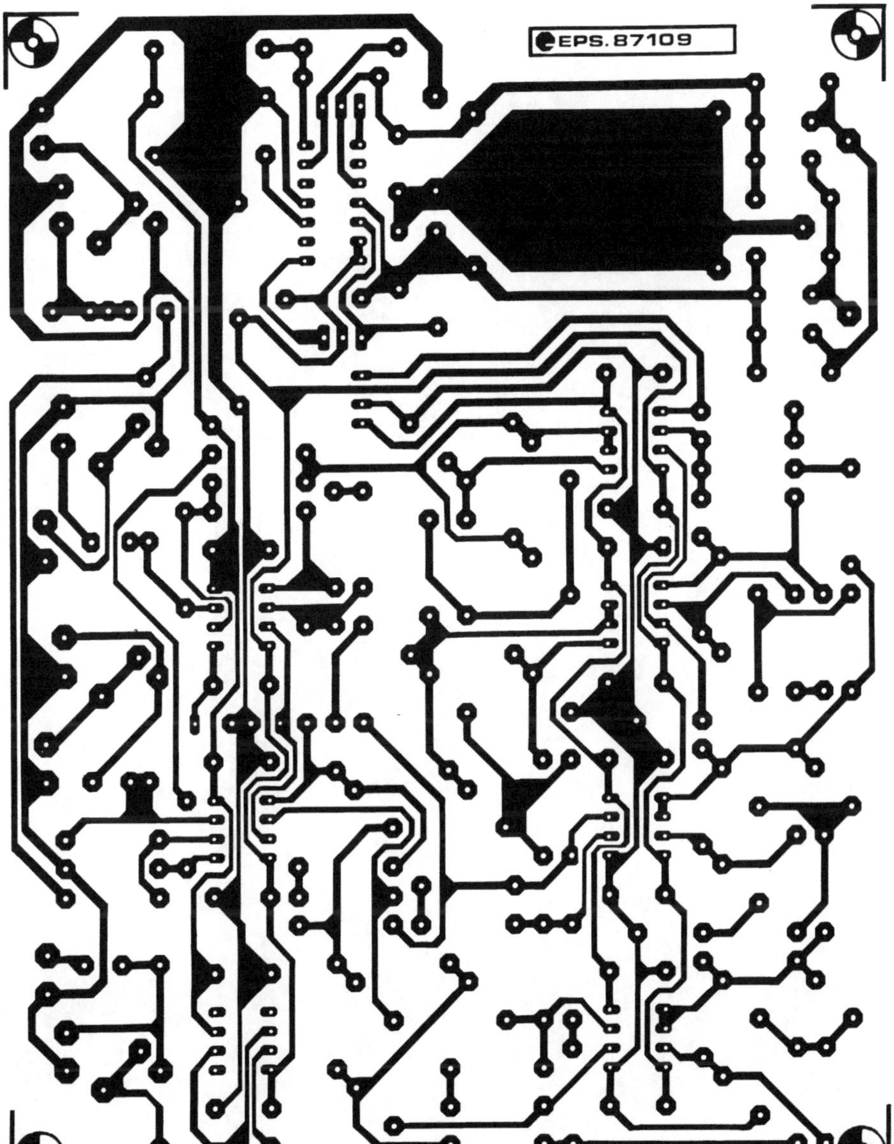

Fig. 14.5.(b). Track layout of the PCB for the phase-linear network.

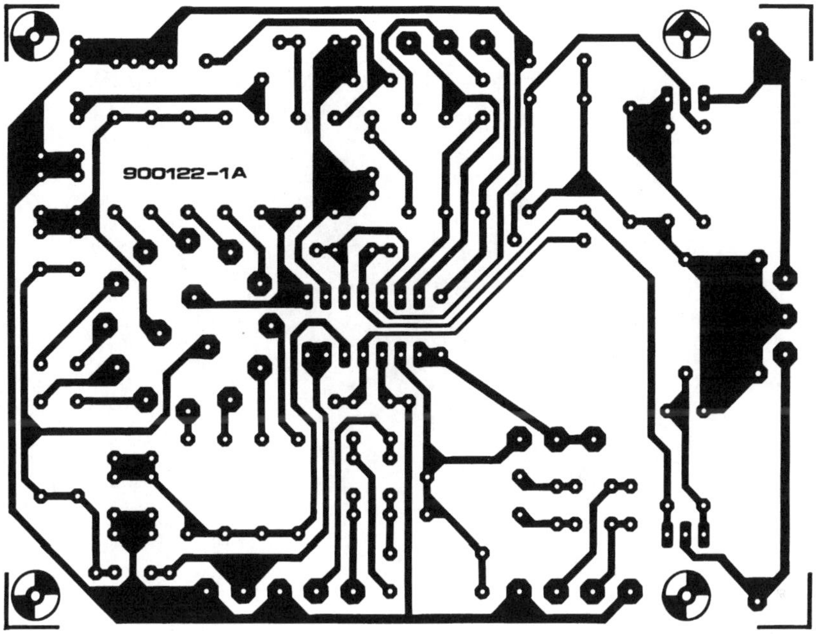

Fig. 15.4.(b). Track layout of the PCB for the correction filter.

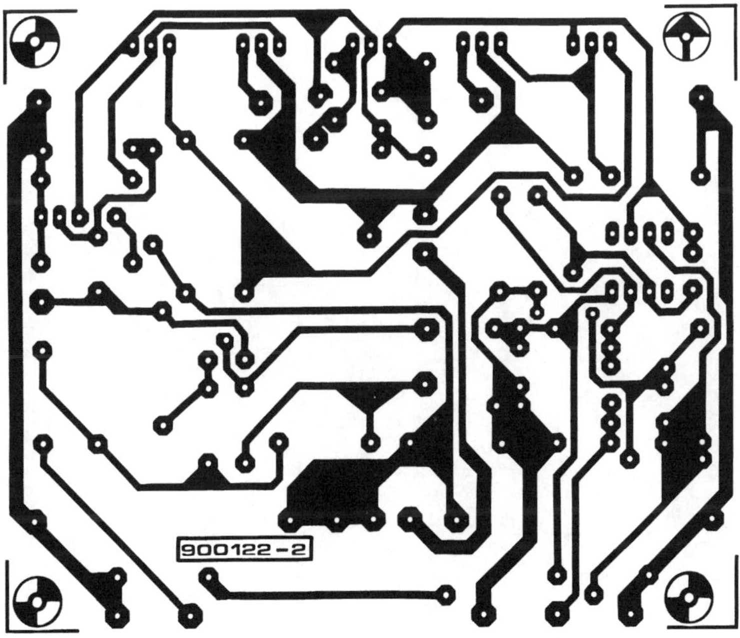

Fig. 15.9.(b). Track layout of the PCB for the output amplifier.

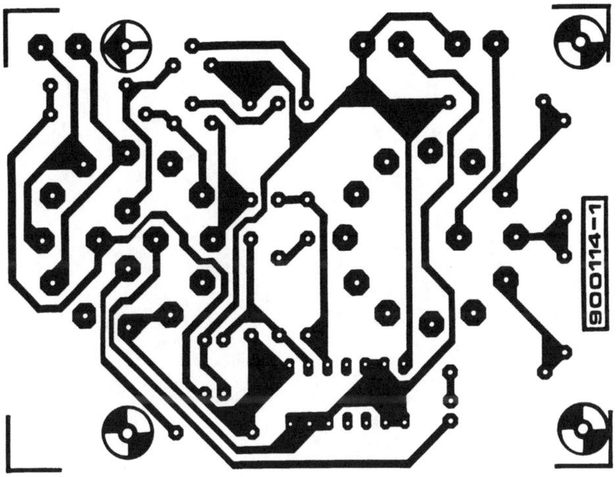

Fig. 16.4.(b). Track layout of the PCB for the pulse transmitter.

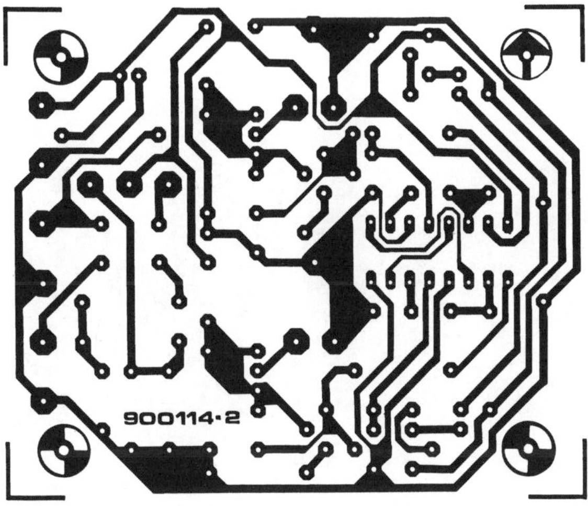

Fig. 16.5.(b). Track layout of the PCB for the pulse receiver.